LINEAR DISCRETE PARABOLIC PROBLEMS

NORTH-HOLLAND MATHEMATICS STUDIES 203

Editor: Jan van Mill
Faculteit der Exacte Wetenschappen
Amsterdam, The Netherlands

ELSEVIER

Amsterdam – Boston – Heidelberg – London – New York – Oxford
Paris – San Diego – San Francisco – Singapore – Sydney – Tokyo

LINEAR DISCRETE PARABOLIC PROBLEMS

NIKOLAI YU. BAKAEV
Moscow, Russia

ELSEVIER

Amsterdam – Boston – Heidelberg – London – New York – Oxford
Paris – San Diego – San Francisco – Singapore – Sydney – Tokyo

ELSEVIER B.V.
Radarweg 29
P.O. Box 211, 1000 AE
Amsterdam, The Netherlands

ELSEVIER Inc.
525 B Street
Suite 1900, San Diego
CA 92101-4495, USA

ELSEVIER Ltd.
The Boulevard
Langford Lane, Kidlington,
Oxford OX5 1GB, UK

ELSEVIER Ltd.
84 Theobalds Road
London WC1X 8RR
UK

First edition 2006

Library of Congress Cataloging in Publication Data
A catalog record is available from the Library of Congress.

British Library Cataloguing in Publication Data
A catalogue record is available from the British Library.

ISBN-13: 978-0-444-52140-8
ISBN-10: 0-444-52140-2
ISSN: 0304-0208 (Series)

∞ The paper used in this publication meets the requirements of ANSI/NISO Z39.48-1992 (Permanence of Paper).
Printed in The Netherlands.

Preface

In this book we give a systematic approach to the study of linear discrete parabolic problems for which we examine the question of well-posedness, in particular, of stability. The developed techniques can also be applied in the analysis of convergence of the corresponding discrete solutions, as demonstrated in Chapter 7. We remark that the use of the concept of evolution equation in discrete time and of some related concepts, which will be explained in the sequel, is of great importance for our investigation. This approach appears natural in view of the fact that our subsequent considerations are restricted to studying the parabolic case only. We expect that the reader will have gotten the impression that the developed discrete theory by no means is a direct analogue of the modern theory of parabolic differential equations in Banach space. It will therefore be seen that the below approach possesses its own distinctive features for which one can find no analogue in the continuous case. On the other hand, in what follows we emphasize frequently the parallels between the present discrete theory and the classical continuous one. Our assertions of general character are further used for examining certain discretization-in-time methods as applied to abstract parabolic equations, and as such we consider Runge-Kutta methods as well as other widespread methods. Note that the achieved results can be used for the study of fully discrete approximations to parabolic partial differential problems that are constructed on the basis of finite difference or finite element discretization in space. However such an investigation would require much more room to describe the necessary details, and we give up the realization of this idea at least in the present book; perhaps it will be the subject for another one.

It is well known that any modern discretization method used in the numerical treatment of non-stationary problems can be equivalently rewritten in the form of a single step method with respect to the time variable, under certain reasonable restrictions. In fact, this circumstance is an essential ingredient of our subsequent considerations. Moreover, it allows us to examine discrete problems at the first stage of analysis without any connection with the starting parabolic problem that has been discretized.

It is generally acknowledged that the questions of stability and convergence are the most essential for numerical analysis of discretizations. In particular, if the method is stable, any numerical errors which arise during the calculating process do not grow in the sequel. The analysis of convergence is closely related to that of stability. Moreover, the so-called equivalence theorems (see, e.g., Richtmyer & Morton [135]) allow one to derive error estimates for both stable and consistent discrete methods, assuming that the data are smooth enough. In dealing with the parabolic case we have the possibility of working with non-smooth data as well. The question of stability/convergence is then settled by using *a priori* inequalities related to the property of strong stability of holomorphic semigroups (cf. Larsson [99]). When working with concrete classes of discretization methods, the questions of existence and uniqueness of a solution in the necessary sense also need to be settled, and then we have to discuss well-posedness instead of stability.

Much work has been devoted to the verification of stability and by now many different tools for its analysis are known. We can mention, among others, the Fourier transform method (see, e.g., Lax & Wendroff [101] and Richtmyer & Morton [135]); the method of separation of variables (see, e.g., Douglas & Gunn [66]); the energy method (see, e.g., Kreiss [98]); the method based on estimating the Green's functions (Brenner & Thomée [51], Schatz, Thomée, and Wahlbin [141], and Serdyukova [143, 145]); and the method based on the maximum principle (see, e.g., Samarskii [139]).

Recently, many authors have successfully dealt with discrete initial-boundary value problems thinking of them as initial value problems for difference-operator equations in Banach (in particular, in Hilbert) spaces. As a consequence, the stability theory of linear discrete non-stationary problems in Hilbert spaces is, at present, well advanced and fairly effective in applications connected with the finite element and finite difference methods as well as with other approaches (see, e.g., Bramble *et al.* [49], Karakashian [90], Keeling [92], Oden & Reddy [121], Samarskii [139], Thomée [161, 162], and references therein). However related analysis within Banach space settings is much more complicated, which results from complexity of the geometry of Banach spaces as compared with that of Hilbert spaces. Most of this work has been devoted to the case of linear differential problems with a constant operator (see, e.g., Bakaev [10, 16, 17, 18, 20, 28, 30, 31, 32, 37], Brenner & Thomée [51, 52], Crouzeix [60], Crouzeix *et al.* [61], Crouzeix, Larsson, and Thomée [62], Lubich & Nevanlinna [104], Nitsche & Wheeler [119], Palencia [127, 128], Schatz, Thomée, and Wahlbin [141], and Thomée [162]).

In applications such a situation corresponds to partial differential initial-boundary value problems with time-independent coefficients. We also remark that certain results for discretized linear integro-differential equations have appeared (see, e.g., Bakaev, Larsson, and Thomée [40, 41], Sloan & Thomée [147], and Zhang [168]).

Less is known for the case of non-stationary problems with a variable (that is, time-dependent) operator in Banach space. The first achievements in this direction have dealt with situations with one spatial dimension (see, e.g., Thomée and Wahlbin [163]) or with concrete families of discrete problems (see, e.g., Gudovich & Terteryan [78] and Sobolevskii [151, 152]). For results of general character, we mention our work [10, 16, 25, 26, 27, 28] carried out in the course of Banach space settings and covering a large variety of applications.[1] Also, problems with a splitting operator have been considered [15, 34, 35], as well as problems with generalized input data [20, 34, 35] and problems with a variable stepsize [44, 30, 36]. Such results allow one to examine, in particular, wide classes of parabolic semidiscrete and fully discrete problems with variable coefficients, constructed on the basis of discretization in time by single step or multistep methods and of finite difference or finite element discretization in space. This research has become possible due to certain resolvent estimates that have been established for elliptic differential, finite difference, and finite element operators (see [156, 14, 29, 38, 43, 39]).

The present book actually gives an unfolded and considerably extended exposition of our results collected in the papers [10] --[13], [15] --[28], and [30] --[36], of which some are quite old and some more or less recent. Most of the above cited works in that direction are either announcements of results or have been published in editions almost unattainable for the readers outside of Russia.

Although in this book we deal only with linear problems, our achievements are significant for studying numerical methods for nonlinear parabolic equations, and we intend to demonstrate in our future work how the assertions in this volume can be applied to the analysis of nonlinear problems as well. Recent results devoted to examining nonlinear discrete parabolic problems in Hilbert and Banach space settings can be found, in particular, in Akrivis & Crouzeix [4], Akrivis, Crouzeix, and Makridakis [5], Akrivis, Karakashian, and Karakatsani [6], Akrivis & Makridakis [7], Beyn & Garay [47], González *et al.* [72], Ostermann & Thalhammer [125], and

[1]Not so long ago González & Palencia [73, 74], using different techniques, showed certain stability assertions that are similar to some of those established within our earlier work.

Ostermann, Thalhammer, and Kirlinger [126].

Our considerations start by examining general questions of stability within the framework of Banach spaces (see Part I). This investigation is, in fact, the core of the book. The main working concept in Part I is that of evolution equation in discrete time. On the one hand, any modern (even implicit and/or multistep) discretization method which is applied to a non-stationary problem necessarily reduces to the Cauchy problem for an evolution equation in discrete time (under reasonable restrictions). On the other hand, rewriting the starting discrete problem in the form of an evolution equation in discrete time allows one to observe its most important features in a more transparent way. As it will be seen below, the discrete theory presented in Part I looks very often rather similar to the modern analysis of abstract differential equations in Banach space (see, e.g., Amann [9] or Lunardi [108]). At the same time it possesses its own distinctive features which have no analogues in continuous theory. Apart from everything else, we hope that our investigation makes a valuable contribution to the theory of discrete semigroups, which is at the moment only at a development stage (see, e.g., Lyubich [110], Nagy & Zemánek [114], and Nevanlinna [116, 117, 118]).

The main problem to be considered is posed in Chapter 1, where we introduce the necessary working concepts for our subsequent use in Chapters 2, 3, and 4.

Furthermore, Chapter 2 is devoted to the analysis of stability of the Cauchy problem for an evolution equation in discrete time. These results are most crucial for the whole subsequent exposition. We distinguish between the autonomous and non-autonomous cases for which we essentially use different techniques. In the former case the main tools of analysis are the operator calculus methods[2] while the study of non-autonomous equations is based on the use of certain discrete versions of Gronwall's lemma. The corresponding stability estimates are then valid for finite time intervals only.

The considerations in Chapter 3 are of particular interest and concern discrete problems with a splitting operator; the ideas employed in the preceding chapter are further developed for examining the stability of such problems. It is significant that our analysis is not based on the assumption that the separate components of the splitting necessarily commute; avoiding this assumption is essential for many important applications. In this chapter we also consider a more involved splitting construction which appears, for

[2]The application of such methods usually deals with integration over suitable contours in the complex plane. Throughout the book the orientation of any contour is taken according to what is accepted in Appendix A.2.

example, when handling second order discretizations.

In Chapter 4 we study from the viewpoint of stability more general equations, that is, discrete evolution equations with a memory term. The techniques used here allow us to show stability on finite intervals only.

The general results of Part I can further be applied to the study of discretizations in time of abstract linear parabolic problems in Banach space. This program of action is realized in Parts II and III. The investigation in these parts is based on reducing the starting discrete problem to the Cauchy problem for an evolution equation in discrete time. We therefore discuss how to perform such a reduction and how to verify the meaning of not clear restrictions on the operator in question if we proceed from a discretization in time of the original differential problem. In Part II we concentrate on the study of single step methods, and as such we consider, in the concrete, the Runge-Kutta methods. However the below techniques can be extended in order to examine stability for other families of modern single step methods (see, e.g., Butcher [53], Hairer, Nørsett, and Wanner [80], Hairer & Wanner [81], or Stetter [155]). Next, in Part III, in order to reveal further possibilities of our approach, we discuss some other ways of discretization, more precisely the θ-method, some operator splitting methods, and the linear multistep methods. In this investigation we stay along the same lines of analysis as above; however our exposition in Part III is not as detailed as that carried out in Part II.

In Chapter 5 we consider Runge-Kutta discretization for the abstract parabolic Cauchy problem. We introduce some working concepts and hypotheses intended for subsequent use, among others, certain conditions of Hölder-continuity; meanwhile some auxiliary assertions are established here as well. Also, we give a suitable classification of the Runge-Kutta methods. However the most important point is that we show, under the sectorialness hypothesis and under one of the Hölder-continuity type conditions, that the starting discrete problem can be equivalently represented as the Cauchy problem for an evolution equation in discrete time and is, therefore, uniquely solvable.

Next, in Chapter 6, we derive stability estimates of the discrete solutions for different classes of Runge-Kutta methods. These results are achieved by showing certain resolvent estimates and Hölder-continuity type results and by subsequently using the general stability results established in Part I.

As concerns convergence estimates, these are presented in Chapter 7 under various restrictions on the Runge-Kutta method in question, distinguishing between the autonomous and non-autonomous cases. Also, we do not pass over the phenomenon of order reduction which may be observed for approximations of equations with an unbounded operator and which shows itself by the fact that the classical order of accuracy is not really achieved.

Completing Part II in Chapter 8, we study variable stepsize Runge-Kutta discretizations and give stability estimates for such discrete problems under some (rather general) restrictions on the families of time partitions used.

In the next three Chapters 9 ––11, constituting Part III, we focus on showing stability only and carry out the investigation under further restrictions and simplifications. This narrowing of scope allows us to concentrate on the most essential aspects of the considered methods.

Our analysis in Chapter 9 concerns the θ-method, and we show that certain refined stability estimates are possible if the operator in question is under stronger restrictions than just the sectorialness hypothesis.

In Chapter 10, under the assumption that the operator in question can be decomposed into the sum of two generally non-commutative operators, we study certain methods with a splitting operator.

In Chapter 11 we deal with some classes of problems constructed by means of linear multistep discretization.

Finally, in Part IV which consists of only one chapter, we discuss the stability of certain approximations to linear integro-differential equations in Banach space by applying the results of Chapter 4.

Our results below will mainly be expressed in terms of stability and convergence estimates in the norm of an abstract Banach space. Throughout the whole exposition there occur constants in assumed, intermediate, and final estimates, whose sizes are unessential for our analysis. They will be denoted by C and c when subject to $C \geq 0$ and $c > 0$, respectively. Both the C's and c's may implicitly depend on other parameters which are fixed in the context;[3] this dependence is never traced. At the same time we take care of giving an explicit mention whenever there may be a danger of confusion.

As already mentioned above, our considerations will be carried out within the framework of Banach spaces. Throughout the book, given an abstract Banach space $\mathfrak{X}$, $\|\cdot\|$ stands for both the vector and operator norm in $\mathfrak{X}$, while $\mathcal{B}(\mathfrak{X})$ is used to denote the space of all linear bounded operators on $\mathfrak{X}$. Furthermore, $\mathfrak{X}^*$ denotes the dual space for $\mathfrak{X}$, $\|\cdot\|_*$ the corresponding norm in $\mathfrak{X}^*$, and $\langle\cdot,\cdot\rangle$ the duality pairing between $\mathfrak{X}^*$ and $\mathfrak{X}$. By I we denote the identity operator both in $\mathfrak{X}$ and in $\mathfrak{X}^*$. In what follows we also deal

[3] A parameter ξ is thought of as fixed without explicit mention if, for example, it appears with the words 'there exists a number ξ ...' or 'for some ξ ...'. The same applies to subsets of the complex plane. On the contrary, if ξ appears with the words 'for all ξ ...' or 'for any ξ ...', it is by no means fixed. In doubtful cases it is explicitly specified if ξ is fixed or not.

with some concrete function Banach spaces; the corresponding notation for their norms will be introduced immediately as required. Furthermore, given a linear (generally unbounded) operator $\mathfrak{L}$, we denote by $\operatorname{Dom}\mathfrak{L}$ and $\operatorname{Ran}\mathfrak{L}$ its domain and range, respectively, while $\mathfrak{L}^*$ denotes the adjoint for $\mathfrak{L}$ (if defined). Next, if $|\cdot|$ is a seminorm, the symbol $\operatorname{Dom}|\cdot|$ is used to denote its domain which is always assumed to be a linear manifold. For future needs we also define some useful classes of unbounded operators. In order to do this, we first denote, given $\varphi \in [0, \pi)$,

$$\Sigma_\varphi = \{\lambda \in \mathbb{C} : \lambda \neq 0, |\arg\lambda| \leq \varphi\} \cup \{0\}.$$

This notation will be used in the sequel as well, without explicit mention.

Definition 0.1. *Let $\mathfrak{L} : \operatorname{Dom}\mathfrak{L} \subseteq \mathfrak{X} \to \mathfrak{X}$ be a closed linear operator. We write $\mathfrak{L} \in \mathcal{S}(\varphi, \sigma)$, with some $\varphi \in (0, \pi/2)$ and $\sigma \in \mathbb{R}$ if the resolvent set of the operator $\mathfrak{L} - \sigma I$ contains the set $-\Sigma_{\pi-\varphi} \setminus \{0\}$ and*

$$\|((\lambda + \sigma)I - \mathfrak{L})^{-1}\| \leq C|\lambda|^{-1} \quad \text{for all } \lambda \in -\Sigma_{\pi-\varphi}. \tag{1}$$

The operator $\mathfrak{L}$ itself is then called (φ, σ)-sectorial *and the resolvent estimate (1) the* sectorial estimate.

It is easily seen that the fact that $\mathfrak{L} \in \mathcal{S}(\varphi, \sigma)$ implies $k\mathfrak{L} \in \mathcal{S}(\varphi, k\sigma)$ for any $k > 0$.

In our subsequent considerations we deal, in fact, with (φ, σ)-sectorial operators for some $0 \leq \varphi < \pi/2$. Differential equations with such operators are usually called parabolic; under certain reasonable restrictions their discrete analogues, in some sense, may be thought of as parabolic as well. We therefore restrict ourselves in this book to the so-called parabolic case which has important distinctive features and admits a rather deep investigation.

Although in this book we restrict ourselves to the study of discretizations of abstract parabolic problems, sometimes with a memory term, our results are applicable, as already mentioned above, to semidiscrete and fully discrete approximations to parabolic partial differential and integro-differential problems. It is an important fact that one is then able to work with multi-dimensional problems in spaces with maximum norm and in related ones, because maximum norm has physical meaning in many applications. At the same time it is worth noting, for comparison, that the work of González & Palencia [73, 74], intended for use when handling problems with time-dependent coefficients, is unadaptable in principle if one cannot be based on *a priori* L_p or Schauder type estimates (see, e.g., Agmon, Douglis, and

Nirenberg [3]) for elliptic partial differential operators and their discrete analogues.

We suppose that the reader is familiar with the fundamental concepts and results of the theory of functions of a complex variable and possesses a basic knowledge of functional analysis. Nevertheless, in order to raise the level of self-containedness of the book, we collect some important facts, needed for our analysis, in an appendix.

In conclusion, the author would like to thank the many colleagues from abroad who kindly sent him reprints of their works that are unavailable in Moscow.

Contents

Part I

EVOLUTION EQUATIONS IN DISCRETE TIME

Chapter 1

Preliminaries

1.1 Preliminaries

Let there be given a complex Banach space $\mathfrak{X}$ and an interval $[0,T]$, $T>0$. In addition, let $\Omega_k = \{t_n : t_n = nk \text{ for } n = 0,\dots,N\}$ be a uniform grid in the interval $[0,T]$, where $k>0$ and $N \in \mathbb{N}$ are connected by the relation $Nk = T$. We allow for the possibility that $T=\infty$; in this case $\Omega_k = \{t_n : t_n = nk \text{ for } n = 0,1,\dots\}$ and $k>0$ is set arbitrarily. The last case is formally arrived at if we put $T=\infty, N=\infty$. We also use the notation $\breve{\Omega}_k = \{t_n : t_n = nk \text{ for } n = 0,\dots,N-1\}$.

In this part of the book all working objects (operators, seminorms, etc.), except the space $\mathfrak{X}$ itself, generally depend on k;[1] this dependence will not mentioned explicitly. At the same time, given k, the dependence on t_n, if generally available, will always be shown explicitly by adding n as a subscript to the corresponding object.[2] In particular, $(\mathfrak{X})$-valued functions of discrete time t_n are used. Given such a function $Z_n : \Omega_k \to \mathfrak{X}$, we denote $\partial Z_n = k^{-1}(Z_{n+1} - Z_n)$, $t_n \in \breve{\Omega}_k$. (This notation will be employed below for scalar discrete functions as well.)

The main object of consideration in the present part of the book will be the discrete Cauchy problem

$$\partial Y_n + \mathfrak{A}_n Y_n = \mathfrak{Q}_n F_n \quad \text{for } t_n \in \breve{\Omega}_k, \qquad Y_0 = \mathfrak{R} y^0, \tag{1.1}$$

for some given $\mathfrak{A}_n, \mathfrak{Q}_n, \mathfrak{R} \in \mathcal{B}(\mathfrak{X})$. The vector $y^0 \in \mathfrak{X}$ and the discrete function $F_n : \breve{\Omega}_k \to \mathfrak{X}$ are considered as the input data of problem (1.1).

[1] In principle, we could let the space $\mathfrak{X}$ depend on k as well, but this is not needed for our subsequent aims.

[2] This means that every object that depends on t_n will have n as a subscript, separated by a semicolon from any additional subscripts.

The discrete function $Y_n : \Omega_k \to \mathfrak{X}$ is to be found. Y_n is therefore thought of as a solution of (1.1).

In the case $T < \infty$, we employ the condition $\mathfrak{A}_N = \mathfrak{A}_{N-1}$. This is not an important restriction, since only the values of $\mathfrak{A}_n$, $n = 0, \ldots, N-1$, are essential, but it will be convenient in our subsequent statements.

Clearly, problem (1.1) can be equivalently rewritten as follows

$$Y_{n+1} = (I - k\mathfrak{A}_n)\, Y_n + k\mathfrak{Q}_n\, F_n \quad \text{for } t_n \in \breve{\Omega}_k, \qquad Y_0 = \mathfrak{R}y^0, \tag{1.2}$$

which is an explicit single step equation with the transition operator

$$\mathfrak{U}_n = I - k\mathfrak{A}_n. \tag{1.3}$$

It is therefore easily seen that problem (1.1) is uniquely solvable and its solution Y_n is given by[3]

$$Y_n = \mathfrak{U}_{n-1,0}\mathfrak{R}y^0 + k\sum_{j=0}^{n-1} \mathfrak{U}_{n-1,j+1}\mathfrak{Q}_j F_j, \tag{1.4}$$

where, for $0 \le j, n \le N$,

$$\mathfrak{U}_{n,j} = \begin{cases} \mathfrak{U}_n\mathfrak{U}_{n-1}\cdots\mathfrak{U}_j & \text{if } \ j \le n, \\ I & \text{if } \ j > n. \end{cases} \tag{1.5}$$

In spite of the fact that the unique solvability of problem (1.1) is evident, it is important to establish stability estimates for the solution of (1.1), which is the main objective in this part of the book.

As far as possible applications are concerned, note that all modern discretization methods (including implicit and/or multistep ones) can, under certain reasonable restrictions, be equivalently written in the form (1.1) or (1.2). In particular, for Runge-Kutta methods this fact is shown in Section 5.3 while for linear multistep methods this observation is pointed out in Section 11.1.

Now we give a series of working definitions.

Definition 1.1. *A difference equation of the form (1.1) is called an* evolution equation in discrete time. *The operator $\mathfrak{U}_{n,j}$ given by (1.5) is called the* discrete evolution operator *for problem (1.1) or, equivalently, for problem (1.2). In the particular case when $\mathfrak{A}_n = \mathfrak{A}$ and $\mathfrak{U}_n = \mathfrak{U}$ do not depend on t_n, $\mathfrak{U}_{n,0} = \mathfrak{U}^n$ is called the* discrete semigroup *generated by the operator $\mathfrak{A}$.*

[3] Throughout the book we use the convention $\sum_{j=l}^{n} \cdots = 0$ if $n < l$.

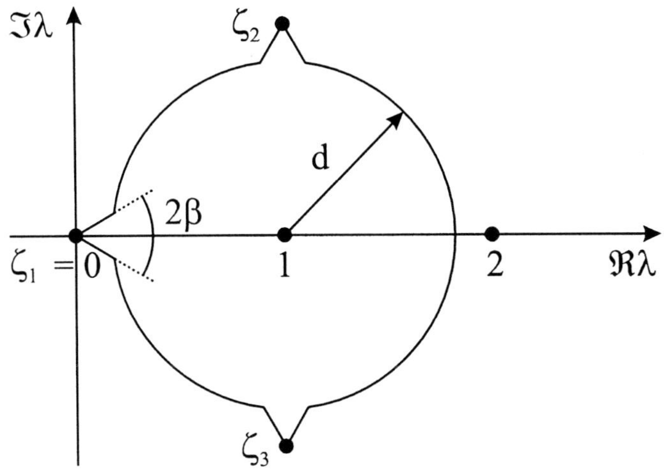

Figure 1.1: **A set of Λ_3-configuration**

When dealing with subsets of the complex plane $\mathbb{C}$, given $V \subseteq \mathbb{C}$, we denote by $\operatorname{Int} V$, $\operatorname{Cl} V$, and ∂V, its interior, closure, and boundary, respectively. Also, we use the notation $\mathcal{D}(z;r) = \{\lambda : |\lambda - z| \leq r\}$ and $\mathcal{D}(r) = \mathcal{D}(0;r)$ for $z \in \mathbb{C}$ and $r \geq 0$.

Definition 1.2. *A set $\Upsilon \subset \mathbb{C}$ is called a* set of Λ_J-configuration *with some integer $J \geq 1$ if Υ is the union of a disk $\mathcal{D}(1;a)$ and of J sets of the form*

$$\sigma_j = \{\lambda : |\arg(\lambda - \zeta_j) - \arg(1 - \zeta_j)| \leq \beta\} \cap \mathcal{D}(\zeta_j; d),\ j = 1, \ldots, J,$$

with some $a \in (0,1)$ and $\beta \in (0, \arcsin a)$, where $d = \cos\beta - (a^2 - \sin^2\beta)^{1/2}$ and $\zeta_1 = 0, \zeta_2, \ldots, \zeta_J$ are different points lying on the circle $\partial\mathcal{D}(1;1)$. The constants a and β are called the parameters *of the set Υ and the points ζ_j, $j = 1, \ldots, J$, the* singular points *of Υ.*

For instance, when $J = 1$, we deal with a set of Λ_1-configuration which has only one singular point: $\zeta_1 = 0$. Such sets will be of particular interest in our subsequent considerations. Figure 1.1 gives an example of a set of Λ_3-configuration.

Definition 1.3. *An operator $\mathfrak{L}_n \in \mathcal{B}(\mathfrak{X})$ is said to satisfy the* Λ_J^k-condition *with some integer $J \geq 1$ if there exist numbers $K > 0$, $\delta \geq 0$, and a set Υ of Λ_J-configuration such that*
(i) for all $k \in (0, K]$, the operator $k\mathfrak{L}_n$ has its spectrum contained in the set $\operatorname{Int} \Upsilon^k \cup \{0\}$, where

$$\Upsilon^k = \cup_{j=1}^J \mathcal{D}(\zeta_j; \delta k) \cup \Upsilon, \tag{1.6}$$

and
(ii) $\mathfrak{L}_n$ *satisfies the resolvent estimate, that is, for all* $k \in (0, K]$, $t_n \in \Omega_k$, *and* $\lambda \notin \operatorname{Int} \Upsilon^k$,

$$\|(\lambda I - k\mathfrak{L}_n)^{-1}\| \leq C \sum_{j=1}^{J} |\lambda - \zeta_j|^{-1}, \tag{1.7}$$

where ζ_j, $j = 1, \ldots, J$, *are the singular points of* Υ. *If (1.7) holds for all* $\lambda \notin \operatorname{Int} \Upsilon$,[4] *the operator* $\mathfrak{L}_n$ *is said to satisfy the* Λ_J-condition. *Any set* Υ *of* Λ_J*-configuration, which may be employed in the above context, is called* Λ_J^k-proper *(*Λ_J-proper*) for the operator* $\mathfrak{L}_n$. *If* Υ *is* Λ_J^k*-proper (*Λ_J*-proper) for* $\mathfrak{L}_n$, *the singular points of* Υ *are also called the* singular points *of the operator* $\mathfrak{L}_n$.

Remark 1.1. *It follows by duality that if an operator* $\mathfrak{L}_n$ *satisfies the* Λ_J^k*-condition (the* Λ_J*-condition) in* $\mathfrak{X}$, *the same is true for its adjoint* $\mathfrak{L}_n^*$ *in* $\mathfrak{X}^*$.

The above concepts are of great importance for our subsequent considerations. One of the main restrictions below on the operator $\mathfrak{A}_n$ in (1.1) will be that $\mathfrak{A}_n$ satisfies the Λ_J- or Λ_J^k-condition with some integer $J \geq 1$, and the situation when $J = 1$ will be of particular interest. At the same time some seminorms, whose properties are related to the above Λ_J- and Λ_J^k- conditions, are used.

Definition 1.4. *Let there be given an* $\mathfrak{L}_n \in \mathcal{B}(\mathfrak{X})$ *and fixed numbers* $J \in \mathbb{N}$, $m \in \mathbb{N}$, $\xi \geq 0$, *with* $m \geq \max\{1, \xi\}$, *and let* $|\!|\!| \cdot |\!|\!|_n$ *and* $|\!|\!| \cdot |\!|\!|_{*;n}$ *be seminorms on* $\mathfrak{X}$ *and* $\mathfrak{X}^*$, *respectively. The pairs* $(|\!|\!| \cdot |\!|\!|_n; \mathfrak{L}_n)$ *and* $(|\!|\!| \cdot |\!|\!|_{*;n}; \mathfrak{L}_n)$ *are called* $\Lambda_J^k(\xi|m)$-concordant *and* $\Lambda_J^k(*; \xi|m)$-concordant, *respectively, if there exist numbers* $K > 0$ *and* $\delta \geq 0$ *and a set* Υ *of* Λ_J*-configuration such that the spectrum of the operator* $k\mathfrak{L}_n$ *is contained in the set* $\operatorname{Int} \Upsilon^k \cup \{0\}$ *whenever* $k \in (0, K]$, *where* Υ^k *is defined by (1.6), and the respective inequalities hold, for all* $k \in (0, K]$, $t_n \in \Omega_k$, $\lambda \notin \operatorname{Int} \Upsilon^k$, $u \in \mathfrak{X}$, *and* $\chi \in \mathfrak{X}^*$,

$$|\!|\!|(\lambda I - k\mathfrak{L}_n)^{-m} u|\!|\!|_n \leq C k^{-\xi} \sum_{j=1}^{J} |\lambda - \zeta_j|^{\xi - m} \|u\| \tag{1.8}$$

and

$$|\!|\!|(\lambda^* I - k\mathfrak{L}_n^*)^{-m} \chi|\!|\!|_{*;n} \leq C k^{-\xi} \sum_{j=1}^{J} |\lambda - \zeta_j|^{\xi - m} \|\chi\|_*, \tag{1.9}$$

where ζ_j, $j = 1, \ldots, J$, *are the singular points of* Υ.[5] *If in addition the*

[4] This means that one can take $\delta = 0$.

[5] By z^* we denote the conjugate comlex number for z.

estimates (1.8) and (1.9) hold for all $\lambda \notin \operatorname{Int}\Upsilon$, *then the pairs* $\left(\|\cdot\|_n; \mathfrak{L}_n\right)$ *and* $\left(\|\cdot\|_{*;n}; \mathfrak{L}_n\right)$ *are called* $\Lambda_J(\xi|m)$-concordant *and* $\Lambda_J(*;\xi|m)$-concordant, *respectively. Any set* Υ *of* Λ_J*-configuration, which may be employed in the above context, is called* $\Lambda_J^k(\ldots)$-proper *(*$\Lambda_J(\ldots)$-proper*) for the corresponding pair. If the size of* m *is not important in the context,* m *does not appear in the above notation.*[6] *Apart from everything else, if the resolvent estimates (1.8) and (1.9) hold for all* $t_n, t_l \in \Omega_k$ *with* $\|\cdot\|_l$ *and* $\|\cdot\|_{*;l}$ *in place of* $\|\cdot\|_n$ *and* $\|\cdot\|_{*;n}$, *respectively, then the pairs* $\left(\|\cdot\|_n; \mathfrak{L}_n\right)$ *and* $\left(\|\cdot\|_{*;n}; \mathfrak{L}_n\right)$ *are called* strictly $\Lambda_J^k(\ldots)$-concordant *(*strictly $\Lambda_J(\ldots)$-concordant*).*

We also need to give some generalizations of the above notions, which are based on the use of certain weighted resolvent estimates.

Definition 1.5. *Let there be given operators* $\mathfrak{L}_n, \mathfrak{V}_n \in \mathcal{B}(\mathfrak{X})$ *and fixed numbers* $J \in \mathbb{N}$, $m \in \mathbb{N}$, $\xi \geq 0$, *with* $m \geq \max\{1, \xi\}$. *In addition, let* $\|\cdot\|_n$ *and* $\|\cdot\|_{*;n}$ *be two seminorms defined on* $\mathfrak{X}$ *and* $\mathfrak{X}^*$, *respectively. The triplets* $\left(\|\cdot\|_n; \mathfrak{L}_n; \mathfrak{V}_n\right)$ *and* $\left(\|\cdot\|_{*;n}; \mathfrak{L}_n; \mathfrak{V}_n\right)$ *are called* $\Lambda_J^k(\xi|m)$*- and* $\Lambda_J^k(*;\xi;|m)$-concordant, *respectively, if there exist numbers* $K > 0$ *and* $\delta \geq 0$ *and a set* Υ *of* Λ_J*-configuration such that the spectrum of the operator* $k\mathfrak{L}_n$ *is contained in the set* $\operatorname{Int}\Upsilon^k \cup \{0\}$ *whenever* $k \in (0, K]$, *where* Υ^k *is defined by (1.6), and the respective inequalities hold, for all* $k \in (0, K]$, $t_n \in \Omega_k$, $\lambda \notin \operatorname{Int}\Upsilon^k$, $u \in \mathfrak{X}$, *and* $\chi \in \mathfrak{X}^*$,

$$\|(\lambda I - k\mathfrak{L}_n)^{-m}\mathfrak{V}_n u\|_n \leq Ck^{-\xi}\sum_{j=1}^{J}|\lambda - \zeta_j|^{\xi - m}\,\|u\| \tag{1.10}$$

and

$$\|(\lambda^* I - k\mathfrak{L}_n^*)^{-m}\mathfrak{V}_n^*\chi\|_{*;n} \leq Ck^{-\xi}\sum_{j=1}^{J}|\lambda - \zeta_j|^{\xi - m}\,\|\chi\|_*, \tag{1.11}$$

where ζ_j, $j = 1, \ldots, J$, *are the singular points of* Υ. *If (1.10) and (1.11) hold for all* $\lambda \notin \operatorname{Int}\Upsilon$, *the triplets* $\left(\|\cdot\|_n; \mathfrak{L}_n; \mathfrak{V}_n\right)$ *and* $\left(\|\cdot\|_{*;n}; \mathfrak{L}_n; \mathfrak{V}_n\right)$ *are called* $\Lambda_J(\xi|m)$*- and* $\Lambda_J(*;\xi|m)$-concordant, *respectively. Any set* Υ *of* Λ_J*-configuration, which may be employed in the above context, is called* $\Lambda_J^k(\ldots)$-proper *(*$\Lambda_J(\ldots)$-proper*) for the corresponding triplet. If the size of* m *is not important in the context,* m *does not appear in the above notation.*

Furthermore, we give a collection of some assertions involving the above-introduced concepts and intended for subsequent reference.

[6]For example, we then write simply $\Lambda_J^k(*;\xi)$ instead of $\Lambda_J^k(*;\xi|m)$.

Lemma 1.1. *If an operator $\mathfrak{L}_n \in \mathcal{B}(\mathfrak{X})$ satisfies the Λ_J^k-condition, it then follows that $\|k\mathfrak{L}_n\| \leq C$ for all $k \in (0, K]$ and $t_n \in \Omega_k$.*

Proof. This fact can easily be shown by using the operator calculus formula (A.2) with $f(\lambda) = \lambda$ and with $k\mathfrak{L}_n$ substituted for $\mathfrak{L}$, where the contour Ξ is taken independent of k and bounded away from the disk $\mathcal{D}(1;1)$. We therefore have

$$k\mathfrak{L}_n = \frac{1}{2\pi i}\int_{\Xi} \lambda(\lambda I - k\mathfrak{L}_n)^{-1} d\lambda,$$

and a trivial estimation shows the claim. □

Lemma 1.2. *Suppose that $\mathfrak{L}_n \in \mathcal{B}(\mathfrak{X})$ while $\|k\mathfrak{L}_n\| \leq C$ for all $k \in (0, K]$ and $t_n \in \Omega_k$. Let also seminorms $|\!|\!|\cdot|\!|\!|_n$ and $|\!|\!|\cdot|\!|\!|_{*;n}$ form respectively $\Lambda_J^k(\gamma|m)$- and $\Lambda_J^k(*;\gamma_*|m)$-concordant pairs with $\mathfrak{L}_n$ for some $J, m \in \mathbb{N}$ and $\gamma, \gamma_* \in [0, m]$. Then, with any fixed $\vartheta \in [0, \gamma]$ and $\vartheta_* \in [0, \gamma_*]$, the seminorms $k^{\vartheta}|\!|\!|\cdot|\!|\!|_n$ and $k^{\vartheta_*}|\!|\!|\cdot|\!|\!|_{*;n}$ will form respectively $\Lambda_J^k(\gamma - \vartheta|m)$- and $\Lambda_J^k(*;\gamma_* - \vartheta_*|m)$-concordant pairs with $\mathfrak{L}_n$. Furthermore, for all $k \in (0, K]$ and $t_n \in \Omega_k$,*

$$k^{\gamma} \sup_{\|u\|=1} |\!|\!|u|\!|\!|_n + k^{\gamma_*} \sup_{\|\chi\|_*=1} |\!|\!|\chi|\!|\!|_{*;n} \leq C. \tag{1.12}$$

Proof. We will show the claim in the case of the seminorm $k^{\vartheta}|\!|\!|\cdot|\!|\!|_n$ only, noting that the needed result for $k^{\vartheta_1}|\!|\!|\cdot|\!|\!|_{*;n}$ can be established in a similar manner. Accordingly, we will prove only that the first term on the left-hand side of (1.12) can be estimated as stated.

Let $R > 0$ be defined by

$$R = 2 \sup_{k\in(0,K],t_n\in\Omega_k} \|k\mathfrak{L}_n\|.$$

Using a standard argument, it follows that for $|\lambda| \geq R$, since $|\lambda|^{-1}\|k\mathfrak{L}_n\| \leq 1/2$,

$$\|(\lambda I - k\mathfrak{L}_n)^{-1}\| = |\lambda|^{-1}\|(I - \lambda^{-1}k\mathfrak{L}_n)^{-1}\| \leq 2|\lambda|^{-1}. \tag{1.13}$$

Furthermore, let Υ be a $\Lambda_J^k(\gamma|m)$-proper set for the pair $\big(|\!|\!|\cdot|\!|\!|_n; \mathfrak{L}_n\big)$, whose singular points are $\zeta_1, \ldots, \zeta_J$, so that with some $\delta \geq 0$ and with Υ^k given by (1.6), for all $\lambda \notin \operatorname{Int}\Upsilon^k$ and $u \in \mathfrak{X}$,

$$|\!|\!|(\lambda I - k\mathfrak{L}_n)^{-m}u|\!|\!|_n \leq Ck^{-\gamma}\sum_{j=1}^{J}|\lambda - \zeta_j|^{\gamma-m}\|u\|. \tag{1.14}$$

The last inequality yields, under the additional restriction $|\lambda| \le R$,

$$k^{\vartheta} \| (\lambda I - k\mathfrak{L})^{-m} u \|_n \le C k^{-(\gamma-\vartheta)} \sum_{j=1}^{J} |\lambda - \zeta_j|^{\gamma-\vartheta-m} \|u\|. \tag{1.15}$$

On the other hand, assuming without loss of generality that $K\delta \le 1$, by (1.13) and (1.14) we have, for $k \in (0, K]$, $\lambda \notin \operatorname{Int} \Upsilon^k$, $|\lambda| \ge R$, and $u \in \mathfrak{X}$, since $-1 \notin \operatorname{Int} \Upsilon^k$ and $\|k\mathfrak{L}_n\| \le C$, that

$$\begin{aligned} k^{\vartheta} \| (\lambda I - k\mathfrak{L})^{-m} u \|_n &= k^{\vartheta} \| (-I - k\mathfrak{L}_n)^{-m} (-I - k\mathfrak{L}_n)^{m} (\lambda I - k\mathfrak{L})^{-m} u \|_n \\ &\le C k^{-(\gamma-\vartheta)} \| (I + k\mathfrak{L}_n)^{m} (\lambda I - k\mathfrak{L})^{-m} u \| \\ &\le C k^{-(\gamma-\vartheta)} |\lambda|^{-m} \|u\| \le C k^{-(\gamma-\vartheta)} \sum_{j=1}^{J} |\lambda - \zeta_j|^{\gamma-\vartheta-m} \|u\|. \end{aligned} \tag{1.16}$$

Now the $\Lambda_J^k(\gamma - \vartheta|m)$-concordance of the pair $\left(k^{\vartheta} \| \cdot \|_n; \mathfrak{L}_n\right)$ follows by combining (1.15) and (1.16).

Using the above argument we also find, for all $k \in (0, K]$ and $u \in \mathfrak{X}$, that

$$\begin{aligned} k^{\gamma} \| u \|_n &= k^{\gamma} \| (-I - k\mathfrak{L}_n)^{-m} (-I - k\mathfrak{L}_n)^{m} u \|_n \\ &\le C \| (I + k\mathfrak{L}_n)^{m} u \| \le C \|u\|, \end{aligned}$$

which shows the desired estimate for the first term on the left-hand side of (1.12). □

Lemma 1.3. *Suppose that $\mathfrak{L}_n \in \mathcal{B}(\mathfrak{X})$ and $\|k\mathfrak{L}_n\| \le C$ for all $k \in (0, K]$ and $t_n \in \Omega_k$. Let also seminorms $\| \cdot \|_n$ and $\| \cdot \|_{*;n}$, defined respectively on $\mathfrak{X}$ and $\mathfrak{X}^*$, satisfy the conditions, with some $\gamma, \gamma_* \in [0, 1]$, $\psi \in (0, \pi/2)$, and $\delta \ge 0$, for all $k \in (0, K]$, $t_n \in \Omega_k$, $\lambda \notin \operatorname{Int}\left(\Sigma_{\psi} \cup \mathcal{D}(\delta)\right)$, $u \in \mathfrak{X}$, and $\chi \in \mathfrak{X}^*$,*

$$\| (\lambda I - \mathfrak{L}_n)^{-1} u \|_n \le C |\lambda|^{\gamma-1} \|u\|$$

and

$$\| (\lambda I - \mathfrak{L}_n^*)^{-1} \chi \|_{*;n} \le C |\lambda|^{\gamma_*-1} \|\chi\|_*.$$

Then, for any fixed $\vartheta \in [0, \gamma]$ and $\vartheta_* \in [0, \gamma_*]$, for all $k \in (0, K]$, $t_n \in \Omega_k$, $\lambda \notin \operatorname{Int}\left(\Sigma_{\psi} \cup \mathcal{D}(\delta)\right)$, $u \in \mathfrak{X}$, and $\chi \in \mathfrak{X}^*$,

$$k^{\vartheta} \| (\lambda I - \mathfrak{L}_n)^{-1} u \|_n \le C |\lambda|^{\gamma-\vartheta-1} \|u\|$$

and

$$k^{\vartheta_*} \| (\lambda I - \mathfrak{L}_n^*)^{-1} \chi \|_{*;n} \le C |\lambda|^{\gamma_*-\vartheta_*-1} \|\chi\|_*.$$

The proof is similar to that of Lemma 1.2.

Lemma 1.4. *Let an operator $\mathfrak{L}_n$ satisfy the Λ_J^k-condition with some $J \geq 1$. Then there exist a $\psi \in (0, \pi/2)$ and a $\mu_1 > 0$ such that for all $k \in (0, K]$, $t_n \in \Omega_k$, $\mu \geq \mu_1$, and $\lambda \notin \operatorname{Int}(\Sigma_\psi \cup \mathcal{D}(\mu))$,*

$$|\lambda| \, \|(\lambda I - \mathfrak{L}_n)^{-1}\| + \|(\mathfrak{L}_n + \mu I)\,(\lambda I - \mathfrak{L}_n)^{-1}\| \leq C.$$

If $\mathfrak{L}_n$ satisfies the Λ_J-condition, then the last estimate holds with $\mu_1 = 0$.

Proof. It is obvious that the first term on the left-hand side of the inequality can be estimated as needed, with some $\psi \in (0, \pi/2)$ and $\mu_1 > 0$. Using this fact together with the identity

$$(\mathfrak{L}_n + \mu I)\,(\lambda I - \mathfrak{L}_n)^{-1} = -I + (\lambda + \mu)\,(\lambda I - \mathfrak{L}_n)^{-1}$$

leads to the desired estimate for the second term as well.

In the case where $\mathfrak{L}_n$ satisfies the Λ_J-condition, it is not hard to see that the same argument also works for $\mu_1 = 0$. □

Lemma 1.5. *Let an operator $\mathfrak{L}_n$ satisfy the Λ_J^k-condition with some integer $J \geq 1$. Then there exists a $\mu_1 > 0$ such that with any fixed $\mu_0 \in \mathbb{R}$, for all $\mu \geq \mu_1$, $k \in (0, K]$, $t_n \in \Omega_k$, $u \in \mathfrak{X}$, and $\chi \in \mathfrak{X}^*$,*

$$\|(\mathfrak{L}_n + \mu_0 I)u\| + \|u\| \leq C\|(\mathfrak{L}_n + \mu I)u\| \tag{1.17}$$

and

$$\|(\mathfrak{L}_n^* + \mu_0 I)\,\chi\|_* + \|\chi\|_* \leq C\|(\mathfrak{L}_n^* + \mu I)\,\chi\|_*.$$

Proof. Since $\mathfrak{L}_n$ satisfies the Λ_J^k-condition, by Lemma 1.4 it follows that with some $\mu_1 > 0$,

$$\|(\mathfrak{L}_n + \mu I)^{-1}\| \leq C|\mu|^{-1} \leq C \qquad \text{for all } \mu \geq \mu_1, \tag{1.18}$$

thus the first term on the left-hand side of (1.17) can be estimated as stated in view of the identity

$$\begin{aligned} \mathfrak{L}_n + \mu_0 I &= (\mathfrak{L}_n + \mu_0 I)\,(\mathfrak{L}_n + \mu I)^{-1}(\mathfrak{L}_n + \mu I) \\ &= \left(I + (\mu_0 - \mu)\,(\mathfrak{L}_n + \mu I)^{-1}\right)\,(\mathfrak{L}_n + \mu I). \end{aligned}$$

It is also easily seen that (1.18) leads to the desired estimate for the second term on the left-hand side of (1.17).

The latter stated estimate now follows in view of Remark 1.1. □

Lemma 1.6. *Let, for some $J \in \mathbb{N}$ and $\xi, \xi_* \in [0,1]$, seminorms $\|\!| \cdot |\!\|_n$ and $\|\!| \cdot |\!\|_{*;n}$ form respectively $\Lambda_J^k(\xi|1)$- and $\Lambda_J^k(*;\xi_*|1)$-concordant pairs with an operator $\mathfrak{L}_n \in \mathcal{B}(\mathfrak{X})$. Then there exist a $\psi \in (0,\pi/2)$ and a $\delta > 0$ such that for all $k \in (0,K]$, $t_n \in \Omega_k$, $\lambda \notin \operatorname{Int}(\Sigma_\psi \cup \mathcal{D}(\delta))$, $u \in \mathfrak{X}$, and $\chi \in \mathfrak{X}^*$,*

$$\|\!|(\lambda I - \mathfrak{L}_n)^{-1} u|\!\|_n \leq C|\lambda|^{\xi-1}\|u\|$$

and

$$\|\!|(\lambda I - \mathfrak{L}_n^*)^{-1} \chi|\!\|_{*;n} \leq C|\lambda|^{\xi_*-1}\|\chi\|_*.$$

The proof is immediate.

Lemma 1.7. *Let an operator $\mathfrak{L}_n$ satisfy the Λ_1^k-condition and let a set Υ be Λ_1^k-proper for $\mathfrak{L}_n$. Then there exists a $\delta > 0$ such that with any fixed $\mu_0 \in \mathbb{R}$, for all $k \in (0,K]$, $t_n \in \Omega_k$, and $\lambda \notin \operatorname{Int}\left(\Upsilon \cup \mathcal{D}(\delta k)\right)$,*

$$\|k(\mathfrak{L}_n + \mu_0 I)\,(\lambda I - k\mathfrak{L}_n)^{-1}\| \leq C.$$

Proof. By the given conditions, the operator $\mathfrak{L}_n$ satisfies (1.7) for $k \in (0,K]$ and $\lambda \notin \operatorname{Int}\Upsilon$, $|\lambda| \geq \delta k$, with $J = 1$ and some $\delta > 0$. With this fact in mind and using the identity

$$k(\mathfrak{L}_n + \mu_0 I)\,(\lambda I - k\mathfrak{L}_n)^{-1} = -I + (\lambda + \mu_0 k)\,(\lambda I - k\mathfrak{L}_n)^{-1},$$

as well as taking into account that $|\lambda| \geq \delta k$ we get the stated result. □

Lemma 1.8. *Suppose that seminorms $\|\!| \cdot |\!\|_n$ and $\|\!| \cdot |\!\|_{*;n}$ form respectively $\Lambda_J^k(\xi|m)$- and $\Lambda_J^k(*;\xi|m)$-concordant pairs with an operator $\mathfrak{L}_n \in \mathcal{B}(\mathfrak{X})$ for some $J \in \mathbb{N}$, $m \in \mathbb{N}$, and $\xi \in [0,m]$. Then, with any fixed $\xi_1 \in [\xi, m]$, the pairs $\left(\|\!| \cdot |\!\|_n; \mathfrak{L}_n\right)$ and $\left(\|\!| \cdot |\!\|_{*;n}; \mathfrak{L}_n\right)$ will be $\Lambda_J^k(\xi_1|m)$- and $\Lambda_J^k(*;\xi_1|m)$-concordant, respectively.*

Proof. It suffices to note that one can take the δ in Definition 1.4 to be strictly positive. □

Observe that if an operator $\mathfrak{L}_n$ fulfils the Λ_J^k-condition, it follows from Lemma 1.4 that with some $\psi \in (0,\pi/2)$ and with any fixed $\mu_0 \geq 0$ sufficiently large, for all $k \in (0,K]$, $t_n \in \Omega_k$, and $\lambda \notin \operatorname{Int}\Sigma_\psi$,

$$\|(\lambda I - (\mathfrak{L}_n + \mu_0 I))^{-1}\| \leq C(1 + |\lambda|)^{-1}.$$

Therefore, if $\mu_0 \geq 0$ is sufficiently large, the operator $\mathfrak{L}_n + \mu_0 I$ is $(\psi, 0)$-sectorial, it has a bounded inverse, and its fractional powers $(\mathfrak{L}_n + \mu_0 I)^\xi$, $-\infty < \xi < \infty$, are defined and bounded on $\mathfrak{X}$ (see Appendix A.3).

Lemma 1.9. *Let an operator $\mathfrak{L}_n$ satisfy the Λ^k_J-condition and let seminorms $\|\!| \cdot \|\!|_n$ and $\|\!| \cdot \|\!|_{*;n}$ form respectively $\Lambda^k_J(\gamma|1)$- and $\Lambda^k_1(*;\gamma_*|1)$-concordant pairs with $\mathfrak{L}_n$, for some $J \in \mathbb{N}$ and $\gamma, \gamma_* \in [0,1)$. Then there exists a $\mu_1 > 0$ such that with any fixed $\gamma_1 > \gamma$ and $\gamma_{*1} > \gamma_*$, for all $k \in (0,K]$, $t_n \in \Omega_k$, $u \in \mathfrak{X}$, and $\chi \in \mathfrak{X}^*$,*

$$\|\!|u\|\!|_n \le C\|(\mathfrak{L}_n + \mu_1 I)^{\gamma_1} u\| \tag{1.19}$$

and

$$\|\!|\chi\|\!|_{*;n} \le C\|(\mathfrak{L}_n^* + \mu I)^{\gamma_{*1}} \chi\|_*.$$

Proof. We will show only (1.19), noting that the second estimate can be proved in a similar way. Also, by the convexity inequality (A.18) it follows that $\|(\mathfrak{L}_n + \mu_1 I)^{-\xi}\| \le C$ for any fixed $\mu_1 > 0$ sufficiently large and for any fixed $\xi \ge 0$. Therefore without loss of generality it can be assumed that $\gamma_1 < 1$.

Let $\psi \in (0, \pi/2)$ and $\delta > 0$ be the same as in the assertion of Lemma 1.6 stated for $\xi = \gamma$, and let $\mu_1 > 0$ be a fixed number such that $|\lambda - \mu_1| \ge \delta$ for $\lambda \notin \operatorname{Int}\Sigma_\psi$. Let furthermore the contour Γ be given by $\Gamma = \partial\Sigma_\psi$. By the operator calculus formula (A.15), we get, for all $u \in \mathfrak{X}$,

$$(\mathfrak{L}_n + \mu_1 I)^{-\gamma_1} u = \frac{1}{2\pi i}\int_\Gamma \lambda^{-\gamma_1}((\lambda - \mu_1)I - \mathfrak{L}_n)^{-1} u \, d\lambda.$$

Since the pair $\big(\|\!| \cdot \|\!|_n; \mathfrak{L}_n\big)$ is $\Lambda^k_1(\gamma)$-concordant and $|\lambda - \mu_1| \ge c(|\lambda| + 1)$ for $\lambda \in \Gamma$, taking next the seminorm $\|\!| \cdot \|\!|_n$ of both sides and using Lemma 1.6 yields

$$\|\!|(\mathfrak{L}_n + \mu_1 I)^{-\gamma_1} u\|\!|_n \le C\|u\| \int_\Gamma |\lambda|^{-\gamma_1}(|\lambda| + 1)^{\gamma - 1} |d\lambda| \le C\|u\|,$$

which shows (1.19). □

Lemma 1.10. *Let $\mathfrak{L}_n \in \mathcal{B}(\mathfrak{X})$, let seminorms $\|\!| \cdot \|\!|_n$ and $\|\!| \cdot \|\!|_{*;n}$ form respectively $\Lambda^k_J(\gamma|1)$- and $\Lambda^k_J(*;\gamma_*|1)$-concordant pairs with $\mathfrak{L}_n$, and let $\mathfrak{M}_n \in \mathcal{B}(\mathfrak{X})$ be a "small" perturbation to $\mathfrak{L}_n$ in the sense that for all $k \in (0,K]$, $t_n \in \Omega_k$, $\chi \in \mathfrak{X}^*$, and $u \in \mathfrak{X}$,*

$$|\langle \chi, \mathfrak{M}_n u\rangle| \le Ck^\vartheta \|\!|\chi\|\!|_{*;n} \|\!|u\|\!|_n, \tag{1.20}$$

for some $J \in \mathbb{N}$, $\gamma, \gamma_* \in [0,1)$, and $\vartheta \in [0,1]$ such that $\gamma + \gamma_* < 1 + \vartheta$. Then there exist a $\psi \in (0, \pi/2)$ and a $\delta > 0$ such that for all $k \in (0,K]$, $t_n \in \Omega_k$, $\lambda \notin \operatorname{Int}\big(\Sigma_\psi \cup \mathcal{D}(\delta)\big)$, $u \in \mathfrak{X}$, and $\chi \in \mathfrak{X}^*$,

$$\|\!|(\lambda I - (\mathfrak{L}_n + \mathfrak{M}_n))^{-1} u\|\!|_n \le C|\lambda|^{\gamma - 1}\|u\|, \tag{1.21}$$

$$\|\!|(\lambda I - (\mathfrak{L}_n^* + \mathfrak{M}_n^*))^{-1}\chi|\!\|_{*;n} \le C|\lambda|^{\gamma_*-1}\|\chi\|_*, \tag{1.22}$$

and

$$\|(\lambda I - (\mathfrak{L}_n + \mathfrak{M}_n))^{-1}\| \le C|\lambda|^{-1}. \tag{1.23}$$

Proof. We will show (1.21) only, noting that the other estimates (1.22) and (1.23) can be proved in a similar manner.

Let γ' and γ'_* be fixed numbers such that $\gamma' \in (\gamma, 1]$, $\gamma'_* \in (\gamma_*, 1]$, and $2\gamma' + \gamma'_* - \gamma < 1 + \vartheta$. By (1.20) and Lemma 1.9, there exists a $\mu_1 > 0$ such that for all $t_n \in \Omega_k$, $u \in \mathfrak{X}$, and $\chi \in \mathfrak{X}^*$,

$$\|\!|u|\!\|_n \le C\|(\mathfrak{L}_n + \mu_1 I)^{\gamma'}u\| \tag{1.24}$$

and

$$\|\!|\chi|\!\|_{*;n} \le C\|(\mathfrak{L}_n^* + \mu_1 I)^{\gamma'_*}\chi\|_*. \tag{1.25}$$

Further, by the convexity inequality (A.18) and Lemma 1.4, we have, with some $\psi \in (0, \pi/2)$ and $\delta_0 > 0$, for all $\lambda \notin \operatorname{Int}\big(\Sigma_\psi \cup \mathcal{D}(\delta_0)\big)$,

$$\begin{aligned}\|(\mathfrak{L}_n + \mu_1 I)^{\gamma'}(\lambda I - \mathfrak{L}_n)^{-1}\| &\le C\|(\mathfrak{L}_n + \mu_1 I)(\lambda I - \mathfrak{L}_n)^{-1}\|^{\gamma'} \\ &\quad \times \|(\lambda I - \mathfrak{L}_n)^{-1}\|^{1-\gamma'} \le C|\lambda|^{\gamma'-1},\end{aligned} \tag{1.26}$$

and likewise, with $\vartheta' = \min\{\vartheta, \gamma' + \gamma'_*\}$, taking into account (A.14), Lemma 1.1, and the fact that $\gamma' + \gamma'_* < 1 + \vartheta$,

$$\begin{aligned}&k^\vartheta\|(\mathfrak{L}_n + \mu_1 I)^{\gamma'}(\lambda I - \mathfrak{L}_n)^{-1}(\mathfrak{L}_n + \mu_1 I)^{\gamma'_*}\| \\ &\qquad \le C\|k^\vartheta(\mathfrak{L}_n + \mu_1 I)^{\vartheta'}\|\,\|(\mathfrak{L}_n + \mu_1 I)^{\gamma'+\gamma'_*-\vartheta'}(\lambda I - \mathfrak{L}_n)^{-1}\| \\ &\qquad \le C|\lambda|^{\gamma'+\gamma'_*-\vartheta'-1}.\end{aligned} \tag{1.27}$$

Let now δ be a fixed number such that $\delta \ge \delta_0$. Multiplying then both sides of the identity

$$(\lambda I - (\mathfrak{L}_n + \mathfrak{M}_n))^{-1} = (\lambda I - \mathfrak{L}_n)^{-1} + (\lambda I - \mathfrak{L}_n)^{-1}\mathfrak{M}_n(\lambda I - (\mathfrak{L}_n + \mathfrak{M}_n))^{-1}, \tag{1.28}$$

from the left by $(\mathfrak{L}_n + \mu_1 I)^{\gamma'}$ and using furthermore a simple estimation by (1.20), (1.24), (1.25), (1.26), and (1.27), we get, for all $\lambda \notin \operatorname{Int}\big(\Sigma_\psi \cup \mathcal{D}(\delta)\big)$, with $w = \|(\mathfrak{L}_n + \mu_1 I)^{\gamma'}(\lambda I - (\mathfrak{L}_n + \mathfrak{M}_n))^{-1}\|$,

$$w \le C|\lambda|^{\gamma'-1} + C|\lambda|^{\gamma'+\gamma'_*-\vartheta'-1}w \le C|\lambda|^{\gamma'-1} + C\delta^{\gamma'+\gamma'_*-\vartheta'-1}w.$$

Note that in this estimate no constant C depends on δ. Taking therefore $\delta > 0$ sufficiently large we obtain

$$\|(\mathfrak{L}_n + \mu_1 I)^{\gamma'}(\lambda I - (\mathfrak{L}_n + \mathfrak{M}_n))^{-1}\| \le C|\lambda|^{\gamma'-1}. \tag{1.29}$$

Applying then the above argument as well as (1.20), (1.24), (1.25), (1.27), and (1.29), it follows that with the same ϑ' as above, for all $\lambda \notin \operatorname{Int}\big(\Sigma_\psi \cup \mathcal{D}(\delta)\big)$ and $u \in \mathfrak{X}$, since $2\gamma' + \gamma'_* - \gamma < 1 + \vartheta'$ and $|\lambda| \geq \delta^{-1}$,

$$\begin{aligned}&\sup_{\|u\|=1} \|\!|(\lambda I - \mathfrak{L}_n)^{-1}\mathfrak{M}_n(\lambda I - (\mathfrak{L}_n + \mathfrak{M}_n))^{-1}u|\!\|_n \\ &\leq Ck^{\vartheta}\|(\mathfrak{L}_n + \mu_1 I)^{\gamma'+\gamma'_*}(\lambda I - \mathfrak{L}_n)^{-1}\| \, \|(\mathfrak{L}_n + \mu_1 I)^{\gamma'}(\lambda I - (\mathfrak{L}_n + \mathfrak{M}_n))^{-1}\| \\ &\qquad \leq C|\lambda|^{2\gamma'+\gamma'_*-\vartheta'-2} = C|\lambda|^{\gamma-1}|\lambda|^{2\gamma'+\gamma'_*-\gamma-\vartheta'-1} \leq C|\lambda|^{\gamma-1}.\end{aligned}$$

With this result in mind, the claim now follows from (1.28) with both sides multiplied from the right by u, using a simple estimation and taking into account Lemma 1.6 and the fact that the pair $\big(|\!\|\cdot|\!\|_n; \mathfrak{L}_n\big)$ is $\Lambda^k_J(\gamma|1)$-concordant. □

For our future needs we examine here also the behaviour of fractional powers when the operator in question is under perturbation.

Lemma 1.11. *Suppose that seminorms $|\!\|\cdot|\!\|_n$ and $|\!\|\cdot|\!\|_{*;n}$ form respectively $\Lambda^k_J(\gamma|1)$- and $\Lambda^k_1(*;\gamma_*|1)$-concordant pairs with $\mathfrak{L}_n \in \mathcal{B}(\mathfrak{X})$, $\mathfrak{L}_n$ itself is under the Λ^k_J-condition, and $\mathfrak{M}_n \in \mathcal{B}(\mathfrak{X})$ satisfies condition (1.20), with some $J \in \mathbb{N}$, $\vartheta \in [0,1]$, and $\gamma, \gamma_* \in [0,1)$ such that $\gamma + \gamma_* < 1 + \vartheta$. Then, with any fixed $\mu_1 > 0$ sufficiently large and any fixed $\xi \in [0, 1 + \vartheta - \gamma_*)$, $\xi_* \in [0, 1 + \vartheta - \gamma)$, for all $k \in (0, K]$, $t_n \in \Omega_k$, $u \in \mathfrak{X}$, and $\chi \in \mathfrak{X}^*$,*

$$\|(\mathfrak{L}_n + \mathfrak{M}_n + \mu_1 I)^{\xi}u\| \leq C\,\|(\mathfrak{L}_n + \mu_1 I)^{\xi}u\| \tag{1.30}$$

and

$$\|(\mathfrak{L}^*_n + \mathfrak{M}^*_n + \mu_1 I)^{\xi_*}\chi\|_* \leq C\,\|(\mathfrak{L}^*_n + \mu_1 I)^{\xi_*}\chi\|_*.$$

Proof. We demonstrate how to prove (1.30) since the second estimate can be derived in a similar manner. Also, note that the case $\xi = 0$ is trivial, so it can be assumed that $\xi > 0$.

Let ψ and δ be chosen as in Lemma 1.10[7] and let γ' be a fixed number such that $\gamma' \in (\gamma, 1)$ and $\gamma' + \gamma_* < 1 + \vartheta$. Note that in the below proof we can use the estimates (1.22) and (1.23), by Lemma 1.10. Also, in the case $\xi \leq \gamma$, by Lemma 1.9 and analogy with (1.27) we have, since $\gamma' - \xi < 1$, with any fixed $\mu_1 > 0$ sufficiently large, for all $\lambda \notin \operatorname{Int}\big(\Sigma_\psi \cup \mathcal{D}(\delta)\big)$ and $u \in \mathfrak{X}$,

$$\begin{aligned}|\!\|(\lambda I - \mathfrak{L}_n)^{-1}u|\!\|_n &\leq C\|(\mathfrak{L}_n + \mu_1 I)^{\gamma'}(\lambda I - \mathfrak{L}_n)^{-1}u\| \\ &\leq C\|(\mathfrak{L}_n + \mu_1 I)^{\gamma'-\xi}(\lambda I - \mathfrak{L}_n)^{-1}\| \, \|(\mathfrak{L}_n + \mu_1 I)^{\xi}u\| \\ &\leq C|\lambda|^{\gamma'-\xi-1}\|(\mathfrak{L}_n + \mu_1 I)^{\xi}u\|. \end{aligned} \tag{1.31}$$

[7] We can assume without loss of generality that ψ and δ are are chosen as in Lemmas 1.3 and 1.4 which will be applied together with Lemma 1.10 in the present proof.

At the same time, if $\xi \in (\gamma, 1+\vartheta-\gamma_*)$, since $\mathfrak{L}_n$ satisfies the Λ_J^k-condition, using again Lemma 1.9 yields, for all $\lambda \notin \operatorname{Int}\left(\Sigma_\psi \cup \mathcal{D}(\delta)\right)$ and $u \in \mathfrak{X}$,

$$\begin{aligned} \|(\lambda I - \mathfrak{L}_n)^{-1}u\|_n &\leq C\|(\mathfrak{L}_n + \mu_1 I)^\xi(\lambda I - \mathfrak{L}_n)^{-1}u\| \\ &\leq C|\lambda|^{-1}\|(\mathfrak{L}_n + \mu_1 I)^\xi u\|. \end{aligned} \tag{1.32}$$

Next note that by Lemma 1.1 we have

$$\|k\mathfrak{L}_n\| \leq C, \tag{1.33}$$

which in turn gives that

$$k^\gamma \sup_{\|u\|=1} \|u\|_n + k^{\gamma_*} \sup_{\|\chi\|_*=1} \|\chi\|_{*;n} \leq C,$$

by Lemma 1.2. Using this result and (1.20) we obtain, since $\gamma+\gamma_* < 1+\vartheta$, that

$$\|k\mathfrak{M}_n\| \leq Ck^{1+\vartheta-\gamma-\gamma_*} \leq C,$$

which, combined with (1.33), obviously implies

$$\|k(\mathfrak{L}_n + \mathfrak{M}_n)\| \leq C.$$

With this estimate in mind and using Lemma 1.3 and (1.22), a simple calculation gives, with $\vartheta' = \min\{\gamma_*, \vartheta\}$, for all $\lambda \notin \operatorname{Int}\left(\Sigma_\psi \cup \mathcal{D}(\delta)\right)$ and $\chi \in \mathfrak{X}^*$,

$$k^{\vartheta'}\|(\lambda I - (\mathfrak{L}_n^* + \mathfrak{M}_n^*))^{-1}\chi\|_{*;n} \leq C|\lambda|^{\gamma_*-\vartheta'-1}\|\chi\|_*. \tag{1.34}$$

From now on we assume that $\mu_1 > 0$ is chosen sufficiently large so that $|\lambda - \mu_1| \geq \delta$ for $\lambda \notin \operatorname{Int}\Sigma_\psi$. We then apply the identity

$$\begin{aligned} &(\lambda I - (\mathfrak{L}_n + \mathfrak{M}_n + \mu_1 I))^{-1} - (\lambda I - (\mathfrak{L}_n + \mu_1 I))^{-1} \\ &\qquad = (\lambda I - (\mathfrak{L}_n + \mathfrak{M}_n + \mu_1 I))^{-1}\mathfrak{M}_n(\lambda I - (\mathfrak{L}_n + \mu_1 I))^{-1}, \end{aligned}$$

and the estimates (1.20), (1.31), (1.32), and (1.34)[8] to obtain, for all $\lambda \notin \operatorname{Int}\Sigma_\psi$ and $u \in \mathfrak{X}$, since $k \leq K$ and $|\lambda - \mu_1| \geq c(|\lambda|+1)$,

$$\begin{aligned} &|\lambda|^\xi \|\left((\lambda I - (\mathfrak{L}_n + \mathfrak{M}_n + \mu_1 I))^{-1} - (\lambda I - (\mathfrak{L}_n + \mu_1 I))^{-1}\right) u\| \\ &\leq \begin{cases} C(|\lambda|+1)^{\gamma_*+\gamma'-\vartheta'-2}\|(\mathfrak{L}_n + \mu_1 I)^\xi u\| & \text{if } \xi \in [0,\gamma], \\ Ck^{\vartheta-\vartheta'}|\lambda|^\xi(|\lambda|+1)^{\gamma_*-\vartheta'-2}\|(\mathfrak{L}_n + \mu_1 I)^\xi u\| & \text{if } \xi \in (\gamma, 1+\vartheta-\gamma_*). \end{cases} \end{aligned} \tag{1.35}$$

[8] In fact, we use the last three estimates with $\lambda - \mu_1$ in place of λ, for $\lambda \notin \operatorname{Int}\Sigma_\psi$.

Next we denote $R = 2\sup_{k\in(0,K],t_n\in\Omega_k}\big(\|k(\mathfrak{L}_n+\mu_1 I)\|+\|k\mathfrak{M}_n\|\big)$. By Lemma 1.4, (1.23), and a standard argument, it follows that for all $\lambda \notin \operatorname{Int}\big(\Sigma_\psi \cap \mathcal{D}(R/k)\big)$,

$$\|(\lambda I-(\mathfrak{L}_n+\mathfrak{M}_n+\mu_1 I))^{-1}\|+\|(\lambda I-(\mathfrak{L}_n+\mu_1 I))^{-1}\| \le C(|\lambda|+1)^{-1}. \quad (1.36)$$

Let now $\Gamma_{(k)}$ be the contour coinciding with $\partial\big(\Sigma_\psi\cap\mathcal{D}(R/k)\big)$. If $\xi<1$, using the operator calculus formula (A.13) with $m=1$, recalling that $\mathfrak{L}_n,\mathfrak{M}_n \in \mathcal{B}(\mathfrak{X})$ and taking (1.36) into account, we get, after a suitable deformation of the path of integration in (A.13), for all $u\in\mathfrak{X}$,

$$\begin{aligned}(\mathfrak{L}_n &+ \mathfrak{M}_n + \mu_1 I)^\xi u - (\mathfrak{L}_n+\mu_1 I)^\xi u \\ &= \frac{1}{2\pi i}\int_{\Gamma_{(k)}} \lambda^{\xi-1}(\mathfrak{L}_n+\mathfrak{M}_n+\mu_1 I)\,(\lambda I-(\mathfrak{L}_n+\mathfrak{M}_n+\mu_1 I))^{-1}u\,d\lambda \\ &\quad - \frac{1}{2\pi i}\int_{\Gamma_{(k)}} \lambda^{\xi-1}(\mathfrak{L}_n+\mu_1 I)\,(\lambda I-(\mathfrak{L}_n+\mu_1 I))^{-1}u\,d\lambda. \end{aligned} \quad (1.37)$$

Furthermore, with the aid of the identity

$$\begin{aligned}(\mathfrak{L}_n+\mathfrak{M}_n+\mu_1 I)\,&(\lambda I-(\mathfrak{L}_n+\mathfrak{M}_n+\mu_1 I))^{-1}-(\mathfrak{L}_n+\mu_1 I)\,(\lambda I-(\mathfrak{L}_n+\mu_1 I))^{-1} \\ &= \lambda\big((\lambda I-(\mathfrak{L}_n+\mathfrak{M}_n+\mu_1 I))^{-1}-(\lambda I-(\mathfrak{L}_n+\mu_1 I))^{-1}\big),\end{aligned}$$

we obtain, instead of (1.37),

$$\begin{aligned}(\mathfrak{L}_n &+ \mathfrak{M}_n + \mu_1 I)^\xi u - (\mathfrak{L}_n+\mu_1 I)^\xi u \\ &= \frac{1}{2\pi i}\int_{\Gamma_{(k)}} \lambda^{\xi}\big((\lambda I-(\mathfrak{L}_n+\mathfrak{M}_n+\mu_1 I))^{-1}-(\lambda I-(\mathfrak{L}_n+\mu_1 I))^{-1}\big)u\,d\lambda.\end{aligned} \quad (1.38)$$

A slightly more delicate calculation, based on applying (A.13) with $m=2$ and using Cauchy's Theorem, shows that (1.38) remains valid for $\xi\in(1,1+\vartheta-\gamma_*)$ as well. Also, in the case $\xi=1$, (1.38) follows directly from (A.2) with $f(\lambda)=\lambda$ and $\Xi=\Gamma_{(k)}$. Since $\gamma'+\gamma_*<1+\vartheta'$, $\xi+\gamma_*<1+\vartheta$, and $k^{\vartheta-\vartheta'}|\lambda|^\xi \le C|\lambda|^{\max\{\xi-\vartheta+\vartheta',0\}}$ for $\lambda\in\Gamma_{(k)}$, the norm of the integral on the right-hand side of (1.38) can be estimated with the aid of (1.35) by $C\|(\mathfrak{L}_n+\mu_1 I)^\xi u\|$ which shows (1.30). □

Remark 1.2. *The reader will have no difficulty in showing that in the particular case where* $|\!|\!|\cdot|\!|\!|_n = \|(\mathfrak{L}_n+\mu_1 I)^\gamma\cdot\|$ *and* $|\!|\!|\cdot|\!|\!|_{*;n} = \|(\mathfrak{L}_n^*+\mu_1 I)^{\gamma_*}\cdot\|$, *with some* $\mu_1\ge 0$ *sufficiently large, the assertions of Lemmas 1.10 and 1.11 are still valid if one requires that* γ,γ_* *belong to* $[0,1]$ *instead of* $[0,1)$.

In the remainder of the section we want to show that under reasonable conditions on the operator in question there do exist seminorms with the properties stated in Definitions 1.4 and 1.5.

Lemma 1.12. *Let $\mathfrak{L}_n$ satisfy the Λ_J^k-condition (Λ_J-condition) with some integer $J \geq 1$ and let ζ_j, $j = 1, \ldots, J$, be the singular points of $\mathfrak{L}_n$. Then, with any fixed $\mu_0 \geq 0$ sufficiently large (with $\mu_0 = 0$ in the case of the Λ_J-condition), with any fixed $m \in \mathbb{N}$, $\eta \in \mathbb{N} \cup \{0\}$, and $\xi \in [0, \min\{m, \eta\}]$, with*

$$\mathfrak{V}_n = \prod_{j=2}^{J} (I - \zeta_j^{-1} k\mathfrak{L}_n)^\eta, \tag{1.39}$$

and with the seminorms $|\!|\!| \cdot |\!|\!|_n$ and $|\!|\!| \cdot |\!|\!|_{*;n}$ given by

$$|\!|\!| u |\!|\!|_n = \|(\mathfrak{L}_n + \mu_0 I)^\xi \mathfrak{V}_n u\| \qquad \text{for } u \in \mathfrak{X}$$

and

$$|\!|\!| \chi |\!|\!|_{*;n} = \|(\mathfrak{L}_n^* + \mu_0 I)^\xi \mathfrak{V}_n^* \chi\|_* \qquad \text{for } \chi \in \mathfrak{X}^*,$$

the pairs $\big(|\!|\!| \cdot |\!|\!|_n; \mathfrak{L}_n\big)$ and $\big(|\!|\!| \cdot |\!|\!|_{*;n}; \mathfrak{L}_n\big)$ will be $\Lambda_J^k(\xi|m)$- and $\Lambda_J^k(*;\xi|m)$-concordant ($\Lambda_J(\xi|m)$- and $\Lambda_J(*;\xi|m)$-concordant), respectively.

Proof. For the sake of brevity we denote $\mathfrak{V}_{j;n} = I - \zeta_j^{-1} k\mathfrak{L}_n$, $j = 2, \ldots, J$, so that we have $\mathfrak{V}_n = \prod_{j=2}^{J} \mathfrak{V}_{j;n}^\eta$.

Our argument is based on using certain smoothing properties of the operators $\mathfrak{V}_{j;n}$, $j = 2, \ldots, J$, and $\mathfrak{V}_n$. Note first that the operators $\mathfrak{V}_{j;n}$ are uniformly bounded

$$\|\mathfrak{V}_{j;n}\| \leq C, \qquad j = 2, \ldots, J, \tag{1.40}$$

because the operator $k\mathfrak{L}_n$ is uniformly bounded by Lemma 1.1. Let furthermore Υ be a Λ_J^k-proper (Λ_J-proper) set for $\mathfrak{L}_n$. With the aid of the identity

$$(\lambda I - k\mathfrak{L}_n)^{-1} \mathfrak{V}_{j;n} = \frac{\zeta_j - \lambda}{\zeta_j} (\lambda I - k\mathfrak{L}_n)^{-1} + \zeta_j^{-1} I,$$

and of the given condition on $\mathfrak{L}_n$, it follows that with some $\delta_1 > 0$ sufficiently small and some $\delta > 0$ ($\delta = 0$ in the case of the Λ_J-condition), for all $j = 2, \ldots, J$ and $\lambda \notin \operatorname{Int} \Upsilon$, $\delta_1 \geq |\lambda - \zeta_j| \geq \delta k$,

$$\|(\lambda I - k\mathfrak{L}_n)^{-1} \mathfrak{V}_{j;n}\| \leq C.$$

Using this result, (1.40) and the fact that $\mathfrak{L}_n$ satisfies the Λ_J^k-condition (Λ_J-condition) yields, for all $\lambda \notin \operatorname{Int}\left(\cup_{j=2}^J \mathcal{D}(\zeta_j;\delta k) \cup \Upsilon\right)$, $\lambda \in \cup_{j=2}^J \mathcal{D}(\zeta_j;\delta_1)$,

$$\|\mathfrak{V}_n(\lambda I - k\mathfrak{L}_n)^{-m}\| \le C \sum_{j=2}^{J} |\lambda - \zeta_j|^{\min\{0,\eta-m\}}. \tag{1.41}$$

Also, by (1.40) it follows that for all $\lambda \notin \operatorname{Int}\left(\cup_{j=1}^J \mathcal{D}(\zeta_j;\delta_1) \cup \Upsilon\right)$,

$$\|\mathfrak{V}_n(\lambda I - k\mathfrak{L}_n)^{-m}\| \le C \sum_{j=2}^{J} |\lambda - \zeta_j|^{-m}. \tag{1.42}$$

Recalling that $\|k\mathfrak{L}_n\| \le C$, by the convexity inequality (A.18) we have

$$\|(\mathfrak{L}_n + \mu_0 I)^{\xi}\| \le Ck^{-\xi}, \tag{1.43}$$

which, combined with (1.41) and (1.42), yields, since $0 \le \xi \le \eta$, for $\lambda \notin \operatorname{Int}\left(\cup_{j=2}^J \mathcal{D}(\zeta_j;\delta k) \cup \Upsilon\right)$, $\delta_1 \le |\lambda|$,

$$\|(\mathfrak{L}_n + \mu_0 I)^{\xi}\mathfrak{V}_n(\lambda I - k\mathfrak{L}_n)^{-m}\| \le Ck^{-\xi} \sum_{j=2}^{J} |\lambda|^{\xi-m}. \tag{1.44}$$

Furthermore, without loss of generality it can be assumed that ψ in the statement of Lemma 1.4 coincides with the parameter β of the set Υ (see Definition 1.2). Therefore, by (A.18), (1.40), and Lemma 1.4 with $\mu = \mu_0$ we get, for all $\lambda \notin \operatorname{Int}\Upsilon$, $\delta_1 \ge |\lambda| \ge \delta k$,

$$\|(\mathfrak{L}_n + \mu_0 I)^{\xi}(\lambda I - k\mathfrak{L}_n)^{-m}\mathfrak{V}_n\| \le Ck^{-\xi}|\lambda|^{\xi-m}.$$

Together with (1.44), this inequality shows that the seminorm $\|\!|\cdot|\!\|_n$ forms a $\Lambda_J^k(\xi|m)$-concordant ($\Lambda_J(\xi|m)$-concordant) pair with $\mathfrak{L}_n$.

Similarly, with the aid of Remark 1.1 it can be shown that the seminorm $\|\!|\cdot|\!\|_{*;n}$ forms a $\Lambda_J^k(*;\xi|m)$-concordant ($\Lambda_J(*;\xi|m)$-concordant) pair with $\mathfrak{L}_n$. □

Lemma 1.13. *Let $\mathfrak{L}_n$ satisfy the Λ_J^k-condition (Λ_J-condition) for some integer $J \ge 1$. Let ζ_j, $j = 1,\dots,J$, be the singular points of $\mathfrak{L}_n$ and let $\mathfrak{V}_n$ be given by (1.39) for some $\eta \in \mathbb{N} \cup \{0\}$. Then, with any fixed $\mu_0 \ge 0$ sufficiently large (with $\mu_0 = 0$ in the case that $\mathfrak{L}_n$ satisfies the Λ_J-condition), with any fixed $m \in \mathbb{N}$ and $\xi \in [0, \min\{m,\eta\}]$, and with the seminorms $\|\!|\cdot|\!\|_n$ and $\|\!|\cdot|\!\|_{*;n}$ given by*

$$\|\!|u|\!\|_n = \|(\mathfrak{L}_n + \mu_0 I)^{\xi}u\| \qquad \textit{for } u \in \mathfrak{X}$$

and

$$|||\chi|||_{*;n} = \|(\mathfrak{L}_n^* + \mu_0 I)^{\xi}\chi\|_* \qquad \textit{for } \chi \in \mathfrak{X}^*,$$

the triplets $\big(|||\cdot|||_n; \mathfrak{L}_n; \mathfrak{V}_n\big)$ *and* $\big(|||\cdot|||_{*;n}; \mathfrak{L}_n; \mathfrak{V}_n\big)$ *are respectively* $\Lambda_J^k(\xi|m)$*- and* $\Lambda_J^k(*;\xi|m)$*-concordant (*$\Lambda_J(\xi|m)$*- and* $\Lambda_J(*;\xi|m)$*-concordant).*

The proof actually reduces to that of Lemma 1.12 because the operators $(\lambda I - k\mathfrak{L}_n)^{-1}$ and $\mathfrak{V}_n$ are permutable.

1.2 Comments and bibliographical remarks

Discrete semigroups are used in Kato [91] and Pazy [131] as a discrete analogue of semigroups of linear bounded operators in operator theory. Evolution equations in discrete time actually appear in [10, 11, 12], but the corresponding terminology is introduced later, in [34]. The Λ_J^k-condition (in particular, the Λ_J-condition) is actually introduced already in [10, 11, 12]. Note that the Λ_J^k-condition may be regarded as a discrete analogue of the sectorialness property in the theory of holomorphic semigroups. The $\Lambda_J^k(\ldots)$-concordant pairs are introduced in this book and are further developed herein for practical purposes.

Chapter 2

Main Results on Stability

This chapter deals with the stability of problem (1.1). In fact, the solution Y_n of this problem is estimated in seminorms forming concordant pairs with the operator $\mathfrak{A}_n$ in (1.1), in the sense of Definitions 1.4 or 1.5. We also show dual estimates which are often used in numerical analysis. Throughout we employ the notation $\mathfrak{U}_n$ and $\mathfrak{U}_{n,j}$ introduced by (1.3) and (1.5) without explicit reference to the corresponding formulas.

2.1 Stability of discrete semigroups

Our first step is examining the stability of discrete semigroups that arise from problem (1.1) considered with frozen operator. In this section we therefore concentrate on the problem of estimating the powers of linear bounded operators. As pointed out above, our analysis will be carried out by using special seminorms. The results of the present section allow us to get stability estimates for problem (1.1) in the autonomous case, that is, where the operator $\mathfrak{A}_n$ does not depend on t_n (see Section 2.4). At the same time it turns out that these estimates are also important for the analysis of the non-autonomous case.

Theorem 2.1. *Let a seminorm $\| \! | \cdot \| \! |_n$ form a $\Lambda_J^k(\xi)$-concordant pair with $\mathfrak{A}_n$ for some $J \in \mathbb{N}$ and $\xi \geq 0$. Then, if $T < \infty$, we have*

$$\| \! | \mathfrak{U}_n^l u \| \! |_n \leq C t_{l+1}^{-\xi} \|u\|,$$

for all $k \in (0, K]$, $t_l, t_n \in \Omega_k$, and $u \in \mathfrak{X}$. Furthermore, if the pair $\left(\| \! | \cdot \| \! |_n ; \mathfrak{A}_n\right)$ is $\Lambda_J(\xi)$-concordant, the last estimate holds also for $T = \infty$, that is, for all $l, n = 0, 1, \ldots$.

Proof. We present the proof only for the case where the pair $\big(|\!|\!| \cdot |\!|\!|_n; \mathfrak{A}_n\big)$ is $\Lambda_J(\xi)$-concordant.[1] Therefore, with this assumed condition, there exists an integer $m \geq \max\{1, \xi\}$ such that the pair $\big(|\!|\!| \cdot |\!|\!|_n; \mathfrak{A}_n\big)$ is $\Lambda_J(\xi|m)$-concordant. Let a set Υ be Λ_J-proper for this pair, let a and β be the parameters of Υ, and let ζ_j, $j = 1, \ldots, J$, be its singular points (see Definition 1.2). Let further d be expressed via a and β as specified in Definition 1.2 and let Γ^l be the contour coinciding with the boundary of the set

$$\cup_{j=1}^{J} D(\zeta_j; a(l+1)^{-1}) \cup \Upsilon, \qquad l = 0, 1, \ldots.$$

By the operator calculus formula (A.2) we have, for all $l = 0, 1, \ldots$ and $u \in \mathfrak{X}$, that

$$\mathfrak{U}_n^l u = (2\pi i)^{-1} \int_{\Gamma^l} (1-\lambda)^l (\lambda I - k\mathfrak{A}_n)^{-1} u \, d\lambda.$$

Thus integrating by parts, we find[2]

$$\mathfrak{U}_n^l u = \frac{(m-1)!}{(2\pi i) \prod_{q=1}^{m-1}(l+q)} \int_{\Gamma^l} (1-\lambda)^{l+m-1} (\lambda I - k\mathfrak{A}_n)^{-m} u \, d\lambda. \tag{2.1}$$

Since the functional $|\!|\!| \cdot |\!|\!|_n$ is a seminorm and the pair $\big(|\!|\!| \cdot |\!|\!|_n; \mathfrak{A}_n\big)$ is $\Lambda_J(\xi|m)$-concordant, it follows from (2.1) that for $l = 0, 1, \ldots$,

$$\begin{aligned} |\!|\!| \mathfrak{U}_n^l u |\!|\!|_n &\leq \frac{(m-1)!}{2\pi \prod_{q=1}^{m-1}(l+q)} \int_{\Gamma^l} |1-\lambda|^{l+m-1} |\!|\!| (\lambda I - k\mathfrak{A}_n)^{-m} u |\!|\!|_n |d\lambda| \\ &\leq Ck^{-\xi}(l+1)^{1-m} \|u\| \int_{\Gamma^l} |1-\lambda|^{l+m-1} G(\lambda)\, |d\lambda|, \end{aligned} \tag{2.2}$$

where we used the notation

$$G(\lambda) = \sum_{j=1}^{J} |\lambda - \zeta_j|^{\xi - m}.$$

Clearly, Γ^l can be decomposed as follows

$$\Gamma^l = \cup_{j=1}^{J} \left(\Gamma_{1,j}^l \cup \Gamma_{2,j}^l\right) \cup \Gamma_3, \qquad l = 0, 1, \ldots, \tag{2.3}$$

---

[1] It will be seen from what follows that our argument can easily be extended to the general case corresponding to the first part of the statement of Theorem 2.1 (see also Remark 2.1).

[2] It is assumed that $\prod_{q=1}^{m-1}(l+q) = 1$ if $m = 1$.

where

$$\begin{aligned}\Gamma^l_{1,j} &= \{\lambda \in \mathbb{C} : |\arg(\zeta_j - \lambda) - \arg(\zeta_j - 1)| \le \beta,\, |\lambda - \zeta_j| = d(l+1)^{-1}\},\\ \Gamma^l_{2,j} &= \{\lambda \in \mathbb{C} : \arg(\lambda - \zeta_j) = \arg(1 - \zeta_j) \pm \beta,\, d/(l+1) \le |\lambda - \zeta_j| \le d\},\end{aligned}$$

and Γ_3 is the part of Γ^l whose points lie on the circle $\partial\mathcal{D}(1;a)$. Further, for all $j = 1, \ldots, J$, $l = 0, 1, \ldots$, and $\lambda \in \Gamma^l_{1,j}$, we have $|1-\lambda|^{l+m-1} \le C$ and $G(\lambda) \le C(l+1)^{m-\xi}$, consequently,

$$\int_{\Gamma^l_{1,j}} |1-\lambda|^{l+m-1} G(\lambda)\,|d\lambda| \le C(l+1)^{m-\xi} \int_{\Gamma^l_{1,j}} |d\lambda| \le C(l+1)^{m-1-\xi}. \quad (2.4)$$

Next, in view of the restriction $\beta < \pi/2$, we find that for all $j = 1, \ldots, J$, $l = 0, 1, \ldots$, and $\lambda \in \Gamma^l_{2,j}$,

$$|1-\lambda|^{l+m-1} \le C \exp(-c(l+1)|\lambda - \zeta_j|).$$

Thus it follows that

$$\begin{aligned}\int_{\Gamma^l_{2,j}} |1-\lambda|^{l+m-1} G(\lambda)\,|d\lambda| &\le C \int_{d/(l+1)}^{\infty} \exp(-(l+1)cx) x^{\xi-m}\,dx\\ &= C(l+1)^{m-1-\xi}. \quad (2.5)\end{aligned}$$

Finally, it is easily seen that $G(\lambda) \le C$ if $\lambda \in \Gamma_3$, which gives, since $a < 1$, for all $l = 0, 1, \ldots$,

$$\int_{\Gamma_3} |1-\lambda|^{l+m-1} G(\lambda)\,|d\lambda| \le C a^{l+m-1} \le C(l+1)^{m-1-\xi}. \quad (2.6)$$

So the claim follows by combining (2.2), (2.3), (2.4), (2.5), and (2.6). □

Remark 2.1. *The general case in Theorem 2.1 where the pair* $(\|\!|\cdot|\!\|_n; \mathfrak{A}_n)$ *is* $\Lambda^k_J(\xi)$*-concordant can be settled similarly by using a slightly different path of integration. The path of integration should deviate from each singular point of the set* Υ*, appearing in the proof of the theorem, by* $\mathcal{O}((l+1)^{-1} + k)$*, while the path follows the boundary of* Υ *as long as the deviation exceeds this value. Note that the assumption* $T < \infty$ *is essential in the general case since we then have to use the fact that* $(1 + \mathcal{O}(k))^l \le C$ *for* $t_l \in \Omega_k$ *in order to bound suitably the quantity* $|1-\lambda|^{l+m-1}$ *for* $\lambda \in \Gamma^l_{1,j}$.

We state as well a dual stability estimate.

Theorem 2.2. *Let a seminorm* $\|\cdot\|_{*;n}$ *form a* $\Lambda_J^k(*;\xi)$*-concordant pair with* $\mathfrak{A}_n$ *for some* $J \in \mathbb{N}$ *and* $\xi \geq 0$. *Then, if* $T < \infty$, *for all* $k \in (0, K]$, $t_l, t_n \in \Omega_k$, *and* $\chi \in \mathfrak{X}^*$,

$$\|(\mathfrak{U}_n^*)^l \chi\|_{*;n} \leq C t_{l+1}^{-\xi} \|\chi\|_*.$$

Furthermore, if the pair $\left(\|\cdot\|_{*;n}; \mathfrak{A}_n\right)$ *is* $\Lambda_J(*;\xi)$*-concordant, the last estimate holds as well for* $T = \infty$, *that is, for all* $l, n = 0, 1, \ldots$.

The proof is similar to that of Theorem 2.1.

Considerations in subsequent parts of this book will use the application of the above Theorems 2.1 and 2.2. We make this statement now in view of its importance.

Theorem 2.3. *Let the operator* $\mathfrak{A}_n$ *satisfy the* Λ_J^k*-condition and let* $\mathfrak{B}_n \in \mathcal{B}(\mathfrak{X})$ *be given by (1.39), with some* $J \in \mathbb{N}$, $\eta \in \mathbb{N} \cup \{0\}$ *and with some* $\mu_0 \geq 0$ *sufficiently large. Then, if* $T < \infty$, *we have, with any fixed* $\xi \in [0, \eta]$, *for all* $k \in (0, K]$ *and* $t_l, t_n \in \Omega_k$, *that*

$$\|(\mathfrak{A}_n + \mu_0 I)^\xi \mathfrak{B}_n \mathfrak{U}_n^l\| \leq C t_{l+1}^{-\xi} \tag{2.7}$$

and

$$\|(\mathfrak{A}_n^* + \mu_0 I)^\xi \mathfrak{B}_n^* (\mathfrak{U}_n^*)^l\|_* \leq C t_{l+1}^{-\xi}. \tag{2.8}$$

If in addition the operator $\mathfrak{A}_n$ *satisfies the* Λ_J*-condition, both estimates are in force for* $T = \infty$, *that is, for* $l, n = 0, 1, \ldots$, *provided that* $\mu_0 = 0$. *In particular, if* $\mathfrak{A}_n$ *satisfies the* Λ_1^k*-condition (*Λ_1*-condition), the above holds with* $\mathfrak{B}_n = I$.

Theorem 2.3 is a direct consequence of Lemma 1.12 combined with Theorems 2.1 and 2.2.

In what follows we also use the following generalizations of the above results.

Theorem 2.4. *Let there be given numbers* $J \in \mathbb{N}$ *and* $\xi \geq 0$, *an operator* $\mathfrak{B}_n \in \mathcal{B}(\mathfrak{X})$, *and a seminorm* $\|\cdot\|_n$ *such that the triplet* $\left(\|\cdot\|_n; \mathfrak{A}_n; \mathfrak{B}_n\right)$ *is* $\Lambda_J^k(\xi)$*-concordant. Then, if* $T < \infty$, *we have, for all* $k \in (0, K]$, $t_l, t_n \in \Omega_k$, *and* $u \in \mathfrak{X}$, *that*

$$\|\mathfrak{U}_n^l \mathfrak{B}_n u\|_n \leq C t_{l+1}^{-\xi} \|u\|.$$

Furthermore, if the triplet $\left(\|\cdot\|_n; \mathfrak{A}_n; \mathfrak{B}_n\right)$ *is* $\Lambda_J(\xi)$*-concordant, the last estimate is in force for* $T = \infty$, *that is, for all* $l, n = 0, 1, \ldots$.

Theorem 2.5. *Let there be given numbers $J \in \mathbb{N}$ and $\xi \geq 0$, an operator $\mathfrak{V}_n \in \mathcal{B}(\mathfrak{X})$, and a seminorm $\|\!| \cdot |\!\|_{*;n}$ such that the triplet $\big(\|\!| \cdot |\!\|_{*;n}; \mathfrak{A}_n; \mathfrak{V}_n\big)$ is $\Lambda_J^k(*;\xi)$-concordant. Then, if $T < \infty$, we have, for all $k \in (0, K]$, $t_l, t_n \in \Omega_k$, and $\chi \in \mathfrak{X}^*$, that*

$$\|\!|(\mathfrak{U}_n^*)^l \mathfrak{V}_n^* \chi|\!\|_{*;n} \leq C t_{l+1}^{-\xi} \|\chi\|_*.$$

Furthermore, if the triplet $\big(\|\!| \cdot |\!\|_{*;n}; \mathfrak{A}_n; \mathfrak{V}_n\big)$ is $\Lambda_J(*;\xi)$-concordant, the last estimate is in force for $T = \infty$, that is, for all $l, n = 0, 1, \ldots$.

The proofs of Theorems 2.4 and 2.5 are similar to that of Theorem 2.1.

Remark 2.2. *Note that Theorem 2.3 can be stated with the order of $\mathfrak{V}_n$ and $\mathfrak{U}_n^l$ reversed in (2.7) and with the one of $\mathfrak{V}_n^*$ and $(\mathfrak{U}_n^*)^l$ reversed in (2.8) because the operators $\mathfrak{U}_n$ and $\mathfrak{V}_n$ commute. Clearly, in this new form Theorem 2.3 may be thought of as a particular case of both Theorems 2.4 and 2.5 collected in one statement.*

Remark 2.3. *The assertions of Theorems 2.1, 2.2, 2.4, and 2.5 remain valid for all $t_n, t_l, t_q \in \Omega_k$, with $\|\!| \cdot |\!\|_q$ and $\|\!| \cdot |\!\|_{*;q}$ in place of $\|\!| \cdot |\!\|_n$ and $\|\!| \cdot |\!\|_{*;n}$, respectively, if instead of the $\Lambda_J^k(\ldots)$-concordance ($\Lambda_J(\ldots)$-concordance) of the corresponding pairs/triplets one requires their strict concordance.*

2.2 Stability of discrete semigroups: some special points

In this section we present certain aspects of the stability analysis of discrete semigroups, which are connected with modifications of our starting hypotheses. Concentrating on special questions of stability, we therefore introduce certain assumptions that facilitate the following presentation and make our analysis more transparent. In particular, we restrict ourselves essentially to the autonomous case,[3] and we consider only the case $T = \infty$. Also, we deal with analogues of the Λ_J-condition only (not of the Λ_J^k-condition).

We begin by posing a slight modification of the above Λ_J-condition.

Definition 2.1. *We say that an operator $\mathfrak{L} \in \mathcal{B}(\mathfrak{X})$ satisfies the $\Lambda_J(M)$-*condition *with some $M \geq 1$ and $J \in \mathbb{N}$ if it satisfies the Λ_J-condition in which the constant C (see (1.7)) is replaced by M. Λ_J-*proper *sets for the operator $\mathfrak{L}$ are then defined in the same way as in Definition 1.3.*

[3] We then write simply $\mathfrak{A}$ and $\mathfrak{U}$ instead of $\mathfrak{A}_n$ and $\mathfrak{U}_n$, respectively.

In fact, employing the $\Lambda_J(M)$-condition means that our results may be used for situations in which the size of the constant in the corresponding resolvent estimate varies considerably.

Now we state an analogue of Theorem 2.3 with $\xi = 0$.[4]

Theorem 2.6. *Let the operator $\mathfrak{A}$ satisfy the $\Lambda_J(M)$-condition with some $J \in \mathbb{N}$ and $M \geq 1$. Then, for all $k \in (0, K]$ and $l = 0, 1, \ldots,$*

$$\|\mathfrak{U}^l\| \leq CM.$$

This estimate follows from (2.1) with $m = 1$ in a manner similar to that used in the proof of Theorem 2.1.

For simplicity we present an analogue of Theorem 2.3 with $\xi > 0$ in the case $J = 1$ only.

Theorem 2.7. *Let the operator $\mathfrak{A}$ satisfy the $\Lambda_1(M)$-condition with some $M \geq 1$. Then, with any fixed $\xi > 0$, for all $k \in (0, K]$ and $l = 0, 1, \ldots,$*

$$\|\mathfrak{A}^\xi \mathfrak{U}^l\| \leq CMt_{l+1}^{-\xi}.$$

Proof. Unfortunately, we cannot act on the complete analogy of the proof of Lemma 1.12 which further implies Theorem 2.3 since in that proof we have used the convexity inequality which does not control the size of the stability constant. We will therefore use instead a slightly different argument which is helpful in proving directly Theorem 2.3 as well.

As in Section 1.1, it follows from Lemma 1.4 that the fractional powers $\mathfrak{A}_n^\xi$, $\xi \geq 0$, are defined and bounded on $\mathfrak{X}$. Further, let a set Υ be Λ_1-proper for $\mathfrak{A}$ and let Γ be the contour coinciding with $\partial\Upsilon$. Using the operator calculus formula (A.17) with $f(\lambda) = (1 - \lambda)^l$, with $k\mathfrak{A}$ in place of $\mathfrak{L}$, and with Γ in place of Ξ_L,[5] we have, for $l = 0, 1, \ldots,$

$$\mathfrak{A}^\xi \mathfrak{U}^l = (2\pi i k^\xi)^{-1} \int_\Gamma \lambda^\xi (1 - \lambda)^l (\lambda I - k\mathfrak{A})^{-1} d\lambda.$$

Now it is not hard to see that the claim follows because of the reasonings already used in the proof of Theorem 2.1. □

Note also that Theorems 2.1, 2.2, 2.4, and 2.5 can be similarly developed. We leave the details for the reader.

In the remainder of the section we intend to get stability estimates based on a slightly different resolvent condition.

---

[4] In this section we present no estimates involving the adjoint operators.

[5] Clearly, the mentioned deformation of the original contour is tolerable by Cauchy's Theorem.

Definition 2.2. *We say that an operator $\mathfrak{L} \in \mathcal{B}(\mathfrak{X})$ satisfies the $ST_J(M)$-condition*[6] *with some $M \geq 1$ and $J \in \mathbb{N}$ if its spectrum is contained in the shifted unit disk $\mathcal{D}(1;1)$ whenever $k \in (0, K]$ and*

$$\|(\lambda I - \mathfrak{L})^{-1}\| \leq M \sum_{j=1}^{J} |\lambda - \zeta_j|^{-1} \qquad \text{for all } k \in (0, K] \text{ and } \lambda \notin \operatorname{Int} \mathcal{D}(1;1),$$

with $\zeta_1 = 0, \zeta_2, \ldots, \zeta_J$ given fixed points lying on the circle $\partial \mathcal{D}(1;1)$.

On the one hand, if $\mathfrak{L}$ satisfies the $\Lambda_J(M)$-condition, then the operator $k\mathfrak{L}$ clearly has satisfied the $ST_J(M)$-condition with the same constant M. On the other hand, it can be shown by analytic continuation (see Theorems A.1 and A.2) that the $ST_J(M)$-condition for the operator $k\mathfrak{L}$ has implied the $\Lambda_J(M_1)$-condition for the operator $\mathfrak{L}$ with some $M_1 > M$. Therefore, if we want to use the above techniques in order to get stability estimates, we have to trace the dependence M_1 on M.

Theorem 2.8. *Let the operator $k\mathfrak{A}$ satisfy the $ST_J(M)$-condition with some $J \in \mathbb{N}$ and $M \geq 1$. Then, for all $k \in (0, K]$ and $l = 0, 1, \ldots,$*

$$\|\mathfrak{U}^l\| \leq CM \log(1 + M).$$

Proof. For the sake of simplicity, we show the result in the particular case $J = 1$.[7] Assuming therefore this condition, let $\varphi = \arcsin 1/(2M)$.

Assume first that $(l+1)^{-1} \leq 2 \sin \varphi$. Let then $\widehat{\Gamma}^{l,M}$ be the contour specified by $\widehat{\Gamma}^{l,M} = \widehat{\Gamma}_1^{l,M} \cup \widehat{\Gamma}_2^{l,M} \cup \widehat{\Gamma}_3^{M}$, where

$$\begin{aligned} \widehat{\Gamma}_1^{l,M} &= \{\lambda \in \mathbb{C} : |\lambda| = 1/(l+1), \, |\arg(-\lambda)| \leq \pi/2 + \varphi\}, \\ \widehat{\Gamma}_2^{l,M} &= \{\lambda \in \mathbb{C} : 1/(l+1) \leq |\lambda| \leq 2 \sin \varphi, \, \arg \lambda = \pm(\pi/2 - \varphi)\}, \\ \widehat{\Gamma}_3^{M} &= \{\lambda \in \mathbb{C} : |\lambda - 1| = 1, \, |\arg \lambda| \leq \pi/2 - \varphi\}. \end{aligned}$$

Since, by the accepted conditions,

$$\|(\lambda I - k\mathfrak{A})^{-1}\| \leq M|\lambda|^{-1} \qquad \text{for all } \lambda \text{ with } \Re\lambda = 0,$$

[6] The abbreviature ST comes of two words: "shifted" and "Tadmor" of which the second is the name of the mathematician who introduced a resolvent condition related to the $ST_J(M)$-condition with $J = 1$ (see Tadmor [158]). In fact, both conditions differ from each other by a shift of the spectrum of the operator in question. Note that Ritt [136] has worked with a resolvent condition which actually is very close to Tadmor's condition.

[7] It is seen from the present proof that our argument can easily be extended to the general case $J \geq 1$.

applying Theorem A.2 yields, for $|\arg(-\lambda)| \leq \pi/2 + \varphi$,

$$\|(\lambda I - k\mathfrak{A})^{-1}\| \leq \frac{M}{1 - M\sin\varphi}|\lambda|^{-1} = 2M|\lambda|^{-1}, \tag{2.9}$$

which shows in particular that

$$\|(\lambda I - k\mathfrak{A})^{-1}\| \leq 2M(l+1) \qquad \text{for } \lambda \in \widehat{\Gamma}_1^{l,M}. \tag{2.10}$$

Also for further reference, observe that the $ST_1(M)$-condition directly implies

$$\|(\lambda I - k\mathfrak{A})^{-1}\| \leq \frac{M}{2\cos(\arg\lambda)} \qquad \text{for all } \lambda \in \partial\mathcal{D}(1;1). \tag{2.11}$$

Using now the operator calculus formula (A.2), we have, for $l = 0, 1, \ldots$,

$$\begin{aligned}
\mathfrak{U}^l &= \frac{1}{2\pi i}\int_{\widehat{\Gamma}^{l,M}} (1-\lambda)^l(\lambda I - k\mathfrak{A})^{-1}\,d\lambda \\
&= \frac{1}{2\pi i}\left(\int_{\widehat{\Gamma}_1^{l,M}} \cdots + \int_{\widehat{\Gamma}_2^{l,M}} \cdots + \int_{\widehat{\Gamma}_3^{M}} \cdots\right) \\
&= \frac{1}{2\pi i}\big(\widehat{I}_1 + \widehat{I}_2 + \widehat{I}_3\big).
\end{aligned} \tag{2.12}$$

Since $|1-\lambda|^l \leq C$ for $\lambda \in \widehat{\Gamma}_1^{l,M}$, with the aid of (2.10) it easily follows that

$$\|\widehat{I}_1\| \leq 2M(l+1)\int_{\widehat{\Gamma}_1^{l,M}} |1-\lambda|^l|d\lambda| \leq CM, \qquad l = 0, 1, \ldots. \tag{2.13}$$

Next, using the estimates, for $l = 0, 1, \ldots$,

$$\begin{aligned}
\big|1 - x\exp\big(i(\pi/2 - \varphi)\big)\big|^l &\leq (1 - x\sin\varphi)^{l/2} \\
&\leq C\exp\big(-x\tfrac{l+1}{2}\sin\varphi\big) \quad \text{if } 0 \leq x \leq \sin\varphi
\end{aligned} \tag{2.14}$$

and

$$\big|1 - x\exp\big(i(\pi/2 - \varphi)\big)\big|^l \leq 1 \quad \text{if } \sin\varphi \leq x \leq 2\sin\varphi,$$

a simple calculation yields

$$\begin{aligned}
&\int_{1/(l+1)}^{2\sin\varphi} \big|1 - x\exp\big(i(\pi/2 - \varphi)\big)\big|^l \frac{dx}{x} \\
&\qquad \leq C\int_{1/(l+1)}^{\sin\varphi} \exp\Big(-x\frac{l+1}{2}\sin\varphi\Big)\frac{dx}{x} + \int_{\sin\varphi}^{2\sin\varphi}\frac{dx}{x} \\
&\qquad \leq C\log(1+M).
\end{aligned}$$

Applying this inequality and (2.9), we get, for $l = 0, 1, \ldots,$

$$\begin{aligned}\|\widehat{I}_2\| &\le 4M \int_{1/(l+1)}^{2\sin\varphi} \left|1 - x\exp\left(i(\pi/2-\varphi)\right)\right|^l \frac{dx}{x} \\ &\le CM\log(1+M). \end{aligned} \tag{2.15}$$

Finally, since $\left|d\big(2\cos\vartheta\exp(i\vartheta)\big)/d\vartheta\right| = 2$ and $|1-\lambda| = 1$ for $\lambda \in \widehat{\Gamma}_3^M$, with the aid of (2.11) we find, for $l = 0, 1, \ldots,$

$$\begin{aligned}\|\widehat{I}_3\| &\le \frac{M}{2}\int_{\widehat{\Gamma}_3^M} \frac{d\lambda}{\cos(\arg\lambda)} \le 2M\int_0^{\pi/2-\varphi} (\cos\vartheta)^{-1} d\vartheta \\ &\le CM\log(1+M). \end{aligned} \tag{2.16}$$

Now it follows from (2.12), (2.13), (2.15), and (2.16) that if $(l+1)^{-1} \le 2\sin\varphi$,

$$\|\mathfrak{U}^l\| \le CM\log(1+M), \qquad l = 0, 1, \ldots.$$

A similar argument in the case where[8] $(l+1)^{-1} \ge 2\sin\varphi$ gives, for $l = 0, 1, \ldots,$

$$\|\mathfrak{U}^l\| \le CM\log(l+2) \le CM\log(1+M).$$

The claim thus follows by combining the last two estimates for $\|\mathfrak{U}^l\|$. □

Theorem 2.9. *Let the operator $k\mathfrak{A}$ satisfy the $ST_1(M)$-condition with some $M \ge 1$. Then, with any fixed $\xi > 0$, for all $l = 0, 1, \ldots,$*

$$\|\mathfrak{A}^\xi \mathfrak{U}^l\| \le CK(M)t_{l+1}^{-\xi}, \tag{2.17}$$

where[9]

$$K(M) = \begin{cases} M^{1+\xi}\log(1+M) & \text{if } \xi = \lfloor\xi\rfloor, \\ M^{1-\lfloor\xi\rfloor+2\xi} & \text{if } \lfloor\xi\rfloor < \xi \le \lfloor\xi\rfloor + \frac{1}{2}, \\ M^{2+\lfloor\xi\rfloor} & \text{if } \lfloor\xi\rfloor + \frac{1}{2} < \xi < \lfloor\xi\rfloor + 1. \end{cases} \tag{2.18}$$

Proof. As above let $\varphi = \arcsin 1/(2M)$ and let $\widetilde{\Gamma}^M$ be the contour specified by $\widetilde{\Gamma}^M = \widetilde{\Gamma}_1^M \cup \widetilde{\Gamma}_2^M \cup \widetilde{\Gamma}_3^M$, where

$$\begin{aligned}\widetilde{\Gamma}_1^M &= \{\lambda \in \partial\Sigma_{\pi/2-\varphi} : 0 \le |\lambda| \le \sin\varphi\}, \\ \widetilde{\Gamma}_2^M &= \{\lambda \in \partial\Sigma_{\pi/2-\varphi} : \sin\varphi \le |\lambda| \le (2-1/M)\sin\varphi\}, \\ \widetilde{\Gamma}_3^M &= \{\lambda \in \Sigma_{\pi/2-\varphi} : \lambda = (2-1/M)\cos(\arg\lambda)\exp(i\arg\lambda)\}.\end{aligned}$$

[8]In this case the path of integration contains no line segments.

[9]$\lfloor\xi\rfloor$ denotes the largest integer not exceeding ξ.

Using now (A.17) with $f(\lambda) = (1-\lambda)^l$ and with $\widetilde{\Gamma}^M$ in place of Ξ_L,[10] we have, for $l = 0, 1, \ldots$,

$$\begin{aligned}
\mathfrak{A}^\xi \mathfrak{U}^l &= (2\pi i k^\xi)^{-1} \int_{\widetilde{\Gamma}^M} \lambda^\xi (1-\lambda)^l (\lambda I - k\mathfrak{A})^{-1}\, d\lambda \\
&= (2\pi i k^\xi)^{-1} \Big(\int_{\widetilde{\Gamma}_1^M} \cdots + \int_{\widetilde{\Gamma}_2^M} \cdots + \int_{\widetilde{\Gamma}_3^M} \cdots \Big) \\
&= (2\pi i k^\xi)^{-1} \big(\widetilde{I}_1 + \widetilde{I}_2 + \widetilde{I}_3\big). \qquad (2.19)
\end{aligned}$$

It thus remains to show suitable estimates for $\widetilde{I}_1$, $\widetilde{I}_2$, and $\widetilde{I}_3$.

A simple estimation which is similar to the one that leads to (2.15) gives, with the aid of (2.14), for $l = 0, 1, \ldots$,

$$\begin{aligned}
\|\widetilde{I}_1\| &\le 4M \int_0^{\sin\varphi} x^{\xi-1} \big|1 - x \exp\big(i(\pi/2 - \varphi)\big)\big|^l \, dx \\
&\le CM \int_0^{\infty} x^{\xi-1} \exp\big(-\frac{x}{2}(l+1)\sin\varphi\big)\, dx \\
&\le CM^{1+\xi}(l+1)^{-\xi}. \qquad (2.20)
\end{aligned}$$

Likewise, with the aid of the estimate

$$\big|1 - x\exp\big(i(\pi/2-\varphi)\big)\big| \le \exp\big(-\sin\varphi\,(\sin\varphi - x/2)\big) \qquad \text{for } \sin\varphi \le x \le 2\sin\varphi,$$

and since $x^{-1}(1-e^{-x}) \le x^{-\xi}$ for $x > 0$, we obtain in the case $0 < \xi \le 1$, for $l = 0, 1, \ldots$,

$$\begin{aligned}
\|\widetilde{I}_2\| &\le 4M \int_{\sin\varphi}^{(2-1/M)\sin\varphi} x^{\xi-1} \big|1 - x\exp\big(i(\pi/2-\varphi)\big)\big|^l \, dx \\
&\le CM(\sin\varphi)^{\xi-1} \int_{\sin\varphi}^{2\sin\varphi} \exp\big(-(l+1)\sin\varphi\,(\sin\varphi - x/2)\big)\, dx \\
&\le CM(\sin\varphi)^{\xi-2}(l+1)^{-1}\Big(1 - \exp\big(-\frac{l+1}{2}\sin^2\varphi\big)\Big) \\
&\le CM^{1+\xi}(l+1)^{-\xi}. \qquad (2.21)
\end{aligned}$$

Next, with (2.11), using the analytic continuation of the resolvent of $k\mathfrak{A}$ from the set $\partial\mathcal{D}(1;1) \setminus \{0\}$ towards the origin within the sector $\Sigma_{\pi/2-\varphi}$ and

---

[10] By Cauchy's Theorem the mentioned deformation of the original path of integration is admissible.

applying Theorem A.1, we come to the following estimate, for all $\lambda \in \widetilde{\Gamma}_3^M$,

$$\|(\lambda I - k\mathfrak{A})^{-1}\| \le \frac{M(2\cos(\arg\lambda))^{-1}}{1 - M\big(2\cos(\arg\lambda)\big)^{-1}M^{-1}\cos(\arg\lambda)} = \frac{M}{\cos(\arg\lambda)}.$$

At the same time we have, for $\lambda \in \widetilde{\Gamma}_3^M$,

$$|1-\lambda|^l = (1-(2-1/M)/M\cos^2\arg\lambda)^{l/2} \le C\exp\big(-\frac{l+1}{2M}\cos^2\arg\lambda\big).$$

Therefore, observing that $\big|d\big((2-1/M)\cos\vartheta\exp(i\vartheta)\big)/d\vartheta\big| \le 2$, applying the last two estimates together with

$$\begin{aligned} J &:= \int_{\sin\varphi}^{\infty} x^{\xi-1}\exp\big(-\frac{l+1}{2M}x^2\big)\,dx + \exp\big(-\frac{l+1}{4M}\big) \\ &\le \int_{\sin\varphi}^{\infty} x^{\xi-1}\exp\big(-x\frac{l+1}{2M}\sin\varphi\big)\,dx + \exp\big(-\frac{l+1}{4M}\big) \\ &\le C\left(\frac{M^2}{l+1}\right)^{\xi}, \end{aligned}$$

and using the above argument, we obtain, for $l = 0, 1, \ldots$,

$$\begin{aligned} \|\widetilde{I}_3\| &\le M\int_{\widetilde{\Gamma}_3^M} |\lambda|^{\xi}\big(\cos(\arg\lambda)\big)^{-1}|1-\lambda|^l\,|d\lambda| \\ &\le CM\int_0^{\pi/2-\varphi} (\cos\vartheta)^{\xi-1}\exp\big(-\frac{l+1}{2M}\cos^2\vartheta\big)\,d\vartheta \\ &\le CMJ \le CM^{1+2\xi}(l+1)^{-\xi}. \end{aligned} \tag{2.22}$$

Combining (2.19), (2.20), (2.21), and (2.22) now yields, if $0 < \xi \le 1$,

$$\|\mathfrak{A}^{\xi}\mathfrak{U}^l\| \le CM^{1+2\xi}t_{l+1}^{-\xi} \qquad \text{for } l = 0, \ldots,$$

which implies the desired result in the case $0 < \xi \le \frac{1}{2}$.

Also, it will be needed to give a slightly different estimate for $\|\widetilde{I}_3\|$. More precisely, using the first three lines in (2.22), we get, for $l + 1 \le M$,

$$\begin{aligned} \|\widetilde{I}_3\| &\le CM\left(\left(\frac{M}{l+1}\right)^{\xi/2}\int_0^{\infty} x^{\xi-1}\exp(-x^2)\,dx + \exp\big(-\frac{l+1}{4M}\big)\right) \\ &\le CM^{1+\xi}(l+1)^{-\xi}. \end{aligned} \tag{2.23}$$

Now we derive one more estimate for $\|\widetilde{I}_2^M + \widetilde{I}_3^M\|$ which will be an important ingredient in the proof of (2.17) for $\frac{1}{2} < \xi \le \frac{3}{2}$. In fact, putting $\lambda_1(\varphi) = \sin\varphi \exp\big(-i(\pi/2-\varphi)\big)$, $\lambda_2(\varphi) = \sin\varphi \exp\big(i(\pi/2-\varphi)\big)$, and

$$H(\lambda) = (l+1)^{-1}(1-\lambda)^{l+1}\lambda^{\xi}(\lambda I - k\mathfrak{A})^{-1},$$

and integrating by parts, we find

$$\begin{aligned}\int_{\widetilde{\Gamma}_2^M\cup\widetilde{\Gamma}_3^M} & \lambda^{\xi}(1-\lambda)^l(\lambda I - k\mathfrak{A})^{-1}\,d\lambda \\ &= \big(-H(\lambda_2(\varphi)) + H(\lambda_1(\varphi))\big) \\ &\quad + \int_{\widetilde{\Gamma}_2^M} \big(\xi\lambda^{-1}H(\lambda) - H(\lambda)\,(\lambda I - k\mathfrak{A})^{-1}\big)\,d\lambda \\ &\quad + \int_{\widetilde{\Gamma}_3^M} \big(\xi\lambda^{-1}H(\lambda) - H(\lambda)\,(\lambda I - k\mathfrak{A})^{-1}\big)\,d\lambda \\ &= H_1 + H_2 + H_3. \end{aligned} \tag{2.24}$$

With the aid of (2.9) and (2.14), dealing with the case $\xi \ge 1$, a direct estimation gives, since we then have $x^{\xi-1}e^{-x} \le C$ for all $x \ge 0$,

$$\|H_1\| \le \begin{cases} CM^{2-\xi}(l+1)^{-1} & \text{if } \ 0<\xi<1, \\ CM^{\xi}(l+1)^{-\xi} & \text{if } \ \xi \ge 1. \end{cases} \tag{2.25}$$

Next, using the argument leading to (2.21) and (2.22), we get

$$\|H_2\| \le \begin{cases} CM^{3-\xi}(l+1)^{-1} & \text{if } \ 0<\xi<1, \\ CM^{1+\xi}(l+1)^{-\xi} & \text{if } \ 1\le\xi\le 2, \end{cases} \tag{2.26}$$

and

$$\|H_3\| \le \begin{cases} CM^{3-\xi}(l+1)^{-1} & \text{if } \ 0<\xi<1, \\ CM^2\log(1+M)\,(l+1)^{-1} & \text{if } \ \xi=1, \\ CM^{2\xi}(l+1)^{-\xi} & \text{if } \ \xi>1. \end{cases} \tag{2.27}$$

Clearly, in the case $0<\xi<1$ it follows from (2.25), (2.26), and (2.27) that

$$\|H_1+H_2+H_3\| \le CM^2(l+1)^{-\xi} \qquad \text{if } M \le l+1.$$

Combining this estimate with (2.19), (2.20), (2.21), (2.23), and (2.24) proves (2.17) for $\frac{1}{2} < \xi < 1$.

It is also seen that if $1 \le \xi \le \frac{3}{2}$, the claim follows by combining (2.19), (2.20), (2.25), (2.26), and (2.27).

Further, to show the result for $\frac{3}{2} < \xi \leq \frac{5}{2}$, we integrate again by parts on the right-hand side of (2.24) and use (2.25) with the above argument. Note that one can find in analogy to (2.23), for $\xi > 1$,

$$\|H_3\| \leq CM^{(3+\xi)/2}(l+1)^{-(1+\xi)/2},$$

which is needed at this stage of the proof.

Clearly, using the above reasonings repeatedly, the claim can be shown for any fixed $\xi > 0$ in a finite number of steps. □

2.3 Stability of discrete evolution operators

In this section we are concerned with the general (non-autonomous) case. This fact means that we do not suppose that $\mathfrak{A}_n$ is necessarily independent of t_n. Instead we work with certain Hölder-continuity type hypotheses imposing restrictions on the behaviour of the difference $\mathfrak{A}_n - \mathfrak{A}_l$. We assume these hypotheses to hold for all $k \in (0, K]$, $t_n, t_l \in \Omega_k$, $u \in \mathfrak{X}$, and $\chi \in \mathfrak{X}^*$, with some $\vartheta, \vartheta_1 \in [0,1]$ and some given seminorms $\|\!| \cdot |\!\|_{\emptyset;n}$, $\|\!| \cdot |\!\|_{I;n}$, $\|\!| \cdot |\!\|_{*\emptyset;n}$, and $\|\!| \cdot |\!\|_{*I;n}$ of which the first two are defined on $\mathfrak{X}$ and the other two on $\mathfrak{X}^*$. More precisely, we will employ the following Hölder-continuity type hypotheses.

$\mathbf{H\ddot{O}L}^{(\rightarrow)}(\vartheta, \vartheta_1; \|\!| \cdot |\!\|_{\mathfrak{m};n}, \mathfrak{m} = \emptyset, I, *\emptyset)$ reads

$$|\langle \chi, (\mathfrak{A}_n - \mathfrak{A}_l)u \rangle| \leq C\left(t_{|n-l|}^{\vartheta} \|\!|u|\!\|_{\emptyset;l} + t_{|n-l|}^{\vartheta_1} \|\!|u|\!\|_{I;l}\right) \|\!|\chi|\!\|_{*\emptyset;n},$$

and

$\mathbf{H\ddot{O}L}^{(\leftarrow)}(\vartheta, \vartheta_1; \|\!| \cdot |\!\|_{\mathfrak{m};n}, \mathfrak{m} = \emptyset, *\emptyset, *I)$ reads

$$|\langle \chi, (\mathfrak{A}_n - \mathfrak{A}_l)u \rangle| \leq C\left(t_{|n-l|}^{\vartheta} \|\!|\chi|\!\|_{*\emptyset;n} + t_{|n-l|}^{\vartheta_1} \|\!|\chi|\!\|_{*I;n}\right) \|\!|u|\!\|_{\emptyset;l}.$$

In the sequel a particular case of both $\mathbf{H\ddot{O}L}^{(\rightarrow)}(\vartheta, \vartheta_1; \|\!| \cdot |\!\|_{\mathfrak{m};n}, \mathfrak{m} = \emptyset, I, *\emptyset)$ and $\mathbf{H\ddot{O}L}^{(\leftarrow)}(\vartheta, \vartheta_1; \|\!| \cdot |\!\|_{\mathfrak{m};n}, \mathfrak{m} = \emptyset, *\emptyset, *I)$ will sometimes be in use, which is now stated as a separate hypothesis as follows:

$\mathbf{H\ddot{O}L}(\vartheta; \|\!| \cdot |\!\|_{\mathfrak{m};n}, \mathfrak{m} = \emptyset, *\emptyset)$ reads

$$|\langle \chi, (\mathfrak{A}_n - \mathfrak{A}_l)u \rangle| \leq Ct_{|n-l|}^{\vartheta} \|\!|u|\!\|_{\emptyset;l} \|\!|\chi|\!\|_{*\emptyset;n}.$$

Throughout the present section we assume that $T < \infty$ since our considerations will essentially be based on a discrete version of Gronwall's lemma. It is stated here in the following form.

Lemma 2.1. *Let $z_{n,l}$ and $w_{n,l}$ be nonnegative solutions of the respective inequalities, for $t_n, t_l \in \Omega_k$, $t_l \leq t_n$,*

$$z_{n,l} \leq Ct_{n-l+1}^{-\zeta_1} + Ck \sum_{q=l}^{n-1} z_{n,q+1} t_{q-l+1}^{-\zeta_2}$$

and

$$w_{n,l} \leq Ct_{n-l+1}^{-\zeta_1} + Ck \sum_{q=l}^{n-1} t_{n-q}^{-\zeta_2} w_{q,l},$$

with some $\zeta_1, \zeta_2 \in [0,1)$. Then we have, for all $t_n, t_l \in \Omega_k$, $t_l \leq t_n$,

$$z_{n,l} \leq Ct_{n-l+1}^{-\zeta_1}$$

and

$$w_{n,l} \leq Ct_{n-l+1}^{-\zeta_1}.$$

In fact, this proposition follows directly from a more general lemma of Gronwall type that is stated in Chapter 8 for the case with a variable stepsize (see Lemma 8.4). On the other hand, Lemma 2.1 can easily be deduced from Lemma 4.1 below.[11]

Our subsequent techniques will be based on using the following telescopic type identities which are trivially verified for all $t_n, t_l \in \Omega_k$ with $t_l \leq t_n$, noting that $\mathfrak{U}_q - \mathfrak{U}_l = k(\mathfrak{A}_l - \mathfrak{A}_q)$,

$$\begin{aligned} \mathfrak{U}_{n-1,l} &= \mathfrak{U}_l^{n-l} + \sum_{q=l+1}^{n-1} \mathfrak{U}_{n-1,q+1} (\mathfrak{U}_q - \mathfrak{U}_l)\, \mathfrak{U}_l^{q-l} \\ &= \mathfrak{U}_l^{n-l} + k \sum_{q=l+1}^{n-1} \mathfrak{U}_{n-1,q+1} (\mathfrak{A}_l - \mathfrak{A}_q)\, \mathfrak{U}_l^{q-l} \end{aligned} \tag{2.28}$$

and

$$\begin{aligned} \mathfrak{U}_{n-1,l} &= \mathfrak{U}_n^{n-l} + \sum_{q=l}^{n-1} \mathfrak{U}_n^{n-q-1} (\mathfrak{U}_q - \mathfrak{U}_n)\, \mathfrak{U}_{q-1,l} \\ &= \mathfrak{U}_n^{n-l} + k \sum_{q=l}^{n-1} \mathfrak{U}_n^{n-q-1} (\mathfrak{A}_n - \mathfrak{A}_q)\, \mathfrak{U}_{q-1,l}. \end{aligned} \tag{2.29}$$

---

[11]Strictly speaking, one can make this assertion about the inequality for $w_{n,l}$ only. Note however that both discrete inequalities are solved in a similar manner.

Sometimes a slight modification of (2.28) will be used, as follows, for all $t_l < t_n$,

$$\mathfrak{U}_{n-1,l+1} = \mathfrak{U}_l^{n-l-1} + k \sum_{q=l+1}^{n-1} \mathfrak{U}_{n-1,q+1}(\mathfrak{A}_l - \mathfrak{A}_q)\,\mathfrak{U}_l^{q-l-1}. \tag{2.30}$$

Denoting $p = \left\lfloor \frac{n+l}{2} \right\rfloor$ and combining (2.28) and (2.29) leads to an additional helpful identity, namely, for all $t_l \le t_n$,

$$\begin{aligned} \mathfrak{U}_{n-1,l} &= k \sum_{q=p}^{n-1} \mathfrak{U}_n^{n-q-1}(\mathfrak{A}_n - \mathfrak{A}_q)\,\mathfrak{U}_{q-1,l} \\ &\quad + k\mathfrak{U}_n^{n-p} \sum_{q=l}^{p-1} \mathfrak{U}_{p-1,q+1}(\mathfrak{A}_l - \mathfrak{A}_q)\mathfrak{U}_l^{q-l} + \mathfrak{U}_n^{n-p}\mathfrak{U}_l^{p-l}, \end{aligned} \tag{2.31}$$

which is in subsequent use as well.

Apart from everything else, we supplement our below results by showing stability estimates in the situation when the operator in question is under perturbation. More precisely, to problem (1.1) we associate the following perturbed problem, with some $\mathfrak{B}_n \in \mathcal{B}(\mathfrak{X})$,

$$\partial Y_n + (\mathfrak{A}_n + \mathfrak{B}_n)\,Y_n = \mathfrak{Q}_n F_n \quad \text{for } t_n \in \breve{\Omega}_k, \qquad Y_0 = \mathfrak{R}y^0. \tag{2.32}$$

For subsequent use we introduce the notation

$$\widehat{\mathfrak{U}}_n = I - k(\mathfrak{A}_n + \mathfrak{B}_n) \text{ and } \widehat{\mathfrak{U}}_{n,j} = \begin{cases} \widehat{\mathfrak{U}}_n\widehat{\mathfrak{U}}_{n-1}\cdots\widehat{\mathfrak{U}}_j & \text{if } j \le n, \\ I & \text{if } j > n. \end{cases}$$

Similarly to (1.4), the solution Y_n of (2.32) is then represented by

$$Y_n = \widehat{\mathfrak{U}}_{n-1,0}\mathfrak{R}y^0 + k \sum_{j=0}^{n-1} \widehat{\mathfrak{U}}_{n-1,j+1}\mathfrak{Q}_j F_j. \tag{2.33}$$

One can easily verify the following identities, for $t_n, t_l \in \Omega_k$ with $t_l \le t_n$,

$$\widehat{\mathfrak{U}}_{n-1,l} = \mathfrak{U}_{n-1,l} - k \sum_{q=l}^{n-1} \mathfrak{U}_{n-1,q+1}\mathfrak{B}_q\widehat{\mathfrak{U}}_{q-1,l} \tag{2.34}$$

and

$$\widehat{\mathfrak{U}}_{n-1,l} = \mathfrak{U}_{n-1,l} - k \sum_{q=l}^{n-1} \widehat{\mathfrak{U}}_{n-1,q+1}\mathfrak{B}_q\mathfrak{U}_{q-1,l}, \tag{2.35}$$

which, in fact, connect the discrect evolution operators $\widehat{\mathfrak{U}}_{n,l}$ and $\mathfrak{U}_{n,l}$ of the problems (2.32) and (1.1), respectively. Also, combining both these equalities yields one more helpful formula, for $t_l \leq t_n$,

$$\begin{aligned}\widehat{\mathfrak{U}}_{n-1,l} &= \mathfrak{U}_{n-1,l} - k\sum_{q=l}^{n-1} \mathfrak{U}_{n-1,q+1}\mathfrak{B}_q\mathfrak{U}_{q-1,l} \\ &\quad + k^2 \sum_{q=l}^{n-1}\sum_{s=q+1}^{n-1} \mathfrak{U}_{n-1,s+1}\mathfrak{B}_s\widehat{\mathfrak{U}}_{s-1,q+1}\mathfrak{B}_q\mathfrak{U}_{q-1,l}. \end{aligned} \tag{2.36}$$

The relations (2.34) −−(2.36) are taken as a basis for applying our perturbation techniques when examining the stability of problem (2.32).

Now we give a series of stability assertions which differ from each other by the assumed Hölder-continuity type hypotheses.

Theorem 2.10. *Let seminorms* $]\![\cdot]\!]_n$, $|\!|\!| \cdot |\!|\!|_n$, *and* $|\!|\!| \cdot |\!|\!|_{1;n}$ *form respectively* $\Lambda_1^k(\xi|1)$-, $\Lambda_1^k(1|m)$-, *and* $\Lambda_1^k(\gamma_1|m)$-*concordant pairs with the operator* $\mathfrak{A}_n$ *for some* $\xi, \gamma_1 \in [0, 1)$ *and* $m \in \mathbb{N}$. *Suppose further that* $\mathfrak{A}_n$ *satisfies the* Λ_1^k-*condition as well as hypothesis* $\mathbf{HÖL}^{(\rightarrow)}(\vartheta, 0; |\!|\!| \cdot |\!|\!|_{\mathfrak{m};n}, \mathfrak{m} = \emptyset, I, *\emptyset)$, *with some* $\vartheta \in (0, 1]$ *and with*

$$|\!|\!| \cdot |\!|\!|_{\mathfrak{m};n} = \begin{cases} |\!|\!| \cdot |\!|\!|_n & \text{if } \ \mathfrak{m} = \emptyset, \\ |\!|\!| \cdot |\!|\!|_{1;n} & \text{if } \ \mathfrak{m} = I, \\ \| \cdot \|_* & \text{if } \ \mathfrak{m} = *\emptyset. \end{cases} \tag{2.37}$$

Finally, let the seminorm $\|\mathfrak{B}_n \cdot \|$, *generated by the operator* $\mathfrak{B}_n$ *in (2.32), form a* $\Lambda_1^k(\gamma_0)$-*concordant pair with* $\mathfrak{A}_n$ *for some* $\gamma_0 \in [0, 1)$. *Then, for all* $k \in (0, K]$, $t_n, t_l \in \Omega_k$ *with* $t_l \leq t_n$, *and* $u \in \mathfrak{X}$,

$$]\![\mathfrak{U}_{n-1,l}u]\!]_n +]\![\widehat{\mathfrak{U}}_{n-1,l}u]\!]_n \leq Ct_{n-l+1}^{-\xi}\|u\|. \tag{2.38}$$

Furthermore, suppose that the operator $\mathfrak{A}_n$ *satisfies in addition hypothesis* $\mathbf{HÖL}(\vartheta; |\!|\!| \cdot |\!|\!|_{\mathfrak{m};n}, \mathfrak{m} = \emptyset, *\emptyset)$ *and we have*

$$\|\mathfrak{B}_n u\| \leq Ck^{\vartheta}|\!|\!|u|\!|\!|_n \quad \textit{for all } u \in \mathfrak{X} \textit{ and } t_n \in \Omega_k, \tag{2.39}$$

with the same ϑ *and* $|\!|\!| \cdot |\!|\!|_{\mathfrak{m};n}, \mathfrak{m} = \emptyset, *\emptyset$, *as above. Then (2.38) is valid for* $\xi \in [0, 1 + \vartheta]$ *provided that the pair* $(]\![\cdot]\!]_n; \mathfrak{A}_n)$ *is still* $\Lambda_1^k(\xi)$-*concordant. However, for* $\xi = 1 + \vartheta$, *the expression* $t_{n-l+1}^{-1}\log(t_{n-l+2}/k)$ *should be added to the factor* $t_{n-l+1}^{-\xi}$ *on the right.*

Proof. We begin by considering the case where $\xi \in [0,1)$, $\mathfrak{A}_n$ satisfies $\mathbf{H\ddot{O}L}^{(\to)}(\vartheta, 0; \|\!|\cdot\|\!|_{\mathfrak{m};n}, \mathfrak{m} = \emptyset, I, *\emptyset)$ with $\|\!|\cdot\|\!|_{\mathfrak{m};n}$, $\mathfrak{m} = \emptyset, I, *\emptyset$, given by (2.37), and the seminorm $\|\mathfrak{B}_n \cdot\|$ forms a $\Lambda_1^k(\gamma_0)$-concordant pair with $\mathfrak{A}_n$. Applying Theorem 2.1, we then obtain, for all $t_l \le t_q$ and $u \in \mathfrak{X}$,

$$\begin{aligned} \|(\mathfrak{A}_l - \mathfrak{A}_q)\,\mathfrak{U}_l^{q-l} u\| &\le C\left(t_{q-l}^{\vartheta} \|\!|\mathfrak{U}_l^{q-l} u\|\!|_l + \|\!|\mathfrak{U}_l^{q-l} u\|\!|_{1;l}\right) \\ &\le C\left(t_{q-l+1}^{\vartheta-1} + t_{q-l+1}^{-\gamma_1}\right) \|u\|. \end{aligned} \tag{2.40}$$

It is not hard to see that there exists a set Υ of Λ_1-configuration which is $\Lambda_1^k(\xi|1)$-proper for $(]\![\cdot]\!]_n; \mathfrak{A}_n)$, $\Lambda_1^k(1|m)$-proper for $(\|\!|\cdot\|\!|_n; \mathfrak{A}_n)$, and $\Lambda_1(\gamma_1|m)$-proper for $(\|\!|\cdot\|\!|_{1;n}; \mathfrak{A}_n)$, simultaneously. Using again the assumed Hölder-continuity type condition on $\mathfrak{A}_n$ and taking into account that $t_{|n-l|}^{\vartheta} \le C$, it follows that with some $\delta > 0$, for all $t_n, t_l \in \Omega_k$, $u \in \mathfrak{X}$, and $\lambda \notin \operatorname{Int}\left(\Upsilon \cup \mathcal{D}(\delta k)\right)$,

$$\begin{aligned} &k\|(\mathfrak{A}_l - \mathfrak{A}_n)\,(\lambda I - k\mathfrak{A}_l)^{-m} u\| \\ &\qquad \le Ck\|\!|(\lambda I - k\mathfrak{A}_l)^{-m} u\|\!|_l + Ck\|\!|(\lambda I - k\mathfrak{A}_l)^{-m} u\|\!|_{1;l} \\ &\qquad \le C\left(1 + k^{1-\gamma_1}|\lambda|^{\gamma_1 - 1}\right)|\lambda|^{1-m}\|u\| \le C|\lambda|^{1-m}\|u\|. \end{aligned}$$

With this fact in mind applying the identity, for $t_n, t_l \in \Omega_k$,

$$\begin{aligned} (\lambda I - k\mathfrak{A}_l)^{-m} &= (\lambda I - k\mathfrak{A}_n)^{-1}(\lambda I - k\mathfrak{A}_l)^{1-m} \\ &\quad + k(\lambda I - k\mathfrak{A}_n)^{-1}(\mathfrak{A}_l - \mathfrak{A}_n)\,(\lambda I - k\mathfrak{A}_l)^{-m}, \end{aligned} \tag{2.41}$$

yields, for all $u \in \mathfrak{X}$ and $\lambda \notin \operatorname{Int}\left(\Upsilon \cup \mathcal{D}(\delta k)\right)$,

$$\begin{aligned}]\![(\lambda I - k\mathfrak{A}_l)^{-m} u]\!]_n &\le]\![(\lambda I - k\mathfrak{A}_n)^{-1}(\lambda I - k\mathfrak{A}_l)^{1-m} u]\!]_n \\ &\quad + k]\![(\lambda I - k\mathfrak{A}_n)^{-1}(\mathfrak{A}_l - \mathfrak{A}_n)\,(\lambda I - k\mathfrak{A}_l)^{-m} u]\!]_n \\ &\le Ck^{-\xi}|\lambda|^{\xi-1}\|(\lambda I - k\mathfrak{A}_l)^{1-m} u\| \\ &\quad + Ck^{1-\xi}|\lambda|^{\xi-1}\|(\mathfrak{A}_l - \mathfrak{A}_n)\,(\lambda I - k\mathfrak{A}_l)^{-m} u\| \\ &\le Ck^{-\xi}|\lambda|^{\xi-m}\|u\|, \end{aligned}$$

because the pair $(]\![\cdot]\!]_n; \mathfrak{A}_n)$ is $\Lambda_1^k(\xi|1)$-concordant and the set Υ is $\Lambda_1^k(\xi|1)$-proper for this pair. The last estimate shows that the pair $(]\![\cdot]\!]_n; \mathfrak{A}_n)$ is, in fact, strictly $\Lambda_1^k(\xi|m)$-concordant. By Theorem 2.1 and Remark 2.3, we then find, for all $t_n, t_l, t_q \in \Omega_k$ and $u \in \mathfrak{X}$,

$$]\![\mathfrak{U}_l^q u]\!]_n \le C t_{q+1}^{-\xi}\|u\|. \tag{2.42}$$

Denoting $z_{n,l} = \sup_{\|u\|=1} [\![\mathfrak{U}_{n-1,l} u]\!]_n$, taking (2.42) and (2.40) into account, and using (2.28), a trivial estimation now gives, for $t_l \le t_n$,

$$\begin{aligned} z_{n,l} &\le \sup_{\|u\|=1} [\![\mathfrak{U}_l^{n-l} u]\!]_n + k \sum_{q=l+1}^{n-1} z_{n,q+1} \left\| \left(\mathfrak{A}_l - \mathfrak{A}_q\right) \mathfrak{U}_l^{q-l} \right\| \\ &\le C t_{n-l+1}^{-\xi} + C k \sum_{q=l+1}^{n-1} z_{n,q+1} \left(t_{q-l+1}^{\vartheta-1} + t_{q-l+1}^{-\gamma_1} \right). \end{aligned}$$

Therefore, with the aid of Lemma 2.1 we conclude, since $\xi < 1$, $1 - \vartheta < 1$, and $\gamma_1 < 1$, that

$$z_{n,l} \le C t_{n-l+1}^{-\xi} \qquad \text{for } t_l \le t_n,$$

which yields that the first term on the left-hand side of (2.38) is estimated as stated. As a consequence of this fact, we note that for all $t_l \le t_n$,

$$\|\mathfrak{U}_{n-1,l}\| \le C \tag{2.43}$$

and

$$\|\mathfrak{B}_n \mathfrak{U}_{n-1,l}\| \le C t_{n-l+1}^{-\gamma_0}, \tag{2.44}$$

since we can set $[\![\cdot]\!]_n = \|\cdot\|$ for $\xi = 0$ (cf. Lemma 1.12) and $[\![\cdot]\!]_n = \|\mathfrak{B}_n \cdot\|$ for $\xi = \gamma_0$.

Applying furthermore (2.34), multiplying from the left by $\mathfrak{B}_n$, and using (2.44), we get, for all $t_l \le t_n$,

$$\|\mathfrak{B}_n \widehat{\mathfrak{U}}_{n-1,l}\| \le C t_{n-l+1}^{-\gamma_0} + C k \sum_{q=l}^{n-1} t_{n-q}^{-\gamma_0} \|\mathfrak{B}_q \widehat{\mathfrak{U}}_{q-1,l}\|.$$

Applying Lemma 2.1 to the last inequality, it follows that

$$\|\mathfrak{B}_n \widehat{\mathfrak{U}}_{n-1,l}\| \le C t_{n-l+1}^{-\gamma_0} \quad \text{for all } t_l \le t_n.$$

With this estimate in mind, multiplying (2.34) from the right by $u \in \mathfrak{X}$ and taking the seminorm $[\![\cdot]\!]_n$ of both sides, an estimation shows the desired second term on the left-hand side of (2.38), since the same estimation for the first term is already shown.

To prove the latter part of the assertion, we assume now that $\xi \in [0, 1+\vartheta]$, the pair $\left([\![\cdot]\!]_n; \mathfrak{A}_n\right)$ is $\Lambda_1^k(\xi)$-concordant, the operator $\mathfrak{A}_n$ satisfies hypothesis $\mathbf{H\ddot{O}L}(\vartheta; \|\!|\cdot|\!\|_{\mathfrak{m};n}, \mathfrak{m} = \emptyset, *\emptyset)$, and (2.39) holds with the same ϑ and $\|\!|\cdot|\!\|_{\mathfrak{m};n}$, $\mathfrak{m} = \emptyset, *\emptyset$, as above. In this case we have, for all $t_l \le t_n$ and $u \in \mathfrak{X}$,

$$\|\left(\mathfrak{A}_n - \mathfrak{A}_l\right) u\| \le C t_{n-l}^{\vartheta} \|\!|u|\!\|_l. \tag{2.45}$$

Denoting then $w_{n,l} = t_{n-l+1} \sup_{\|u\|=1} \|\mathfrak{U}_{n-1,l}u\|_n$ and using (2.43), (2.45), and Theorem 2.1, we obtain from (2.31), after a simple estimation, for $t_l \leq t_n$,

$$w_{n,l} \leq Ck \sum_{q=p}^{n-1} t_{n-q}^{\vartheta-1} \frac{n-l+1}{q-l+1} w_{q,l} + C\frac{n-l+1}{n-p+1}\left(k\sum_{q=l}^{p-1} t_{q-l+1}^{\vartheta-1} + 1\right). \quad (2.46)$$

Noting next that $\frac{n-l+1}{n-p+1} \leq 2$ and $\frac{n-l+1}{q-l+1} \leq 2$ for $t_n, t_l, t_q \in \Omega_k$ such that $t_l \leq t_n$ and $t_p \leq t_q$, it follows from (2.46) that

$$w_{n,l} \leq Ck \sum_{q=p}^{n-1} t_{n-q}^{\vartheta-1} w_{q,l} + C,$$

which further implies

$$w_{n,l} \leq C + Ck \sum_{q=l}^{n-1} t_{n-q}^{\vartheta-1} w_{q,l}.$$

Observing that $1-\vartheta < 1$ and applying Lemma 2.1 with $\zeta_1 = 0$, we find that $w_{n,l} \leq C$ for $t_l \leq t_n$, which gives, for all $t_l \leq t_n$ and $u \in \mathfrak{X}$,

$$\|\mathfrak{U}_{n-1,l}u\|_n \leq Ct_{n-l+1}^{-1}\|u\|. \quad (2.47)$$

Now we are ready to show (2.38) in the case $\xi \in [0, 1+\vartheta]$. In fact, noting that $\xi - \vartheta < 1$ and using (2.31), (2.47), Theorem 2.1, and the above argument, we have, for $t_l \leq t_n$ and $u \in \mathfrak{X}$,

$$\begin{aligned}]\!\![\mathfrak{U}_{n-1,l}u]\!\!]_n &\leq Ck\sum_{q=p}^{n-1} t_{n-q}^{\vartheta-\xi}\|\mathfrak{U}_{q-1,l}u\|_q \\ &\quad + Ct_{n-l+1}^{-\xi}\left(k\sum_{q=l}^{p-1} t_{q-l+1}^{\vartheta-1} + 1\right)\|u\| \\ &\leq C\left(t_{n-l+1}^{-\xi} + k\sum_{q=p}^{n-1} t_{n-q}^{\vartheta-\xi} t_{q-l+1}^{-1}\right)\|u\|, \end{aligned}$$

which shows that the quantity $]\!\![\mathfrak{U}_{n-1,l}u]\!\!]_n$ is estimated as in (2.38). Further, by Lemma 1.2 the seminorm $k^{\vartheta/2}\|\cdot\|_n$ forms a $\Lambda_1^k(1-\vartheta/2)$-concordant pair with $\mathfrak{A}_n$. Therefore, noting that the former part of the theorem is already proved, we find, for all $t_l \leq t_n$ and $u \in \mathfrak{X}$,

$$\|\mathfrak{B}_n\widehat{\mathfrak{U}}_{n-1,l}u\| \leq Ck^{\vartheta}\|\widehat{\mathfrak{U}}_{n-1,l}u\|_n \leq Ck^{\vartheta/2}t_{n-l+1}^{\vartheta/2-1}\|u\|. \quad (2.48)$$

At the same time we have, for all $t_l \le t_n$ and $u \in \mathfrak{X}$, that

$$\|\mathfrak{B}_n \mathfrak{U}_{n-1,l} u\| \le C k^{\vartheta} \|\mathfrak{U}_{n-1,l} u\|_n \le C k^{\vartheta} t_{n-l+1}^{-1} \|u\|, \tag{2.49}$$

since (2.38) for $]\![\mathfrak{U}_{n-1,l} u]\!]_n$ is already shown. Using again the fact that the quantity $]\![\mathfrak{U}_{n-1,l} u]\!]_n$ is under the desired estimate and with the aid of (2.48) and (2.49), a trivial estimation on the basis of (2.36) yields, for all $t_l \le t_n$ and $u \in \mathfrak{X}$,

$$\begin{aligned}
]\![\widehat{\mathfrak{U}}_{n-1,l} u]\!]_n &\le C t_{n-l+1}^{-\xi} \|u\| + C k \sum_{q=l}^{n-1} t_{n-q}^{-\xi} \|\mathfrak{B}_q U_{q-1,l}\| \, \|u\| \\
&\quad + C k^2 \sum_{q=l}^{n-1} \sum_{s=q+1}^{n-1} t_{n-s}^{-\xi} \|\mathfrak{B}_s \widehat{U}_{s-1,q+1}\| \, \|\mathfrak{B}_q U_{q-1,l}\| \, \|u\| \\
&\le C t_{n-l+1}^{-\xi} \|u\| + C k^{1+\vartheta} \sum_{q=l}^{n-1} t_{n-q}^{-\xi} t_{q-l+1}^{-1} \|u\| \\
&\quad + C k^{2+\frac{3}{2}\vartheta} \sum_{q=l}^{n-1} \sum_{s=q+1}^{n-1} t_{n-s}^{-\xi} t_{s-q-1}^{\vartheta/2-1} t_{q-l+1}^{-1} \|u\|,
\end{aligned}$$

which easily leads to the needed estimate for the quantity $]\![\widehat{\mathfrak{U}}_{n-1,l} u]\!]_n$ as well.

So the proof is complete. □

Theorem 2.11. *Let a seminorm* $]\![\cdot]\!]_{*;n}$ *form a* $\Lambda_1^k(*;\xi_*)$*-concordant pair with* $\mathfrak{A}_n$ *for some* $\xi_* \ge 0$ *and let seminorms* $\|\!|\cdot|\!\|_n$ *and* $\|\!|\cdot|\!\|_{1;n}$ *be under the conditions of Theorem 2.10 for some* $\gamma_1 \in [0,1)$ *and* $m \in \mathbb{N}$. *Suppose further that the operator* $\mathfrak{A}_n$ *satisfies the* Λ_1^k*-condition as well as hypothesis* $\mathbf{H\ddot{O}L}^{(\to)}(\vartheta, 0; \|\!|\cdot|\!\|_{\mathfrak{m};n}, \mathfrak{m} = \emptyset, I, *\emptyset)$ *and that the seminorm* $\|\mathfrak{B}_n \cdot\|$ *is* $\Lambda_1^k(\gamma_0)$*-concordant with* $\mathfrak{A}_n$, *with some* $\vartheta \in (0,1]$, $\gamma_0 \in [0,1)$ *and with* $\|\!|\cdot|\!\|_{\mathfrak{m};n}$, $\mathfrak{m} = \emptyset, I, *\emptyset$, *given by (2.37). Then, if* $0 \le \xi_* < \min\{\vartheta, 1-\gamma_1\}$, *we have, for* $k \in (0,K]$, $t_n, t_l \in \Omega_k$ *such that* $t_l < t_n$, *and* $\chi \in \mathfrak{X}^*$, *that*

$$]\![\mathfrak{U}_{n-1,l+1}^* \chi]\!]_{*;l} \le C t_{n-l}^{-\xi_*} \|\chi\|_*. \tag{2.50}$$

In the case $\xi_* = \min\{\vartheta, 1-\gamma_1\}$, *the last estimate remains true if one adds* $\log(t_{n-l+2}/k)$ *to the factor* $t_{n-l+1}^{-\xi_*}$ *on the right. Furthermore, if* $0 \le \xi_* \le \min\{\vartheta, 1-\gamma_1, 1-\gamma_0\}$, *(2.50) holds with* $\widehat{\mathfrak{U}}_{n-1,l}$ *substituted for* $\mathfrak{U}_{n-1,l}$ *provided that one adds* $\log(t_{n-l+2}/k)$ *to the factor* $t_{n-l+1}^{-\xi_*}$ *on the right if* $\xi_* = \min\{\vartheta, 1-\gamma_1, 1-\gamma_0\}$.

Proof. By duality it follows from (2.43) that for all $t_n, t_l \in \Omega_k$ with $t_l \le t_n$,

$$\|\mathfrak{U}^*_{n-1,l}\|_* \le C \tag{2.51}$$

and

$$\|\widehat{\mathfrak{U}}^*_{n-1,l}\|_* + t^{\gamma_0}_{n-l+1}\|\mathfrak{U}^*_{n-1,l}\mathfrak{B}^*_n\|_* \le C. \tag{2.52}$$

By (2.51) the desired estimate for $]\![\mathfrak{U}^*_{n-1,l+1}\chi]\!]_{*;l}$ now easily follows when using (2.30), with the operators on both sides converted into their adjoints, and applying further Theorem 2.2 and the above argument. Next, based on the fact just shown, the quantity $]\![\widehat{\mathfrak{U}}^*_{n-1,l+1}\chi]\!]_{*;l}$ is estimated with the aid of (2.52), using (2.35) with the operators on both sides converted beforehand into their adjoints. □

Now we investigate other situations which are also important for applications.

Theorem 2.12. *Let seminorms* $|\!|\!|\cdot|\!|\!|_n$, $|\!|\!|\cdot|\!|\!|_{1;n}$, $|\!|\!|\cdot|\!|\!|_{*;n}$, $|\!|\!|\cdot|\!|\!|_{0;n}$, *and* $|\!|\!|\cdot|\!|\!|_{*0;n}$ *form respectively* $\Lambda^k_J(\gamma)$-, $\Lambda^k_J(\gamma_1)$-, $\Lambda^k_J(*;\gamma_*)$-, $\Lambda^k_J(\gamma_0)$-, *and* $\Lambda^k_J(*;\gamma_{*0})$-*concordant pairs with the operator* $\mathfrak{A}_n$ *for some* $\gamma, \gamma_1, \gamma_*, \gamma_0, \gamma_{*0} \in [0,1)$, $J \in \mathbb{N}$. *Suppose further that* $\mathfrak{A}_n$ *is under hypothesis* $\mathbf{H\ddot{O}L}^{(\to)}(\vartheta, \vartheta_1; |\!|\!|\cdot|\!|\!|_{\mathfrak{m};n}, \mathfrak{m} = \emptyset, I, *\emptyset)$ *with some* $\vartheta, \vartheta_1 \in [0,1]$ *such that* $\gamma_* < \xi_0 = 1 + \min\{\vartheta - \gamma, \vartheta_1 - \gamma_1\}$ *and with*

$$|\!|\!|\cdot|\!|\!|_{\mathfrak{m};n} = \begin{cases} |\!|\!|\cdot|\!|\!|_n & \text{if } \mathfrak{m} = \emptyset, \\ |\!|\!|\cdot|\!|\!|_{1;n} & \text{if } \mathfrak{m} = I, \\ |\!|\!|\cdot|\!|\!|_{*;n} & \text{if } \mathfrak{m} = *\emptyset, \end{cases} \tag{2.53}$$

while the operator $\mathfrak{B}_n$ *in (2.32) satisfies the condition, with some* $\vartheta_0 \in [0,1]$ *subject to* $\gamma_0 + \gamma_{*0} < 1 + \vartheta_0$, *for all* $k \in (0, K]$, $t_n \in \Omega_k$, $u \in \mathfrak{X}$, *and* $\chi \in \mathfrak{X}^*$,

$$|\langle \chi, \mathfrak{B}_n u\rangle| \le Ck^{\vartheta_0}|\!|\!|\chi|\!|\!|_{*0;n}|\!|\!|u|\!|\!|_{0;n}. \tag{2.54}$$

Then, if a seminorm $]\![\cdot]\!]_{*;n}$ *forms a* $\Lambda^k_J(*;\xi_*)$-*concordant pair with* $\mathfrak{A}_n$ *for some* $\xi_* \in [0, \xi_0]$, *we have, with* $\epsilon = 0$, *for all* $k \in (0, K]$, $t_n, t_l \in \Omega_k$ *with* $t_l < t_n$, *and* $\chi \in \mathfrak{X}^*$, *that*

$$]\![\mathfrak{U}^*_{n-1,l+1}\chi]\!]_{*;l} + \epsilon]\![\widehat{\mathfrak{U}}^*_{n-1,l+1}\chi]\!]_{*;l} \le \begin{cases} Ct^{-\xi_*}_{n-l}\|\chi\|_* & \text{if } \xi_* < \xi_0, \\ C\left(t^{-\xi_*}_{n-l} + t^{-\gamma_*}_{n-l}\log(t_{n-l+1}/k)\right)\|\chi\|_* & \text{if } \xi_* = \xi_0. \end{cases} \tag{2.55}$$

If in addition $\gamma_0 - \vartheta_0 < \xi_1 = 1 - \gamma_* + \min\{\vartheta, \vartheta_1\}$, $\gamma_{*0} < \xi_0$, *and* $\xi_* \leq 1 + \vartheta_0 - \gamma_0$, *then (2.55) holds for* $\epsilon = 1$, *provided that the factor* $t_{n-l}^{-\xi_*}$ *is replaced by* $t_{n-l}^{-\xi_*} + t_{n-l}^{-\gamma_{*0}} \log(t_{n-l+1}/k)$ *in the case* $\xi_* = 1 + \vartheta_0 - \gamma_0$.

Also, if a seminorm $]\![\cdot]\!]_n$ *forms a* $\Lambda_J^k(\xi)$*-concordant pair with* $\mathfrak{A}_n$ *for some* $\xi \in [0, \xi_1]$ *and if* $\max\{\gamma, \gamma_1\} < \xi_1$, *then we have, with* $\epsilon = 0$, *for all* $k \in (0, K]$, $t_n, t_l \in \Omega_k$ *with* $t_l \leq t_n$, *and* $u \in \mathfrak{X}$, *that*

$$
]\![\mathfrak{U}_{n-1,l}u]\!]_n + \epsilon]\![\widehat{\mathfrak{U}}_{n-1,l}u]\!]_n \leq \begin{cases} C t_{n-l+1}^{-\xi} \|u\| & \text{if } \xi < \xi_1, \\ C\left(t_{n-l+1}^{-\xi} + t_{n-l+1}^{-\gamma_2} \log(t_{n-l+2}/k)\right) \|u\| & \text{if } \xi = \xi_1, \end{cases} \tag{2.56}
$$

where

$$
\gamma_2 = \begin{cases} \gamma & \text{if } \vartheta < \vartheta_1, \\ \gamma_1 & \text{if } \vartheta > \vartheta_1, \\ \max\{\gamma, \gamma_1\} & \text{if } \vartheta = \vartheta_1. \end{cases} \tag{2.57}
$$

If in addition $\gamma_0 < \xi_1$, $\gamma_{*0} < \xi_0 + \vartheta_0$, *and* $\xi \leq 1 + \vartheta_0 - \gamma_{*0}$, *then (2.56) holds for* $\epsilon = 1$, *provided that the factor* $t_{n-l+1}^{-\xi}$ *is replaced by* $t_{n-l+1}^{-\xi} + t_{n-l+1}^{-\gamma_0} \log(t_{n-l+2}/k)$ *in the case* $\xi = 1 + \vartheta_0 - \gamma_{*0}$.

Proof. Assume first that $\gamma + \gamma_* - \vartheta < 1$ and $\gamma_1 + \gamma_* - \vartheta_1 < 1$. Using Theorems 2.1, 2.2, and the fact that the operator $\mathfrak{A}_n$ satisfies hypothesis $\mathbf{H\ddot{O}L}^{(\rightarrow)}(\vartheta, \vartheta_1; \|\!|\cdot\|\!|_{\mathfrak{m};n}, \mathfrak{m} = \emptyset, I, *\emptyset)$ with $\|\!|\cdot\|\!|_{\mathfrak{m}}$, $\mathfrak{m} = \emptyset, I, *\emptyset$, given by (2.53), we get, for all $t_l, t_q, t_n \in \Omega_k$ and $\chi \in \mathfrak{X}^*$, that

$$
\|\!|(\mathfrak{U}_l^*)^q \chi\|\!|_{*;l} \leq C t_{q+1}^{-\gamma_*} \|\chi\|_*, \tag{2.58}
$$

$$
]\![(\mathfrak{U}_l^*)^q \chi]\!]_{*;l} \leq C t_{q+1}^{-\xi_*} \|\chi\|_*, \tag{2.59}
$$

and

$$
\begin{aligned}
\|(\mathfrak{U}_l^*)^n (\mathfrak{A}_l^* - \mathfrak{A}_q^*)\chi\|_* &\leq C \|\!|\chi\|\!|_{*;q}\, t_{|q-l|}^{\vartheta} \sup_{\|u\|=1} \|\!|\mathfrak{U}_l^n u\|\!|_l \\
&\quad + C \|\!|\chi\|\!|_{*;q}\, t_{|q-l|}^{\vartheta_1} \sup_{\|u\|=1} \|\!|\mathfrak{U}_l^n u\|\!|_{1;l} \\
&\leq C\left(t_{|q-l|}^{\vartheta} t_{n+1}^{-\gamma} + t_{|q-l|}^{\vartheta_1} t_{n+1}^{-\gamma_1}\right) \|\!|\chi\|\!|_{*;q}.
\end{aligned} \tag{2.60}
$$

Combining (2.58), (2.59), and (2.60) therefore yields, with $p = \lfloor (q-l-1)/2 \rfloor$, for all $t_q, t_l \in \Omega_k$ with $t_l < t_q$ and $\chi \in \mathfrak{X}^*$,

$$
\begin{aligned}
\|\!|(\mathfrak{U}_l^*)^{q-l-1} (\mathfrak{A}_l^* - \mathfrak{A}_q^*)\chi\|\!|_{*;l} &\leq C t_{p+1}^{-\gamma_*} \|(\mathfrak{U}_l^*)^{q-l-1-p} (\mathfrak{A}_l^* - \mathfrak{A}_q^*)\chi\|_* \\
&\leq C\left(t_{q-l}^{\vartheta-\gamma-\gamma_*} + t_{q-l}^{\vartheta_1-\gamma_1-\gamma_*}\right) \|\!|\chi\|\!|_{*;q},
\end{aligned} \tag{2.61}
$$

and likewise,

$$]\!\![(\mathfrak{U}_l^*)^{q-l-1}(\mathfrak{A}_l^* - \mathfrak{A}_q^*)\,\chi]\!\![_{*;l} \le C\left(t_{q-l}^{\vartheta-\gamma-\xi_*} + t_{q-l}^{\vartheta_1-\gamma_1-\xi_*}\right) |\!|\!|\chi|\!|\!|_{*;q}. \qquad (2.62)$$

Taking now the seminorm $|\!|\!|\cdot|\!|\!|_{*;l}$ of both sides of the identity (2.30), with the operators converted beforehand into their adjoints and multiplied from the right by χ, and using further Theorem 2.2 and (2.61), we get, with $z_{n-1,l} = \sup_{\|\chi\|_*=1} |\!|\!|\mathfrak{U}_{n-1,l+1}^*\chi|\!|\!|_{*;l}$,

$$z_{n-1,l} \le Ct_{n-l}^{-\gamma_*} + Ck\sum_{q=l+1}^{n-1}\left(t_{q-l}^{\vartheta-\gamma-\gamma_*} + t_{q-l}^{\vartheta_1-\gamma_1-\gamma_*}\right) z_{n-1,q}, \quad t_l < t_n.$$

Since $\gamma + \gamma_* - \vartheta < 1$ and $\gamma_1 + \gamma_* - \vartheta_1 < 1$, applying Lemma 2.1, with $n-1$ substituted for n, to this inequality immediately yields

$$z_{n-1,l} \le Ct_{n-l}^{-\gamma_*} \qquad \text{for } t_l < t_n. \qquad (2.63)$$

Taking next the seminorm $]\!\![\cdot]\!\![_{*;l}$ of both sides of (2.30), with the operators converted into their adjoints and multiplied from the right by χ, and applying then (2.62), (2.63), and Theorem 2.2, we obtain, for $t_l < t_n$,

$$\sup_{\|\chi\|_*=1}]\!\![\mathfrak{U}_{n-1,l+1}^*\chi]\!\![_{*;l} \le Ct_{n-l}^{-\xi_*} + Ck\sum_{q=l+1}^{n-1}\left(t_{q-l}^{\vartheta-\gamma-\xi_*} + t_{q-l}^{\vartheta_1-\gamma_1-\xi_*}\right) t_{n-q-1}^{-\gamma_*},$$

which directly implies (2.55) with $\epsilon = 0$ since $t_{n-l}^{1+\vartheta-\gamma-\gamma_*} + t_{n-l}^{1+\vartheta_1-\gamma_1-\gamma_*} \le C$.

Let now $\max\{\gamma,\gamma_1\} + \gamma_* < 1 + \min\{\vartheta,\vartheta_1\}$. Since $\mathfrak{A}_n$ is under hypothesis $\mathbf{H\ddot{O}L}^{(\rightarrow)}(\vartheta,\vartheta_1;|\!|\!|\cdot|\!|\!|_{\mathfrak{m};n}, \mathfrak{m} = \emptyset, I, *\emptyset)$ and the pairs $(]\!\![\cdot]\!\![_n;\mathfrak{A}_n)$, $(|\!|\!|\cdot|\!|\!|_n;\mathfrak{A}_n)$, and $(|\!|\!|\cdot|\!|\!|_{1;n};\mathfrak{A}_n)$ are $\Lambda_J^k(\xi|1)$-, $\Lambda_J^k(\gamma|1)$-, and $\Lambda_J^k(\gamma_1|1)$-concordant, respectively, using Theorems 2.1 and 2.2 and reasonings similar to those leading to (2.61), it follows that for all $u \in \mathfrak{X}$ and $t_l, t_n, t_q \in \Omega_k$ such that $t_q \le t_n$,

$$t_{l+1}^{\gamma}|\!|\!|\mathfrak{U}_n^l u|\!|\!|_n + t_{l+1}^{\gamma_1}|\!|\!|\mathfrak{U}_n^l u|\!|\!|_{1;n} \le C\|u\| \qquad (2.64)$$

and

$$]\!\![\mathfrak{U}_n^l u]\!\![_n \le Ct_{l+1}^{-\xi}\|u\|, \qquad (2.65)$$

as well as

$$t_{n-q}^{\gamma}|\!|\!|\mathfrak{U}_n^{n-q-1}(\mathfrak{A}_n - \mathfrak{A}_q)\,u|\!|\!|_n + t_{n-q}^{\gamma_1}|\!|\!|\mathfrak{U}_n^{n-q-1}(\mathfrak{A}_n - \mathfrak{A}_q)\,u|\!|\!|_{1;n}$$
$$\le C\left(t_{n-q}^{\vartheta-\gamma_*}|\!|\!|u|\!|\!|_n + t_{n-q}^{\vartheta_1-\gamma_*}|\!|\!|u|\!|\!|_{1;n}\right) \qquad (2.66)$$

and

$$]\!\![\mathfrak{U}_n^{n-q-1}(\mathfrak{A}_n - \mathfrak{A}_q)\, u]\!\![_n \leq C\left(t_{n-q}^{-\xi-\gamma_*+\vartheta}\|\!|u|\!\|_n + t_{n-q}^{-\xi-\gamma_*+\vartheta_1}\|\!|u|\!\|_{1;n}\right). \qquad (2.67)$$

Applying (2.64) and (2.66), we take in turn the seminorms $\|\!|\cdot|\!\|_n$ and $\|\!|\cdot|\!\|_{1;n}$ of both sides of (2.29), with the expressions multiplied from the right by u, and we use the above argument to obtain

$$w_{n,l} \leq Ct_{n-l+1}^{-\gamma} + Ck\sum_{q=l}^{n-1}\left(t_{n-q}^{-\gamma+\vartheta-\gamma_*}w_{q,l} + t_{n-q}^{-\gamma+\vartheta_1-\gamma_*}w_{q,l}^{(1)}\right) \qquad (2.68)$$

and

$$w_{n,l}^{(1)} \leq Ct_{n-l+1}^{-\gamma_1} + Ck\sum_{q=l}^{n-1}\left(t_{n-q}^{-\gamma_1+\vartheta-\gamma_*}w_{q,l} + t_{n-q}^{-\gamma_1+\vartheta_1-\gamma_*}w_{q,l}^{(1)}\right), \qquad (2.69)$$

for all $t_n, t_l \in \Omega_k$ with $t_l \leq t_n$, with $w_{n,l} = \sup_{\|u\|=1}\|\!|\mathfrak{U}_{n-1,l}u|\!\|_n$ and $w_{n,l}^{(1)} = \sup_{\|u\|=1}\|\!|\mathfrak{U}_{n-1,l}u|\!\|_{1;n}$. As an easy consequence of (2.68) and (2.69), we have, with $\gamma_3 = \max\{\gamma, \gamma_1\}$ and $\gamma_4 = \max\{\gamma_3 - \vartheta + \gamma_*, \gamma_3 - \vartheta_1 + \gamma_*\}$,

$$w_{n,l} + w_{n,l}^{(1)} \leq Ct_{n-l+1}^{-\gamma_3} + Ck\sum_{q=l}^{n-1} t_{n-q}^{-\gamma_4}\left(w_{q,l} + w_{q,l}^{(1)}\right).$$

Thus, by Lemma 2.1 we get, since $\gamma_3, \gamma_4 < 1$,

$$w_{n,l} + w_{n,l}^{(1)} \leq Ct_{n-l+1}^{-\gamma_3}.$$

Combining this inequality with (2.68) (if $\gamma \leq \gamma_1$) or with (2.69) (if $\gamma \geq \gamma_1$) yields

$$w_{n,l} \leq Ct_{n-l+1}^{-\gamma} \qquad (2.70)$$

and

$$w_{n,l}^{(1)} \leq Ct_{n-l+1}^{-\gamma_1}. \qquad (2.71)$$

Using therefore (2.65), (2.67), (2.70), and (2.71) leads to (2.56) with $\epsilon = 0$, after a simple estimation based on applying (2.29) with both sides multiplied from the right by u.

It remains to be proved that under the corresponding conditions, (2.55) and (2.56) hold for $\epsilon = 1$ as well. Since both estimates can be established in a similar manner, we will focus on how to show (2.56) for $\epsilon = 1$.

To that end, assuming that $\gamma_0 < \xi_1$, $\gamma_{*0} < \xi_0+\vartheta_0$ and $\xi \le 1+\vartheta_0-\gamma_{*0}$, we use (2.34) and take (2.54) into account to obtain, after a simple estimation,

$$\begin{aligned}\sup_{\|u\|=1} |\!|\!|\widehat{\mathfrak{U}}_{n-1,l}u|\!|\!|_{0;n} \le \sup_{\|u\|=1} |\!|\!|\mathfrak{U}_{n-1,l}u|\!|\!|_{0;n} \\ + Ck^{1+\vartheta_0}\sum_{q=l}^{n-1} \sup_{\|u\|=1} |\!|\!|\mathfrak{U}_{n-1,q'+1}u|\!|\!|_{0;n} \\ \times \sup_{\|\chi\|_*=1} |\!|\!|\mathfrak{U}^*_{q',q+1}\chi|\!|\!|_{*0;q} \sup_{\|u\|=1} |\!|\!|\widehat{\mathfrak{U}}_{q-1,l}u|\!|\!|_{0;q},\end{aligned}$$

with $q' = \lfloor (n-1+q)/2 \rfloor$, for $t_n, t_l \in \Omega_k$ such that $t_l \le t_n$. Further, seeing from Lemma 1.2 that the seminorm $k^{\vartheta_0}|\!|\!|\cdot|\!|\!|_{*0;n}$ forms a $\Lambda^k_J(*;\gamma_{*0}-\vartheta_0)$-concordant pair with $\mathfrak{A}_n$ and using (2.55) and (2.56), both with $\epsilon = 0$, we have instead

$$\sup_{\|u\|=1} |\!|\!|\widehat{\mathfrak{U}}_{n-1,l}u|\!|\!|_{0;n} \le Ct_{n-l+1}^{-\gamma_0} + Ck\sum_{q=l}^{n-1} t_{n-q}^{\vartheta_0-\gamma_0-\gamma_{*0}} \sup_{\|u\|=1} |\!|\!|\widehat{\mathfrak{U}}_{q-1,l}u|\!|\!|_{0;q}.$$

By Lemma 2.1, since $\gamma_0 < 1$ and $\gamma_0+\gamma_{*0} < 1+\vartheta_0$, this immediately implies, for $t_l \le t_n$,

$$\sup_{\|u\|=1} |\!|\!|\widehat{\mathfrak{U}}_{n-1,l}u|\!|\!|_{0;n} \le Ct_{n-l+1}^{-\gamma_0}.$$

With this fact in mind, estimating similarly $\sup_{\|u\|=1}]\![\widehat{\mathfrak{U}}_{n-1,l}u]\!]_n$, again on the basis of (2.34), it is readily seen that (2.56) holds with $\epsilon = 1$. □

Using the same method one can prove

Theorem 2.13. *Suppose that for some* $\gamma, \gamma_*, \gamma_{*1}, \gamma_0, \gamma_{*0} \in [0,1)$, *and* $J \in \mathbb{N}$, *seminorms* $|\!|\!|\cdot|\!|\!|_n$, $|\!|\!|\cdot|\!|\!|_{*;n}$, $|\!|\!|\cdot|\!|\!|_{*1;n}$, $|\!|\!|\cdot|\!|\!|_{0;n}$, *and* $|\!|\!|\cdot|\!|\!|_{*0;n}$ *form respectively* $\Lambda^k_J(\gamma)$-, $\Lambda^k_J(*;\gamma_*)$-, $\Lambda^k_J(*;\gamma_{*1})$-, $\Lambda^k_J(\gamma_0)$-, *and* $\Lambda^k_J(*;\gamma_{*0})$*-concordant pairs with the operator* $\mathfrak{A}_n$. *Furthermore, let* $\mathfrak{A}_n$ *itself be under hypothesis* $\mathbf{H\ddot{O}L}^{(\leftarrow)}(\vartheta,\vartheta_1;|\!|\!|\cdot|\!|\!|_{\mathfrak{m};n}, \mathfrak{m} = \emptyset, *\emptyset, *I)$ *with some* $\vartheta, \vartheta_1 \in [0,1]$ *such that* $\gamma < \xi_0 = 1+\min\{\vartheta-\gamma_*, \vartheta_1-\gamma_{*1}\}$ *and with*

$$|\!|\!|\cdot|\!|\!|_{\mathfrak{m};n} = \begin{cases} |\!|\!|\cdot|\!|\!|_n & \textit{if } \ \mathfrak{m} = \emptyset, \\ |\!|\!|\cdot|\!|\!|_{*;n} & \textit{if } \ \mathfrak{m} = *\emptyset, \\ |\!|\!|\cdot|\!|\!|_{*1;n} & \textit{if } \ \mathfrak{m} = *I, \end{cases}$$

and let the operator $\mathfrak{B}_n$ *in (2.32) satisfy condition (2.54) with some* $\vartheta_0 \in [0,1]$ *subject to* $\gamma_0+\gamma_{*0} < 1+\vartheta_0$. *Then, if a seminorm* $]\![\cdot]\!]_n$ *forms a* $\Lambda^k_J(\xi)$*-concordant pair with* $\mathfrak{A}_n$ *for some* $\xi \in [0,\xi_0]$, *we have, with* $\epsilon = 0$, *for all*

$k \in (0, K]$, $t_n, t_l \in \Omega_k$ with $t_l \leq t_n$, and $u \in \mathfrak{X}$, that

$$\rrbracket \mathfrak{U}_{n-1,l} u \llbracket_n + \epsilon \rrbracket \widehat{\mathfrak{U}}_{n-1,l} u \llbracket_n \leq \begin{cases} C t_{n-l+1}^{-\xi} \|u\| & \text{if } \xi < \xi_0, \\ C\left(t_{n-l+1}^{-\xi} + t_{n-l+1}^{-\gamma} \log(t_{n-l+2}/k)\right) \|u\| & \text{if } \xi = \xi_0. \end{cases} \tag{2.72}$$

If in addition $\gamma_{*0} - \vartheta_0 < \xi_1 = 1 - \gamma + \min\{\vartheta, \vartheta_1\}$, $\gamma_0 < \xi_0$, and $\xi \leq 1 + \vartheta_0 - \gamma_{*0}$, then (2.72) holds for $\epsilon = 1$ provided that the factor $t_{n-l+1}^{-\xi}$ is replaced by $t_{n-l+1}^{-\xi} + t_{n-l+1}^{-\gamma_0} \log(t_{n-l+2}/k)$ in the case $\xi = 1 + \vartheta_0 - \gamma_{*0}$.

Also, if a seminorm $\rrbracket \cdot \llbracket_{*;n}$ forms a $\Lambda_J^k(*; \xi_*)$-concordant pair with $\mathfrak{A}_n$ for some $\xi_* \in [0, \xi_1]$ and if $\max\{\gamma_*, \gamma_{*1}\} < \xi_1$, it then holds, with $\epsilon = 0$, for all $k \in (0, K]$, $t_n, t_l \in \Omega_k$ with $t_l < t_n$, and $\chi \in \mathfrak{X}^*$,

$$|\!|\!| \mathfrak{U}_{n-1,l+1}^* \chi |\!|\!|_{*;l} + \epsilon |\!|\!| \widehat{\mathfrak{U}}_{n-1,l+1}^* \chi |\!|\!|_{*;l} \leq \begin{cases} C t_{n-l}^{-\xi_*} \|\chi\|_* & \text{if } \xi_* < \xi_1, \\ C\left(t_{n-l}^{-\xi_*} + C t_{n-l}^{-\gamma_{*2}} \log(t_{n-l+1}/k)\right) \|\chi\|_* & \text{if } \xi_* = \xi_1, \end{cases} \tag{2.73}$$

where

$$\gamma_{*2} = \begin{cases} \gamma_* & \text{if } \vartheta < \vartheta_1, \\ \gamma_{*1} & \text{if } \vartheta > \vartheta_1, \\ \max\{\gamma_*, \gamma_{*1}\} & \text{if } \vartheta = \vartheta_1. \end{cases} \tag{2.74}$$

If in addition $\gamma_{*0} < \xi_1$, $\gamma_0 < \xi_0 + \vartheta_0$, and $\xi_* \leq 1 + \vartheta_0 - \gamma_0$, then (2.73) holds for $\epsilon = 1$, provided that $t_{n-l}^{-\xi_*}$ is replaced by $t_{n-l+1}^{-\xi_*} + t_{n-l+1}^{-\gamma_{*0}} \log(t_{n-l+2}/k)$ in the case $\xi_* = 1 + \vartheta_0 - \gamma_0$.

Some further developments are contained in the next two assertions.

Theorem 2.14. *Let there be an operator $\mathfrak{V}_n \in \mathcal{B}(\mathfrak{X})$ and seminorms $|\!|\!| \cdot |\!|\!|_n$ and $|\!|\!| \cdot |\!|\!|_{1;n}$ such that the triplets $\left(|\!|\!| \cdot |\!|\!|_n; \mathfrak{A}_n; \mathfrak{V}_n\right)$ and $\left(|\!|\!| \cdot |\!|\!|_{1;n}; \mathfrak{A}_n; \mathfrak{V}_n\right)$ are $\Lambda_J^k(1)$- and $\Lambda_J^k(\gamma_1)$-concordant, respectively, for some $J \in \mathbb{N}$ and $\gamma_1 \in [0, 1)$. Let also $\mathfrak{A}_n$ satisfy hypothesis $\mathbf{H\ddot{O}L}^{(\rightarrow)}(\vartheta, 0; |\!|\!| \cdot |\!|\!|_{\mathfrak{m};n}, \mathfrak{m} = \emptyset, I, *\emptyset)$ and let, for all $t_n \in \Omega_k$, $u \in \mathfrak{X}$, and $\chi \in \mathfrak{X}^*$,*

$$|\langle \chi, \mathfrak{B}_n u \rangle| \leq C |\!|\!| \chi |\!|\!|_{*\emptyset;n} \left(k^{\vartheta} |\!|\!| u |\!|\!|_{\emptyset;n} + |\!|\!| u |\!|\!|_{I;n}\right),$$

with some $\vartheta \in (0, 1]$ and with

$$|\!|\!| \cdot |\!|\!|_{\mathfrak{m};n} = \begin{cases} |\!|\!| \cdot |\!|\!|_n & \text{if } \mathfrak{m} = \emptyset, \\ |\!|\!| \cdot |\!|\!|_{1;n} & \text{if } \mathfrak{m} = I, \\ \|\mathfrak{V}_n^* \cdot \|_* & \text{if } \mathfrak{m} = *\emptyset. \end{cases}$$

Finally, suppose that for some $\mathfrak{G}_n \in \mathcal{B}(\mathfrak{X})$, *with the seminorm* $]\![\cdot]\!]_n$ *given by*

$$]\![u]\!]_n = \|\mathfrak{G}_n u\| \quad \text{for } u \in \mathfrak{X}, \tag{2.75}$$

the triplet $\left(]\![\cdot]\!]_n; \mathfrak{A}_n; \mathfrak{V}_n\right)$ *is strictly* $\Lambda_J^k(\xi)$*-concordant for some* $\xi \in [0,1)$. *Then, putting* $\mathfrak{V}_{-1} = \mathfrak{V}_0$, *we have*

$$]\![\mathfrak{U}_{n-1,l}\mathfrak{V}_{l-1}u]\!]_n +]\![\widehat{\mathfrak{U}}_{n-1,l}\mathfrak{V}_{l-1}u]\!]_n \le Ct_{n-l+1}^{-\xi}\|u\|, \tag{2.76}$$

for all $k \in (0,K]$, $t_n, t_l \in \Omega_k$ *with* $t_l \le t_n$, *and* $u \in \mathfrak{X}$.

Proof. For the sake of convenience we put $\mathfrak{U}_{-1} = \mathfrak{U}_0$ and $\mathfrak{A}_{-1} = \mathfrak{A}_0$. It follows from Theorem 2.4 and Remark 2.3 that for all $t_n, t_l \in \Omega_k$ with $t_l \le t_n$ and $u \in \mathfrak{X}$,

$$]\![\mathfrak{U}_{l-1}^{n-l}\mathfrak{V}_{l-1}u]\!]_n \le Ct_{n-l+1}^{-\xi}\|u\|. \tag{2.77}$$

Using further the accepted Hölder-continuity type condition on $\mathfrak{A}_n$, taking into account the properties of the seminorms $|\!|\!|\cdot|\!|\!|_n$, $|\!|\!|\cdot|\!|\!|_{1;n}$, and applying Theorem 2.1, we find, for $t_q, t_l \in \Omega_k$ with $t_l \le t_q$, $u \in \mathfrak{X}$, and $\chi \in \mathfrak{X}^*$,

$$|\langle \chi, (\mathfrak{A}_{l-1} - \mathfrak{A}_q)\,\mathfrak{U}_{l-1}^{q-l}\mathfrak{V}_{l-1}u\rangle| \le C\left(t_{q-l+1}^{\vartheta-1} + t_{q-l+1}^{-\gamma_1}\right)\|\mathfrak{V}_q^*\chi\|_*\|u\|.$$

Since the seminorm $]\![\cdot]\!]_n$ is of a specific construction, the last estimate furthermore implies, for all $t_n, t_q, t_l \in \Omega_k$ with $t_l \le t_q < t_n$,

$$\begin{aligned}\sup_{\|u\|=1}]\![\mathfrak{U}_{n-1,q+1}(\mathfrak{A}_{l-1} - \mathfrak{A}_q)\,\mathfrak{U}_{l-1}^{q-l}\mathfrak{V}_{l-1}u]\!]_n \\ \le C\left(t_{q-l+1}^{\vartheta-1} + t_{q-l+1}^{-\gamma_1}\right)\sup_{\|u\|=1}]\![\mathfrak{U}_{n-1,q+1}\mathfrak{V}_q u]\!]_n.\end{aligned} \tag{2.78}$$

Applying next (2.30), with $l-1$ substituted for l and with both sides multiplied from the right by $\mathfrak{V}_{l-1}u$, and using (2.77) and (2.78), a simple estimation gives

$$z_{n,l} \le Ct_{n-l+1}^{-\xi} + Ck\sum_{q=l}^{n-1} z_{n,q+1}\left(t_{q-l+1}^{\vartheta-1} + t_{q-l+1}^{-\gamma_1}\right),$$

for all $t_n, t_l \in \Omega_k$ such that $t_l \le t_n$, with $z_{n,l} = \sup_{\|u\|=1}]\![\mathfrak{U}_{n-1,l}\mathfrak{V}_{l-1}u]\!]_n$. Since $1-\vartheta < 1$, $\gamma_1 < 1$, and $\xi < 1$, the needed estimate for the first term on the left-hand side of (2.76) now follows by Lemma 2.1. With this fact in mind, the second term is then estimated by using (2.35) and the above argument. □

Remark 2.4. *In the conditions of Theorem 2.14 the requirement that the seminorm* $]\![\cdot]\!]_n$ *must be represented in the form (2.75) can be avoided if, instead of hypothesis* $\mathbf{H\ddot{O}L}^{(\to)}(\vartheta, 0; \| \cdot \|_{\mathfrak{m};n}, \mathfrak{m} = \emptyset, I, *\emptyset)$, *it is accepted that* $\mathfrak{A}_n - \mathfrak{A}_l = \mathfrak{V}_l \mathfrak{H}_{n,l}$ *with some* $\mathfrak{H}_{n,l} \in \mathcal{B}(\mathfrak{X})$, *where, for all* $t_n, t_l \in \Omega_k$ *and* $u \in \mathfrak{X}$,

$$\|\mathfrak{H}_{n,l}u\| \leq C\left(t^{\vartheta}_{|n-l|}\|u\|_n + \|u\|_{1;n}\right),$$

with the same ϑ, $\| \cdot \|_n$, $\| \cdot \|_{1;n}$, *and* $\mathfrak{V}_n$ *as in the statement of the theorem.*

This result can easily be shown by using a slight modification of the above proof.

Theorem 2.15. *Suppose that the conditions of Theorem 2.14 are satisfied for some* $J \in \mathbb{N}$, $\vartheta \in (0, 1]$, *and* $\gamma_1, \xi \in [0, 1)$, *with the difference that we now accept the* $\Lambda^k_J(\xi|1)$*-concordance of the triplet* $\left(]\![\cdot]\!]_n; \mathfrak{A}_n; \mathfrak{V}_n\right)$ *instead of its strict* $\Lambda^k_J(\xi)$*-concordance. In addition we suppose that* $\operatorname{Ran}\mathfrak{V}_n$ *is independent of* t_n *and the operator* $\mathfrak{V}_n$ *has an inverse* $\mathfrak{V}_n^{-1}$ *such that for all* $k \in (0, K]$ *and* $t_n, t_l \in \Omega_k$,

$$\|\mathfrak{V}_n^{-1}\mathfrak{V}_l\| \leq C. \tag{2.79}$$

Then the stability estimate (2.76) holds.

Proof. Using (2.41) with $m = 1$, with both sides multiplied beforehand from the right by $\mathfrak{V}_l u$, and applying furthermore (2.79) and the above argument (cf. the proof of Theorem 2.10), we conclude that the triplet $\left(]\![\cdot]\!]_n; \mathfrak{A}_n; \mathfrak{V}_n\right)$ is, in fact, strictly $\Lambda^k_J(\xi)$-concordant. It thus remains to apply Theorem 2.14. □

2.4 Stability of discrete Cauchy problems

Now we apply the above results for showing stability estimates for problem (1.1).

Theorem 2.16. *Let* $T < \infty$, *let seminorms* $]\![\cdot]\!]_n$, $\| \cdot \|_n$, *and* $\| \cdot \|_{1;n}$ *and the operators* $\mathfrak{A}_n$, $\mathfrak{B}_n$ *satisfy the conditions posed in the former part of the assertion of Theorem 2.10, with some* $\vartheta \in (0, 1]$ *and* $\xi, \gamma_1, \gamma_0 \in [0, 1)$, *and let*

$$\|\mathfrak{Q}_n\| + \|\mathfrak{R}\| \leq C \qquad \text{for all } k \in (0, K] \text{ and } t_n \in \breve{\Omega}_k. \tag{2.80}$$

Then the solution Y_n of both (1.1) and (2.32) satisfies the estimate, for all $k \in (0, K]$ and $t_n \in \Omega_k$,

$$\rrbracket Y_n \llbracket_n \le C \left(t_{n+1}^{-\xi} \|y^0\| + k \sum_{l=0}^{n-1} t_{n-l}^{-\xi} \|F_l\| \right). \tag{2.81}$$

If in addition the operators $\mathfrak{A}_n$, $\mathfrak{B}_n$ and the seminorm $\rrbracket \cdot \llbracket_n$ satisfy the conditions imposed in the latter part of the assertion of Theorem 2.10, with the same ϑ as above and with some $\xi \in [0, 1+\vartheta]$, then the estimate (2.81) remains valid provided that the expression $t_q^{-1} \log(t_{q+1}/k)$ is added to the factor $t_q^{-\xi}$ on the right, for $q = n+1$ and $q = n - l$, whenever $\xi = 1 + \vartheta$.

This assertion is a direct consequence if Theorem 2.10 is applied to (1.4) and (2.33).

Theorem 2.17. *Let the operator $\mathfrak{A}_n$ and seminorms $\rrbracket \cdot \llbracket_n$, $\| \cdot \|_n$, $\| \cdot \|_{*;n}$, and $\| \cdot \|_{1;n}$ be under the conditions of Theorem 2.12 for some $J \in \mathbb{N}$, $\xi \ge 0$, $\gamma, \gamma_*, \gamma_1 \in [0, 1)$, and $\vartheta, \vartheta_1 \in [0, 1]$ such that $\max\{\gamma, \gamma_1\} < \xi_1 = 1 - \gamma_* + \min\{\vartheta, \vartheta_1\}$, $\xi \le \xi_1$, and let (2.80) hold. Then, if $T < \infty$, the solution Y_n of (1.1) satisfies the stability estimate (2.81) in which, if $\xi = \xi_1$, $t_q^{-\xi}$ should be replaced by $t_q^{-\xi} + t_q^{-\gamma_2} \log(t_{q+1}/k)$ for $q = n + 1$ and $q = n - l$, where γ_2 is defined by (2.57).*

Furthermore, suppose in addition that the operator $\mathfrak{B}_n$ and seminorms $\| \cdot \|_{0;n}$, $\| \cdot \|_{*0;n}$ are subject to the conditions of Theorem 2.12 for some $\gamma_0 < \xi_1$ and $\gamma_{*0} < 1 + \vartheta_0 + \min\{\vartheta - \gamma, \vartheta_1 - \gamma_1\}$. Then, keeping all the above restrictions and assuming in addition that $\xi < 1 + \vartheta_0 - \gamma_{*0}$, (2.81) also holds for the solution of the perturbed problem (2.32). The last assertion remains valid for $\xi = 1 + \vartheta_0 - \gamma_{*0}$ as well provided that the factor $t_q^{-\xi}$ is replaced by $t_q^{-\xi} + t_q^{-\gamma_0} \log(t_{q+1}/k)$ for $q = n + 1$ and $q = n - l$.

In the particular case where $\mathfrak{A}_n = \mathfrak{A}$ is independent of t_n, (2.81) is valid for the solution of (1.1) for any fixed $\xi \ge 0$. If in addition the pair $(\rrbracket \cdot \llbracket_n; \mathfrak{A})$ is $\Lambda_J(\xi)$-concordant, then (2.81) for the solution of (1.1) holds in the case $T = \infty$, that is, for $n = 0, 1, \ldots$.

The claim follows by Theorems 2.12 and 2.1 and with the aid of (1.4) and (2.33).

Theorem 2.18. *Let the operator $\mathfrak{A}_n$ and seminorms $\rrbracket \cdot \llbracket_n$, $\| \cdot \|_n$, $\| \cdot \|_{*;n}$, and $\| \cdot \|_{*1;n}$ be under the conditions of Theorem 2.13 for some $J \in \mathbb{N}$, $\xi \ge 0$, $\gamma, \gamma_*, \gamma_{*1} \in [0, 1)$, and $\vartheta, \vartheta_1 \in [0, 1]$ such that $\gamma < \xi_0 = 1 + \min\{\vartheta - \gamma_*, \vartheta_1 - \gamma_{*1}\}$, $\xi \le \xi_0$, and let (2.80) hold. Then, if $T < \infty$, the solution Y_n of*

(1.1) satisfies the stability estimate (2.81) in which, if $\xi = \xi_0$, $t_q^{-\xi}$ *should be replaced by* $t_q^{-\xi} + t_q^{-\gamma}\log(t_{q+1}/k)$ *for* $q = n+1$ *and* $q = n-l$.

Furthermore, suppose in addition that the operator $\mathfrak{B}_n$ *and seminorms* $\|\!| \cdot |\!\|_{0;n}$, $\|\!| \cdot |\!\|_{*0;n}$ *are subject to the conditions of Theorem 2.13 for some* $\gamma_{*0} < 1 + \vartheta_0 - \gamma + \min\{\vartheta, \vartheta_1\}$ *and* $\gamma_0 < \xi_0$. *Then, keeping all the above restrictions and assuming in addition that* $\xi \le 1 + \vartheta_0 - \gamma_{*0}$, *(2.81) also holds for the solution of the perturbed problem (2.32). The last assertion remains valid for* $\xi = 1 + \vartheta_0 - \gamma_{*0}$ *provided that the factor* $t_q^{-\xi}$ *is replaced by* $t_q^{-\xi} + t_q^{-\gamma_0}\log(t_{q+1}/k)$ *for* $q = n+1$ *and* $q = n-l$.

This result is a consequence if Theorem 2.13 is applied to (1.4) and (2.33).

Theorem 2.19. *Let an operator* $\mathfrak{V}_n \in \mathcal{B}(\mathfrak{X})$, *seminorms* $[\![\cdot]\!]_n$, $\|\!| \cdot |\!\|_n$, *and* $\|\!| \cdot |\!\|_{1;n}$, *and the operators* $\mathfrak{A}_n$, $\mathfrak{B}_n$ *satisfy the conditions of Theorem 2.14 or 2.15 with some* $J \in \mathbb{N}$, $\xi \ge 0$, $\gamma_1 \in [0,1)$, *and* $\vartheta \in (0,1]$, *and let, for all* $t_n \in \breve{\Omega}_k$, *and* $\chi \in \mathfrak{X}^*$,

$$\|\mathfrak{Q}_n^*\chi\|_* \le C\|\mathfrak{V}_n^*\chi\|_* \text{ and } \|\mathfrak{R}^*\chi\|_* \le C\|\mathfrak{V}_0^*\chi\|_*. \tag{2.82}$$

Then, if $T < \infty$ *and* $\xi \in [0,1)$, *the solution of both (1.1) and (2.32) satisfies (2.81). If in addition the operator* $\mathfrak{A}_n = \mathfrak{A}$ *is independent of* t_n *and the triplet* $([\![\cdot]\!]_n; \mathfrak{A}; \mathfrak{V}_n)$ *is* $\Lambda_J^k(\xi)$*-concordant, then (2.81) for the solution of (1.1) is valid for any fixed* $\xi \ge 0$. *Moreover, in the case where the triplet* $([\![\cdot]\!]_n; \mathfrak{A}; \mathfrak{V}_n)$ *is* $\Lambda_J(\xi)$*-concordant, the last assertion holds with* $T = \infty$, *that is, for all* $n = 0, 1, \ldots$.

The result immediately follows when applying Theorems 2.14 or 2.15 and Theorem 2.4 to (1.1) and to (2.32) and taking (2.82) into account.

In conclusion we consider an application of the results of Section 2.2, concerning the behaviour of stability constants. With the same restrictions as in Section 2.2 we are able to deal with the case $T = \infty$.

Theorem 2.20. *Let the operator* $\mathfrak{A}_n = \mathfrak{A}$ *be independent of* t_n, *let it satisfy the* $\Lambda_J(M)$*-condition with some* $J \in \mathbb{N}$ *and* $M \ge 1$, *and let (2.80) hold. Suppose further that* ξ *is a fixed number such that* $\xi \ge 0$ *if* $J = 1$ *and* $\xi = 0$ *if* $J \ge 2$. *Then the solution* Y_n *of (1.1) satisfies the estimate, for all* $k \in (0, K]$ *and* $n = 0, 1, \ldots$,

$$\|\mathfrak{A}^\xi Y_n\| \le CM\left(t_{n+1}^{-\xi}\|y^0\| + k\sum_{l=0}^{n-1} t_{n-l}^{-\xi}\|F_l\|\right).$$

If instead the operator $k\mathfrak{A}$ satisfies the $ST_J(M)$-condition, then the last estimate holds with $K(M)$ given by (2.18) in place of M.

This result follows directly as an application of Theorems 2.6, 2.7, 2.8, and 2.9.

2.5 Comments and bibliographical remarks

Section 2.1: The problem of stability of discrete semigroups in Banach space settings has been extensively examined in the literature by using different resolvent conditions on the operator in question. For such results in finitely many dimensions, we mention the work of Kreiss [97], Le Veque & Trefethen [103], Miller & Strang [112], Morton [113], Spijker [153], and Tadmor [157, 158]. As concerns the infinite-dimensional case, we refer to our work [10, 11, 12, 17, 33, 34] as well as to the work of El-Fallah & Ransford [67], Kalton *et al.* [89], Lubich & Nevanlinna [104], Nagy & Zemánek [114], Nevanlinna [116, 117, 118], Spijker & Straetemans [154], and Vitse [165] (see also references therein). These results are often intended for use in rather general situations while in the present book we restrict ourselves to the parabolic case for which it is natural, as pointed out above, to work with the notions defined in Chapter 1. In particular, Theorem 2.3, as stated under the Λ_J^k-condition, actually arises from our earlier work [10, 11] (see also [17]). However, for the aims of this book we prefer to present our main results as stated in Theorems 2.1, 2.2, 2.4, and 2.5 which further generalize Theorem 2.3 and are useful for various applications of the theory.

Also, we remark that the above estimate (2.7) has a counterpart in the theory of holomorphic semigroups. More precisely, for a closed, densily defined operator A such that $A \in \mathcal{S}(\varphi, 0)$ with some $\varphi \in (0, \pi/2)$, it follows that A generates a holomorphic bounded semigroup e^{-tA} and, with any fixed $\xi \geq 0$,[12]

$$\|A^{\xi} e^{-tA}\| \leq Ct^{-\xi} \qquad \text{for } t > 0.$$

(Cf. Komatsu [93, Theorem 12.2].)

Section 2.2: In our opinion, the $\Lambda_J(M)$-condition for an operator $\mathfrak{L}$ looks more natural than the $ST_J(M)$-condition for the operator $k\mathfrak{L}$, so that the $ST_J(M)$-condition is of more theoretical interest. As it follows from the discussion in Section 2.2, in fact, the $ST_J(M)$-condition as stated for $k\mathfrak{L}$ implies the $\Lambda_J(M)$-condition for $\mathfrak{L}$, with a larger constant M_1 in place of

[12]The fractional powers A^{ξ}, $\xi \geq 0$, are defined under the stated conditions.

M. This fact means that problem (1.1) is naturally parabolic if the operator $k\mathfrak{A}$ is under the $ST_J(M)$-condition. On the other hand, as it will be seen subsequently in the book (see Remark 6.4), the $\Lambda_J(M)$-condition appears naturally and should therefore be thought of as original, not as a produce of the $ST_J(M_0)$-condition with some $M_0 < M$ in place of M. Nevertheless, the $ST_J(M)$-condition is examined here for the reason that it is actively discussed in the literature (see, e.g., [48] and [118]), where the question arises in which way the size of the constant M appears in stability estimates. It is worth noting that it has been shown in Borovykh, Drissi, and Spijker [48] (cf. also El-Fallah & Ransford [67] and Vitse [165]), under a condition which is equivalent to the $ST_1(M)$-condition, that $\|\mathfrak{U}^n\| \leq CM^2$, while the above Theorem 2.8 gives $\|\mathfrak{U}^n\| \leq CM\log(1+M)$. Also, under the same restriction, Yuan [167] has found an estimate which is equivalent to the fact that $\|\mathfrak{A}\mathfrak{U}^n\| \leq CM^3 t_{n+1}^{-1}$, while it follows from Theorem 2.9 that $\|\mathfrak{A}\mathfrak{U}^n\| \leq CM^2\log(1+M)\, t_{n+1}^{-1}$. Apart from everything else, Theorem 2.9 gives an estimation of the quantity $\|\mathfrak{A}^\xi\mathfrak{U}^n\|$ for fractional values of $\xi \geq 0$.
Section 2.3: Discrete versions of Gronwall's lemma is a widespread tool of analysis in discrete problems. As a rule, inequalities like those appearing in the statement of Lemma 2.1 have been dealt with in the case $\zeta_1 = \zeta_2 = 0$ (see, e.g., Sloan & Thomée [147] or Zhang [168]). For our needs we are forced to treat the general case where $0 \leq \zeta_1, \zeta_2 < 1$.

As concerns the stability assertions collected in the present section, we are not informed of any related results, except our own work [10, 11, 34] based on assumptions in a more concrete form.

Chapter 3

Operator Splitting Problems

In this chapter we examine discrete problems that are constructed on the basis of some ideas of operator splitting. For this class of problems the operator in question is decomposed in a special way and our starting assumptions below are stated in terms of restrictions on each separate component of the original operator. Our investigation here is mainly devoted to the analysis of one of the simplest constructions based on operator splitting.[1] More precisely, in this chapter we mainly deal with problem (1.1) in which $\mathfrak{A}_n$ is represented as follows

$$\mathfrak{A}_n = \mathfrak{A}_{1;n} + \mathfrak{A}_{2;n} - k\mathfrak{A}_{2;n}\mathfrak{A}_{1;n}, \tag{3.1}$$

with some given operators $\mathfrak{A}_{j;n} \in \mathcal{B}(\mathfrak{X})$, $j = 1, 2$. To settle even this case one has to employ essentially new techniques in comparison with the analysis carried out in Chapter 2. In order to avoid extra difficulties and to make our presentation readable enough, we restrict ourselves to the case where $\mathfrak{A}_n$ satisfies the Λ_1^k-condition, which does happen, as it will be seen later, if both $\mathfrak{A}_{1;n}$ and $\mathfrak{A}_{2;n}$ satisfy the Λ_1^k-condition.[2] For the same reason we assume that the condition[3] $\mathfrak{Q}_n = \mathfrak{R} = I$ holds and we use no seminorms which form concordant pairs with the operators $\mathfrak{A}_n$ or $\mathfrak{A}_{j;n}$, $j = 1, 2$, and which would differ from $\|(\mathfrak{A}_n + \mu_0 I)^\xi \cdot \|$ or $\|(\mathfrak{A}_{j;n} + \mu_0 I)^\xi \cdot \|$, $\xi \geq 0$. At the same time our argument is further extended in Section 3.3 in order to analyse a more involved construction which admits second order solvers. However the analysis becomes more involved in the case of general higher

[1] In applications this kind of splitting leads, at most, to first order solvers.

[2] The case with the Λ_J^k-condition, $J \geq 2$, in principle can be considered but it turns out to be rather technical and it is not discussed here.

[3] We could suppose (2.80) instead.

order solvers, so we leave the corresponding details outside of our present scope.

In the considered case $\mathfrak{Q}_n = \mathfrak{R} = I$, problem (1.1), (3.1) can be equivalently represented as a time stepping method

$$\begin{aligned} Y_{n+1} &= (I - k\mathfrak{A}_n)Y_n + kF_n \\ &= (I - k\mathfrak{A}_{2;n})\,(I - k\mathfrak{A}_{1;n})Y_n + kF_n \qquad \text{for } t_n \in \breve{\Omega}_k, \\ Y_0 &= y^0, \end{aligned}$$

for which the transition operator $\mathfrak{U}_n = I - k\mathfrak{A}_n$ is therefore split into two factors.

As a matter of fact, the analysis of problems with a splitting operator becomes rather difficult if the operators $\mathfrak{A}_{1;n}$ and $\mathfrak{A}_{2;n}$ in the splitting construction do not commute, as we will see in the case below. In this situation the concept of commutator will play an important role in what follows. More precisely, given linear operators $\mathfrak{L}_1$ and $\mathfrak{L}_2$, their commutator Com $(\mathfrak{L}_1|\mathfrak{L}_2)$ is defined as

$$\text{Com }(\mathfrak{L}_1|\mathfrak{L}_2) = \mathfrak{L}_1\mathfrak{L}_2 - \mathfrak{L}_2\mathfrak{L}_1,$$

whenever the expression on the right makes sense. (It is certainly the case if both $\mathfrak{L}_1$ and $\mathfrak{L}_2$ are bounded.) In the following considerations we base ourselves on two hypotheses which involve the commutator Com $(\mathfrak{A}_{1;n}|\mathfrak{A}_{2;n})$, as follows, with some $\kappa \in [0,1)$ and $\mu_0 \geq 0$, for all $k \in (0,K]$, $t_n \in \Omega_k$, $\chi \in \mathfrak{X}^*$, and $u \in \mathfrak{X}$:

COM1$[\kappa]$ which stands for

$$\begin{aligned} |\langle \chi, \text{Com }(\mathfrak{A}_{1;n}|\mathfrak{A}_{2;n})\,u\rangle| &\leq C\|(\mathfrak{A}^*_{j;n} + \mu_0 I)\,\chi\|_*^{\kappa}\|\chi\|_*^{1-\kappa}\|(\mathfrak{A}_{3-j;n} + \mu_0 I)\,u\| \\ &+ C\|(\mathfrak{A}^*_{j;n} + \mu_0 I)\,\chi\|_*\|(\mathfrak{A}_{3-j;n} + \mu_0 I)\,u\|^{\kappa}\|u\|^{1-\kappa}, \quad j = 1,2, \quad (3.2) \end{aligned}$$

and

COM2$[\kappa]$ which stands for

$$|\langle \chi, \text{Com }(\mathfrak{A}_{1;n}|\mathfrak{A}_{2;n})\,u\rangle| \leq C\sum_{j=1}^{2}\|(\mathfrak{A}^*_{j;n} + \mu_0 I)^{\kappa}\chi\|_*\|(\mathfrak{A}_{3-j;n} + \mu_0 I)\,u\|. \quad (3.3)$$

Accepting these conditions means that even though the operators $\mathfrak{A}_{1;n}$ and $\mathfrak{A}_{2;n}$ may not commute, they "almost commute" in the sense that their commutator is "weaker" than their product, as stated in (3.2) and (3.3). Also, we remark that making use of functions of ordered non-commutative operators is an essential ingredient of this chapter's analysis. The necessary knowledge of such functions and the corresponding notation are contained in Appendix A.4.

Note that due to the non-commutativity of the operators $\mathfrak{A}_{j;n}$, $j = 1, 2$, even if both satisfy the Λ_1-condition, the spectrum of the original operator $\mathfrak{A}_n$ may contain points with negative real part. This fact leads to an exponential growth in stability estimates at large time values. For this reason we restrict ourselves to the case of finite time intervals, that is, where $T < \infty$. The last condition therefore is assumed throughout the chapter.

At the end of the chapter we will briefly discuss problem (1.1) with a more intricate operator splitting construction involving the operators $\mathfrak{A}_{j;n}$, $j = 1, 2$, under the same restrictions on $\mathfrak{A}_{j;n}$ as above. The main motivation for studying such methods is that they allow one, unlike the above method (1.1), (3.1), to deal with second order approximations to the starting differential equation in the case where $\mathfrak{A}_{1;n}$ and $\mathfrak{A}_{2;n}$ do not commute.

3.1 Estimates for discrete semigroups

In the present section we examine the stability of the discrete semigroup $\mathfrak{U}_n^l = (I - k\mathfrak{A}_n)^l$, uniformly with respect to $t_n, t_l \in \Omega_k$. These results lead directly to stability estimates for the solution of problem (1.1), (3.1) in the autonomous case, that is, where $\mathfrak{A}_n$ is independent of t_n. In the non-autonomous case the below assertions are important as well when combined with suitable frozen-operator techniques which are similar to those already used in Section 2.3.

This first assertion is an analogue of Theorem 2.3.

Theorem 3.1. *Suppose that the operators $\mathfrak{A}_{j;n}$, $j = 1, 2$, both satisfy the Λ_1^k-condition and that hypothesis* **COM1**$[\kappa]$ *holds for some $\kappa \in [0, 1)$. Then, with any fixed $\mu_0 \geq 0$ sufficiently large and with any fixed $\xi \geq 0$, for all $k \in (0, K]$ and $t_l, t_n \in \Omega_k$,*

$$\|(\mathfrak{A}_n + \mu_0 I)^\xi \mathfrak{U}_n^l\| \leq C t_{l+1}^{-\xi}, \tag{3.4}$$

where the operator $\mathfrak{A}_n$ is specified by $\mathfrak{A}_{j;n}$, $j = 1, 2$, through (3.1).

Proof. It is convenient to denote[4]

$$\mathcal{L} = \left[\!\!\left[\left(\lambda I - \left(k \overset{1}{\mathfrak{A}}_{1;n} + k \overset{2}{\mathfrak{A}}_{2;n} - k^2 \overset{2}{\mathfrak{A}}_{2;n} \overset{1}{\mathfrak{A}}_{1;n} \right) \right)^{-1} \right]\!\!\right],$$

[4]The newly used notation is explained in Appendix A.4.

and

$$\mathcal{U} = (I - k\mathfrak{A}_{2;n})\,\mathrm{Com}\,(k\mathfrak{A}_{1;n}|\mathcal{L}). \tag{3.5}$$

Our further argument will be based on the use of the operator equation[5]

$$\mathcal{R} = \mathcal{L} + \mathcal{R}\mathcal{U}, \tag{3.6}$$

which is solved for the operator $\mathcal{R} \in \mathcal{B}(\mathfrak{X})$.

Let now Γ_λ be a contour[6] which encircles the spectrum of the operator $k\mathfrak{A}_{n;1}$. Next, using (A.19) with Γ_λ in place of Ξ_1, the operator $\mathcal{L}$ is represented as follows, with $G_0(\lambda; z) = ((\lambda - z)I - (1-z)k\mathfrak{A}_{2;n})^{-1}$,

$$\mathcal{L} = \frac{1}{2\pi i} \int_{\Gamma_\lambda} G_0(\lambda; z)\,(zI - k\mathfrak{A}_{1;n})^{-1}\,dz. \tag{3.7}$$

Clearly, there exist a set Υ of Λ_1-configuration, which is Λ_1^k-proper for both $\mathfrak{A}_{1;n}$ and $\mathfrak{A}_{2;n}$, and a set Υ' of Λ_1-configuration such that

$$\nu = \nu_1 + \nu_2 - \nu_1\nu_2 \in \Upsilon' \qquad \text{for all } \nu_j \in \Upsilon,\ j = 1, 2.$$

Further, select a set Υ'' of Λ_1-configuration such that $\Upsilon' \subset \operatorname{Int}\Upsilon'' \cup \{0\}$. Our aim now is to bound the quantity $\|\mathcal{L}\|$ for all $\lambda \notin \operatorname{Int}\big(\Upsilon'' \cup \mathcal{D}(\delta_0 k)\big)$ with some $\delta_0 > 0$ which will be chosen later. In order to perform this action we distinguish between two cases: $|\lambda| \le \delta_1$ and $|\lambda| \ge \delta_1$, with $\delta_1 = \delta_0 K$, where K appears in the statement of the theorem and is the upper bound for k. Further, we let the contour Γ_λ in (3.7) coincide with the boundary of the set $\Upsilon_\lambda = \Upsilon \cup \mathcal{D}(\min\{\delta_2|\lambda|, \delta_1\delta_2\})$, with some $\delta_2 > 0$ chosen sufficiently small in order to satisfy the inclusion

$$\Upsilon'' \cup \mathcal{D}(|\lambda|/2) \supset \{z \in \mathbb{C} :\ z = \nu_1 + \nu_2 - \nu_1\nu_2,\ \nu_1 \in \Upsilon_\lambda,\ \nu_2 \in \Upsilon\}.$$

Since, with some $\delta > 0$,

$$\sum_{j=1}^{2} \|(zI - k\mathfrak{A}_{j;n})^{-1}\| \le C|z|^{-1} \qquad \text{for } z \notin \operatorname{Int}\Upsilon,\ |z| \ge \delta k,$$

we find, with $\delta_0 = \delta/\delta_2$, for all $\lambda \notin \operatorname{Int}\Upsilon''$, $\delta_1 \ge |\lambda| \ge \delta_0 k$ and $z \in \Gamma_\lambda$,

$$\|G_0(\lambda; z)\| \le C(|\lambda| + |z|)^{-1}, \tag{3.8}$$

[5]The idea to consider such equations arises from Sobolevskii [150].

[6]It will be chosen later, depending on λ.

and

$$\|(zI - k\mathfrak{A}_{1;n})^{-1}\| \le C|z|^{-1}. \tag{3.9}$$

It then follows from (3.7), (3.8), and (3.9) that if $\lambda \notin \operatorname{Int}\Upsilon''$ and $\delta_1 \ge |\lambda| \ge \delta_0 k$, then

$$\|\mathcal{L}\| \le C \int_{\Gamma_\lambda} (|\lambda| + |z|)^{-1}|z|^{-1}\,|dz| \le C|\lambda|^{-1}. \tag{3.10}$$

Also, using similar reasonings we see that (3.8) and (3.9) are valid for $\lambda \notin \operatorname{Int}\big(\Upsilon'' \cup \mathcal{D}(\delta_1)\big)$ as well, so that we find

$$\|\mathcal{L}\| \le C(1 + |\lambda|)^{-1} \qquad \text{if } \lambda \notin \operatorname{Int}\Upsilon'' \text{ and } |\lambda| \ge \delta_1. \tag{3.11}$$

Combining (3.10) and (3.11) thus yields

$$\|\mathcal{L}\| \le C|\lambda|^{-1} \qquad \text{for } \lambda \notin \operatorname{Int}\Upsilon'',\ |\lambda| \ge \delta_0 k. \tag{3.12}$$

Next, since both $\mathfrak{A}_{1;n}$ and $\mathfrak{A}_{2;n}$ satisfy the Λ_1^k-condition, by Lemma 1.1 it follows that

$$\|k\mathfrak{A}_{j;n}\| \le C, \qquad j = 1, 2, \tag{3.13}$$

which further yields, combined with (3.5),

$$\|\mathcal{U}\| \le C\|\mathrm{Com}\ (k\mathfrak{A}_{1;n}|\mathcal{L})\|. \tag{3.14}$$

To bound suitably the right-hand side of (3.14), we use the representation

$$\mathrm{Com}\ (k\mathfrak{A}_{1;n}|\mathcal{L}) = \frac{1}{2\pi i}\int_{\Gamma_\lambda} G(\lambda; z)\,dz, \tag{3.15}$$

with

$$G(\lambda; z) = k^2 G_0(\lambda; z)\,(1 - z)\mathrm{Com}\ (\mathfrak{A}_{1;n}|\mathfrak{A}_{2;n})\,G_0(\lambda; z)\,(zI - k\mathfrak{A}_{1;n})^{-1},$$

and with $G_0(\lambda; z)$ just the same as above. In fact, (3.15) is a consequence of (3.7) and of the following identity, for $\mathfrak{L}_1, \mathfrak{L}_2 \in \mathcal{B}(\mathfrak{X})$,

$$\mathrm{Com}\ (\mathfrak{L}_1|(\lambda I - \mathfrak{L}_2)^{-1}) = (\lambda I - \mathfrak{L}_2)^{-1}\,\mathrm{Com}\ (\mathfrak{L}_1|\mathfrak{L}_2)\,(\lambda I - \mathfrak{L}_2)^{-1}. \tag{3.16}$$

For our below needs, the parameter $\delta_0 > 0$ should be chosen large enough.[7] Observe however that if δ_0 is enlarged, the estimates (3.8) and (3.9) remain

[7]Clearly, δ_0 can be taken as large as needed choosing the above $\delta > 0$ sufficiently large. Also, we keep the parameter K in conformity with δ_0 via the relation $\delta_1 = \delta_0 K$, where δ_1 is always unchanged.

valid[8] with the same C's as for the former size of δ_0. Moreover, we have, for $k \in (0, K]$, $\lambda \notin \mathrm{Int}\,(\Upsilon'' \cup \mathcal{D}(\delta_0 k))$, and $z \in \Gamma_\lambda$, since the operators $\mathfrak{A}_{j;n}$, $j = 1, 2$, both satisfy the Λ_1^k-condition and $|z| \geq \delta_2\delta_0 k$, that

$$\|G_0(\lambda; z) k(\mathfrak{A}_{2;n} + \delta_0 I)\| \leq C \tag{3.17}$$

and

$$\|k(\mathfrak{A}_{1;n} + \delta_0 I)\,(zI - k\mathfrak{A}_{1;n})^{-1}\| \leq C, \tag{3.18}$$

where no constant C depends on δ_0. Also, note that by Lemma 1.5 we can apply hypothesis **COM1**$[\kappa]$ with $\mu_0 = \delta_0$ while no constant C in (3.2) depends on δ_0, whichever size it has.[9] Using the above observations, we obtain by hypothesis **COM1**$[\kappa]$ and (3.8), (3.9), (3.17), (3.18), for $k \in (0, K]$, $\lambda \notin \mathrm{Int}\,(\Upsilon'' \cup \mathcal{D}(\delta_0 k))$, and $z \in \Gamma_\lambda$,

$$\begin{aligned}\|G(\lambda; z)\| &\leq Ck^{1-\kappa}(|\lambda| + |z|)^{-(1-\kappa)}\|G_1(\lambda; z)\| \\ &\quad + Ck^{1-\kappa}\|G_0(\lambda; z)\,(zI - k\mathfrak{A}_{1;n})^{-1}\|^{1-\kappa}\|G_1(\lambda; z)\|^{\kappa},\end{aligned} \tag{3.19}$$

where

$$G_1(\lambda; z) = k(\mathfrak{A}_{1;n} + \delta_0 I)\,G_0(\lambda; z)\,(zI - k\mathfrak{A}_{n;1})^{-1}.$$

In order to bound the norm of $G_1(\lambda, z)$ in a suitable way, we proceed from the identity

$$\begin{aligned}G_1(\lambda; z) &= kG_0(\lambda; z)\,(\mathfrak{A}_{n;1} + \delta_0 I)\,(zI - k\mathfrak{A}_{1;n})^{-1} \\ &\quad + k(1 - z)\,G_0(\lambda; z)\mathrm{Com}\,(\mathfrak{A}_{2;n}|\mathfrak{A}_{1;n})\,G_0(\lambda; z)\,(zI - k\mathfrak{A}_{1;n})^{-1},\end{aligned}$$

which is a consequence of (3.16). Taking norms of both sides of this equation and using next **COM1**$[\kappa]$, (3.8), (3.9), (3.17), (3.18), and the trivial estimate, for all $u \in \mathfrak{X}$,

$$\begin{aligned}\|(\mathfrak{A}_{1;n} + \delta_0 I)u\|^{\kappa}\|u\|^{1-\kappa} &= \|(\mathfrak{A}_{1;n} + \delta_0 I)u\|^{\kappa} \\ &\quad \times \|(\mathfrak{A}_{1;n} + \delta_0 I)^{-1}(\mathfrak{A}_{1;n} + \delta_0 I)u\|^{1-\kappa} \\ &\leq \|(\mathfrak{A}_{1;n} + \delta_0 I)^{-1}\|^{1-\kappa}\|(\mathfrak{A}_{1;n} + \delta_0 I)u\|,\end{aligned}$$

we obtain, for $k \in (0, K]$, $\lambda \notin \mathrm{Int}\,(\Upsilon'' \cup \mathcal{D}(\delta_0 k))$, and $z \in \Gamma_\lambda$,

$$\begin{aligned}\|G_1(\lambda; z)\| &\leq C(|\lambda| + |z|)^{-1} + C\left((k/|\lambda|)^{1-\kappa} + \delta_0^{\kappa-1}\right)\|G_1(\lambda; z)\| \\ &\leq C(|\lambda| + |z|)^{-1} + C\delta_0^{\kappa-1}\|G_1(\lambda; z)\|,\end{aligned}$$

[8] As pointed out above, the restriction $|\lambda| \leq \delta_1$ need not, in fact, be imposed for these estimates to hold.

[9] The size of δ_0 should be sufficiently large.

where no constant C depends on δ_0. Taking $\delta_0 > 0$ sufficiently large we now have

$$\|G_1(\lambda; z)\| \le C(|\lambda| + |z|)^{-1}. \tag{3.20}$$

Combining this equation with (3.19) leads to the following estimate, for $k \in (0, K]$, $\lambda \notin \text{Int}\,\big(\Upsilon'' \cup \mathcal{D}(\delta_0 k)\big)$, and $z \in \Gamma_\lambda$,

$$\|G(\lambda; z)\| \le Ck^{1-\kappa}(|\lambda| + |z|)^{-1}|z|^{-(1-\kappa)},$$

which further implies, after a simple estimation on the basis of (3.15), for $k \in (0, K]$ and $\lambda \notin \text{Int}\,\big(\Upsilon'' \cup \mathcal{D}(\delta_0 k)\big)$, that

$$\|\text{Com}\,(k\mathfrak{A}_{1;n}|\mathcal{L})\| \le Ck^{1-\kappa}|\lambda|^{\kappa-1} \le C\delta_0^{\kappa-1},$$

where $\delta_0 > 0$ may be taken as large as needed while no constant C depends on δ_0. Inserted in (3.15), this result gives, with δ_0 sufficiently large, since $\kappa < 1$,

$$\|\mathcal{U}\| \le C\delta_0^{\kappa-1} \le \frac{1}{2},$$

thus, for $\lambda \notin \text{Int}\,\big(\Upsilon'' \cup \mathcal{D}(\delta_0 k)\big)$,

$$\|(I - \mathcal{U})^{-1}\| \le (1 - \|\mathcal{U}\|)^{-1} \le \frac{1}{2}. \tag{3.21}$$

It follows from (3.12) and (3.21) that equation (3.6) is uniquely solvable and, for $\lambda \notin \text{Int}\,\big(\Upsilon'' \cup \mathcal{D}(\delta_0 k)\big)$,

$$\|\mathcal{R}\| \le C|\lambda|^{-1}. \tag{3.22}$$

Furthermore, a simple calculation shows that

$$\begin{aligned}(\lambda I - k\mathfrak{A}_{2;n})\, G_0(\lambda; z)\,(zI - k\mathfrak{A}_{1;n})^{-1}&\\ - (I - k\mathfrak{A}_{2;n})\, G_0(\lambda; z)\,(zI - k\mathfrak{A}_{1;n})^{-1}\, k\mathfrak{A}_{1;n}&\\ = (zI - k\mathfrak{A}_{1;n})^{-1} + (I - k\mathfrak{A}_{2;n})\, G_0(\lambda; z).&\end{aligned}$$

With this fact in mind and again using (3.7) we get, for $\lambda \notin \text{Int}\,\big(\Upsilon'' \cup \mathcal{D}(\delta_0 k)\big)$,

$$\begin{aligned}&(\lambda I - k\mathfrak{A}_{2;n})\,\mathcal{L} - (I - k\mathfrak{A}_{2;n})\,\mathcal{L}\, k\mathfrak{A}_{1;n}\\ &= \frac{1}{2\pi i}\int_{\Gamma_\lambda} (zI - k\mathfrak{A}_{1;n})\, dz + \frac{1}{2\pi i}\int_{\Gamma_\lambda} (I - k\mathfrak{A}_{2;n})\, G_0(\lambda; z)\, dz\\ &= I,\end{aligned}$$

since the second integral vanishes. It is then not hard to see that by the definition of $\mathcal{L}$ and $\mathcal{U}$ we have, for $\lambda \notin \text{Int}\,(\Upsilon'' \cup \mathcal{D}(\delta_0 k))$,

$$(\lambda I - k\mathfrak{A}_n)\,\mathcal{R} = I. \tag{3.23}$$

Similarly, it can be shown that there exists an operator $\widetilde{\mathcal{R}} \in \mathcal{B}(\mathfrak{X})$ (depending on λ) such that with $\delta_0 > 0$ sufficiently large, for $\lambda \notin \text{Int}\,(\Upsilon'' \cup \mathcal{D}(\delta_0 k))$,

$$\widetilde{\mathcal{R}}\,(\lambda I - k\mathfrak{A}_n) = I. \tag{3.24}$$

Now we conclude by (3.23) and (3.24) (cf. Hille & Phillips [84, Paragraph 1.13]) that $\mathcal{R} = \widetilde{\mathcal{R}} = (\lambda I - k\mathfrak{A}_n)^{-1}$ for $\lambda \notin \text{Int}\,(\Upsilon'' \cup \mathcal{D}(\delta_0 k))$, which together with (3.22) shows that the operator $\mathfrak{A}_n$ satisfies the Λ_1^k-condition and the claim thus follows by Theorem 2.3. □

Next we show a slightly different stability estimate in comparison with (3.4). This new estimate is of independent interest and is used in the stability analysis of problem (1.1), (3.1) in the non-autonomous case.

Theorem 3.2. *Let the conditions of Theorem 3.1 be satisfied and let in addition hypothesis* **COM1**$[\kappa]$ *hold for some* $\kappa \in [0,1)$. *Then, with any fixed* $\mu_0 > 0$ *sufficiently large and with any fixed* $\xi \in [0,1]$, *for all* $k \in (0, K]$ *and* $t_l, t_n \in \Omega_k$, *we have*

$$\sum_{j=1}^{2} \|(\mathfrak{A}_{j;n} + \mu_0 I)^{\xi} \mathfrak{U}_n^l\| \le C t_{l+1}^{-\xi}. \tag{3.25}$$

Proof. Let $\mu_0 > 0$ be chosen sufficiently large in order to make our below argument correct. In particular, this fact yields that the fractional powers $(\mathfrak{A}_{j;n} + \mu_0 I)^{\xi}$, $\xi \ge 0$, of the operators $(\mathfrak{A}_{j;n} + \mu_0 I)$, $j = 1, 2$, are defined and bounded on $\mathfrak{X}$. Also, we fix arbitrarily an $\eta \in [0,1)$.

Let further $\mathcal{L}$, Υ'', Γ_λ, and $G_0(\lambda; z)$ be defined in the same way as in the proof of Theorem 3.1. Using then (3.8), (3.9), (3.17), (3.18), the convexity inequality (A.18), and the above argument, we get, for $\lambda \notin \text{Int}\,(\Upsilon'' \cup \mathcal{D}(\mu_0 k))$, and $z \in \Gamma_\lambda$,

$$\|k^{\eta} (\mathfrak{A}_{1;n} + \mu_0 I)^{\eta} (zI - k\mathfrak{A}_{1;n})^{-1}\| \le C|z|^{\eta - 1} \tag{3.26}$$

and

$$\|k^{\eta} (\mathfrak{A}_{2;n} + \mu I)^{\eta} G_0(\lambda; z)\| \le C(|\lambda| + |z|)^{\eta - 1}. \tag{3.27}$$

Also, by (3.27) and (3.7), acting as in the preceding proof and noting that $\eta < 1$, we find, for $\lambda \notin \text{Int}\left(\Upsilon'' \cup \mathcal{D}(\mu_0 k)\right)$,

$$\|k^\eta(\mathfrak{A}_{2;n} + \mu_0 I)^\eta \mathcal{L}\| \leq C|\lambda|^{\eta-1}. \tag{3.28}$$

Similarly, changing the order of the resolvents on the right-hand side of (2.4) and further using the following identity, for $\mathfrak{L}_1, \mathfrak{L}_2 \in \mathcal{B}(\mathfrak{X})$,

$$\begin{aligned}\text{Com }&\left((\lambda I - \mathfrak{L}_2)^{-1} | (\mu I - \mathfrak{L}_1)^{-1}\right) \\ &= (\mu I - \mathfrak{L}_1)^{-1}(\lambda I - \mathfrak{L}_2)^{-1}\text{Com }(\mathfrak{L}_2|\mathfrak{L}_1)\,(\lambda I - \mathfrak{L}_2)^{-1}(\mu I - \mathfrak{L}_1)^{-1},\end{aligned}$$

it is not hard to see, with the aid of (3.26), **COM1**$[\kappa]$, (3.8), (3.17), and (3.20), that for $\lambda \notin \text{Int}\left(\Upsilon'' \cup \mathcal{D}(\mu_0 k)\right)$,

$$\|k^\eta(\mathfrak{A}_{1;n} + \mu_0 I)^\eta \mathcal{L}\| \leq C|\lambda|^{\eta-1}. \tag{3.29}$$

Therefore, by virtue of (3.28), (3.29), and (3.21) and using (3.6) we have, for $\lambda \notin \text{Int}\left(\Upsilon'' \cup \mathcal{D}(\mu_0 k)\right)$,

$$\|k^\eta(\mathfrak{A}_{j;n} + \mu_0 I)^\eta(\lambda I - k\mathfrak{A}_n)^{-1}\| \leq C|\lambda|^{\eta-1}, \quad j = 1, 2. \tag{3.30}$$

We need to derive some different resolvent estimates as well. With $\mathcal{L}$ just the same as above and with $\mathcal{V}$ given by

$$\mathcal{V} = \text{Com }(\mathcal{L}|k\mathfrak{A}_{2;n})\,(I - k\mathfrak{A}_{1;n}),$$

using the operator equation, solved for $\widetilde{\mathcal{R}} \in \mathcal{B}(\mathfrak{X})$,

$$\widetilde{\mathcal{R}} = \mathcal{L} + \mathcal{V}\widetilde{\mathcal{R}},$$

and acting as above (see the proof of Theorem 3.1 and the preceding part of the present proof), we now obtain, for $\lambda \notin \text{Int}\left(\Upsilon''\mathcal{D}(\mu_0 k)\right)$,

$$\|(\lambda I - k\mathfrak{A}_n)^{-1}k^\eta(\mathfrak{A}_{j;n} + \mu_0 I)^\eta\| \leq C|\lambda|^{\eta-1}, \qquad j = 1, 2. \tag{3.31}$$

Next we denote

$$\mathcal{H}_j = k(\mathfrak{A}_{j;n} + \mu_0 I)\,(\lambda I - k\mathfrak{A}_n)^{-2},$$

and

$$\widetilde{\mathcal{H}}_j = (\lambda I - k\mathfrak{A}_n)^{-1}k(\mathfrak{A}_{j;n} + \mu_0 I)\,(\lambda I - k\mathfrak{A}_n)^{-1}.$$

As a trivial consequence of (3.30) and (3.31), both taken with $\eta = 1/2$, we have, for $\lambda \notin \text{Int}\left(\Upsilon'' \cup \mathcal{D}(\mu_0 k)\right)$,

$$\|\widetilde{\mathcal{H}}_j\| \leq C|\lambda|^{-1}, \qquad j = 1, 2. \tag{3.32}$$

Since $\mathfrak{A}_{1;n}$ and $\mathfrak{A}_{2;n}$ both satisfy the Λ_1^k-condition, it follows by Lemma 1.1 that $\|k\mathfrak{A}_{j;n}\| \le C$, $j = 1, 2$. Using Lemma 1.5 with $\mathfrak{A}_{j;n}$, $j = 1, 2$, substituted for $\mathfrak{L}_n$ we have, for all $u \in \mathfrak{X}$,

$$\begin{aligned}\sum_{j=1}^{2} \|(\mathfrak{A}_{j;n} + \mu_0 I)(I - k\mathfrak{A}_{1;n})u\| &\le C\left(\sum_{j=1}^{2} \|\mathfrak{A}_{j;n}u\| + \|u\|\right) \\ &\le C\sum_{j=1}^{2} \|(\mathfrak{A}_{j;n} + \mu_0 I)u\|. \qquad (3.33)\end{aligned}$$

We now apply (3.16) to obtain

$$\mathcal{H}_1 = \widetilde{\mathcal{H}}_1 + (\lambda I - k\mathfrak{A}_n)^{-1}k^2 \mathrm{Com}\,(\mathfrak{A}_{1;n}|\mathfrak{A}_{2;n})(I - k\mathfrak{A}_{1;n})(\lambda I - k\mathfrak{A}_n)^{-2},$$

which implies, with the aid of (3.32), **COM2**$[\kappa]$,[10] (3.31), and (3.33), for $\lambda \notin \mathrm{Int}\,(\Upsilon'' \cup \mathcal{D}(\mu_0 k))$,

$$\begin{aligned}\|\mathcal{H}_1\| &\le C|\lambda|^{-1} + Ck^{1-\kappa}|\lambda|^{\kappa-1} \\ &\quad \times \sum_{j=1}^{2} \|k(\mathfrak{A}_{j;n} + \mu_0 I)(I - k\mathfrak{A}_{1;n})(\lambda I - k\mathfrak{A}_n)^{-2}\| \\ &\le C|\lambda|^{-1} + C\mu_0^{\kappa-1}\sum_{j=1}^{2} \|\mathcal{H}_j\|, \qquad (3.34)\end{aligned}$$

where no constant C depends on μ_0. Similarly we get, for $\lambda \notin \mathrm{Int}\,(\Upsilon'' \cup \mathcal{D}(\mu_0 k))$,

$$\|\mathcal{H}_2\| \le C|\lambda|^{-1} + C\mu_0^{\kappa-1}\sum_{j=1}^{2} \|\mathcal{H}_j\|. \qquad (3.35)$$

Therefore, summing the inequalities (3.34) and (3.35) and taking $\mu_0 > 0$ sufficiently large, we find, for $\lambda \notin \mathrm{Int}\,(\Upsilon'' \cup \mathcal{D}(\mu_0 k))$,

$$\sum_{j=1}^{2} \|\mathcal{H}_j\| \le C|\lambda|^{-1}.$$

The last estimate yields that the pairs $(\|(\mathfrak{A}_{j;n} + \mu_0 I)\cdot\|; \mathfrak{A}_n)$, $j = 1, 2$, are $\Lambda_1^k(1)$-concordant and therefore (3.25) with $\xi = 1$ immediately follows, by Theorem 2.1. Using this fact and noting that (3.25) for $\xi = 0$ results, in fact, from Theorem 3.1, the claim in the general case $0 \le \xi \le 1$ follows by the convexity inequality (A.18). □

[10]By Lemma 1.5 we can take for μ_0 in (3.3) the same value as in the present proof.

3.2 Stability of operator splitting problems

Now we are in a position to state our results on stability for problem (1.1), (3.1).[11]

In the autonomous case, using straightforwardly Theorems 3.1 and 3.2, we have the following.

Theorem 3.3. *Suppose that the operators $\mathfrak{A}_{j;n}$, $j = 1, 2$, are both independent of t_n and each of them satisfies the Λ_1^k-condition. Let also the hypotheses* **COM1**$[\kappa]$ *and* **COM2**$[\kappa]$ *hold for some $\kappa \in [0, 1)$. Then, with any fixed $\mu_0 \geq 0$ sufficiently large and with any fixed $\xi \in [0, 1]$, the solution Y_n of (1.1), (3.1) satisfies the estimate, for all $k \in (0, K]$ and $t_n \in \Omega_k$,*

$$\|(\mathfrak{A}_n + \mu_0 I)^\xi Y_n\| + \sum_{j=1}^{2} \|(\mathfrak{A}_{j;n} + \mu_0 I)^\xi Y_n\| \leq C \left(t_{n+1}^{-\xi} \|y^0\| + k \sum_{l=0}^{n-1} t_{n-l}^{-\xi} \|F_l\| \right). \tag{3.36}$$

At the same time the first term on the left can be estimated as stated for any fixed $\xi \geq 0$, even though we have not assumed **COM2**$[\kappa]$.

Further, we turn to the general case where the operators $\mathfrak{A}_{j;n}$, $j = 1, 2$, may depend on t_n.

Together with problem (1.1), (3.1) we will examine the following perturbed problem, with some $\mathfrak{B}_{j;n} \in \mathcal{B}(\mathfrak{X})$, $j = 1, 2$,

$$\partial Y_n + \widehat{\mathfrak{A}}_n Y_n = F_n \quad \text{for } t_n \in \breve{\Omega}_k, \qquad Y_0 = y^0, \tag{3.37}$$

where

$$\widehat{\mathfrak{A}}_n = \widehat{\mathfrak{A}}_{1;n} + \widehat{\mathfrak{A}}_{2;n} - k\widehat{\mathfrak{A}}_{2;n}\widehat{\mathfrak{A}}_{1;n} \text{ and } \widehat{\mathfrak{A}}_{j;n} = \mathfrak{A}_{j;n} + \mathfrak{B}_{j;n},\ j = 1, 2. \tag{3.38}$$

Clearly, denoting

$$\widehat{\mathfrak{U}}_n = (I - k\widehat{\mathfrak{A}}_{2;n})\,(I - k\widehat{\mathfrak{A}}_{1;n}) \text{ and } \widehat{\mathfrak{U}}_{n,l} = \begin{cases} \widehat{\mathfrak{U}}_n \cdots \widehat{\mathfrak{U}}_l & \text{if } n \geq l, \\ I & \text{if } n < l, \end{cases}$$

the solution Y_n of (3.37), (3.38) admits the representation

$$Y_n = \widehat{\mathfrak{U}}_{n-1,0} y^0 + k \sum_{l=0}^{n-1} \widehat{\mathfrak{U}}_{n-1,l+1} F_l, \quad t_n \in \Omega_k. \tag{3.39}$$

This formula will be used in our subsequent analysis.

[11] We recall that it is assumed that $\mathfrak{Q}_n = \mathfrak{R} = I$.

Theorem 3.4. *Let the operators $\mathfrak{A}_{j;n}$, $j = 1, 2$, both obey the Λ_1^k-condition and let the hypotheses* **COM1**$[\kappa]$ *and* **COM2**$[\kappa]$ *be satisfied for some $\kappa \in [0, 1)$. Let further the following Hölder-continuity type condition hold, with some $\mu_1 \geq 0, \vartheta \in (0, 1]$, and $\gamma_1 \in [0, 1)$, for all $j = 1, 2$, $k \in (0, K]$, $t_n, t_l \in \Omega_k$, and $u \in \mathfrak{X}$,*

$$\|(\mathfrak{A}_{j;n} - \mathfrak{A}_{j;l})\, u\| \leq C t_{|n-l|}^{\vartheta} \|(\mathfrak{A}_{j;n} + \mu_1 I)\, u\| + C \|(\mathfrak{A}_{j;n} + \mu_1 I)\, u\|^{\gamma_1} \|u\|^{1-\gamma_1}. \tag{3.40}$$

Apart from everything else, suppose that with some $\gamma_0 \in [0, 1)$, for all $k \in (0, K]$, $t_n \in \Omega_k$, and $u \in \mathfrak{X}$,

$$\|\mathfrak{B}_{j;n} u\| \leq C \|(\mathfrak{A}_{j;n} + \mu_1 I)^{\gamma_0} u\|, \quad j = 1, 2. \tag{3.41}$$

Then, for any fixed $\mu_0 > 0$ sufficiently large and for any fixed $\xi \in [0, 1)$, the solution Y_n of both problem (1.1), (3.1) and problem (3.37), (3.38) satisfies the estimate (3.36).

Proof. Let there be fixed $\mu_0 \geq 0$ sufficiently large and $\xi \in [0, 1)$. By (A.18), (3.40), Lemma 1.5,[12] and Theorem 3.2, it follows that for $j = 1, 2$ and $t_n, t_l \in \Omega_k$ such that $t_l \leq t_n$,

$$\begin{aligned} \|(\mathfrak{A}_{j;n} + \mu_0 I)^{\xi} \mathfrak{U}_l^{n-l}\| &\leq C \|(\mathfrak{A}_{j;n} + \mu_0 I) \mathfrak{U}_l^{n-l}\|^{\xi} \|\mathfrak{U}_l^{n-l}\|^{1-\xi} \\ &\leq C \left(\|(\mathfrak{A}_{j;l} + \mu_0 I) \mathfrak{U}_l^{n-l}\| + \|(\mathfrak{A}_{j;l} - \mathfrak{A}_{j;n})\, \mathfrak{U}_l^{n-l}\| \right)^{\xi} \\ &\leq C \|(\mathfrak{A}_{j;l} + \mu_0 I) \mathfrak{U}_l^{n-l}\|^{\xi} \leq C t_{n-l+1}^{-\xi}. \end{aligned} \tag{3.42}$$

Similarly, applying the trivial inequality, for all $t_n, t_l \in \Omega_k$ and $u \in \mathfrak{X}$,

$$\begin{aligned} \|(\mathfrak{A}_n - \mathfrak{A}_l)\, u\| &\leq \sum_{j=1}^{2} \|(\mathfrak{A}_{j;n} - \mathfrak{A}_{j;l})\, u\| + \|k \mathfrak{A}_{2;n} (\mathfrak{A}_{1;n} - \mathfrak{A}_{1;l})\, u\| \\ &\quad + \|k (\mathfrak{A}_{2;n} - \mathfrak{A}_{2;l})\, \mathfrak{A}_{1;n} u\|, \end{aligned} \tag{3.43}$$

with the aid of (3.13) we get, for $t_l \leq t_n$,

$$\|(\mathfrak{A}_n + \mu_0 I)^{\xi} \mathfrak{U}_l^{n-l}\| \leq C t_{n-l+1}^{-\xi}. \tag{3.44}$$

Next note that by virtue of (3.13) and (3.40), the above estimate (3.43) implies as well, for all $t_n, t_l \in \Omega_k$ and $u \in \mathfrak{X}$,

$$\begin{aligned} \|(\mathfrak{A}_n - \mathfrak{A}_l)\, u\| &\leq C t_{|n-l|}^{\vartheta} \sum_{j=1}^{2} \|(\mathfrak{A}_{j;n} + \mu_0 I)\, u\| \\ &\quad + C \sum_{j=1}^{2} \|(\mathfrak{A}_{j;n} + \mu_0 I)\, u\|^{\gamma_1} \|u\|^{1-\gamma_1}. \end{aligned} \tag{3.45}$$

[12] If the lemma is applied with $\mathfrak{A}_{j;n}$ in place of $\mathfrak{A}_n$.

Using now (3.42), (3.44), (3.45), and the identity (2.28), with both sides multiplied from the left consecutively by $(\mathfrak{A}_n + \mu_0 I)^\xi$ and by $(\mathfrak{A}_{n;j} + \mu_0 I)^\xi$, and applying finally Lemma 2.1 yields that (3.40) holds for the solution of (1.1), (3.1), in view of (1.4). In particular, this fact implies that for $t_l \leq t_n$,

$$\|(\mathfrak{A}_n + \mu_0 I)^\xi \mathfrak{U}_{n-1,l}\| + \sum_{j=1}^{2} \|(\mathfrak{A}_{j;n} + \mu_0 I)^\xi \mathfrak{U}_{n-1,l}\| \leq C t_{n-l+1}^{-\xi}, \tag{3.46}$$

which is applied in the sequel.

Furthermore, note that the discrete evolution operators $\widehat{\mathfrak{U}}_{n,l}$ and $\mathfrak{U}_{n,l}$ are connected by the identity, for $0 \leq t_l \leq t_n$,

$$\widehat{\mathfrak{U}}_{n-1,l} = \mathfrak{U}_{n-1,l} + \sum_{q=l}^{n-1} \widehat{\mathfrak{U}}_{n-1,q+1}(\widehat{\mathfrak{U}}_q - \mathfrak{U}_q)\mathfrak{U}_{q-1,l}. \tag{3.47}$$

Since

$$\widehat{\mathfrak{U}}_q - \mathfrak{U}_q = -k\mathfrak{B}_{2;n}(I - k\mathfrak{A}_{1;n}) - (I - k\mathfrak{A}_{2;n})k\mathfrak{B}_{1;n},$$

with the aid of (3.41) it follows that for $t_l \leq t_q$,

$$\begin{aligned}\|(\widehat{\mathfrak{U}}_q - \mathfrak{U}_q)\mathfrak{U}_{q-1,l}\| &\leq Ck\sum_{j=1}^{2} \|\mathfrak{B}_{j;q}\mathfrak{U}_{q-1;l}\| + Ck^2\|\mathfrak{B}_{2;q}\mathfrak{A}_{1;q}\mathfrak{U}_{q-1,l}\| \\ &\leq Ck\sum_{j=1}^{2} \|(\mathfrak{A}_{j;q} + \mu_1 I)^{\gamma_0}\mathfrak{U}_{q-1;l}\| + Ck^2\|(\mathfrak{A}_{2;q} + \mu_1 I)^{\gamma_0}\mathfrak{A}_{1;q}\mathfrak{U}_{q-1,l}\|. \end{aligned}\tag{3.48}$$

Recalling now that both $\mathfrak{A}_{1;n}$ and $\mathfrak{A}_{2;n}$ satisfy the Λ_1^k-condition, by Lemmas 1.1 and 1.5 we have, for all $u \in \mathfrak{X}$,

$$\begin{aligned} k\|(\mathfrak{A}_{2;q} + \mu_1 I)^{\gamma_0}\mathfrak{A}_{1;q}u\| &\leq Ck^{1-\gamma_0}\|\mathfrak{A}_{1;q}u\| \leq Ck^{1-\gamma_0}\|(\mathfrak{A}_{1;q} + \mu_1 I)u\| \\ &\leq C\|(\mathfrak{A}_{1;q} + \mu_1 I)^{\gamma_0}u\|. \end{aligned}$$

Using this estimate as well as (3.46) with γ_0 substituted for ξ, instead of (3.48) we get, for $t_l \leq t_q$,

$$\|(\widehat{\mathfrak{U}}_q - \mathfrak{U}_q)\mathfrak{U}_{q-1,l}\| \leq Ck\sum_{j=1}^{2} \|(\mathfrak{A}_{j;q} + \mu_1 I)^{\gamma_0}\mathfrak{U}_{q-1;l}\| \leq Ckt_{q-l+1}^{-\gamma_0}.$$

Applying this inequality together with (3.46) for an estimation on the basis of (3.47), with both sides multiplied from the left by $(\mathfrak{A}_n + \mu_0 I)^\xi$, we obtain, for $0 \le t_l \le t_n$,

$$\|(\mathfrak{A}_n + \mu_0 I)^{\xi} \widehat{\mathfrak{U}}_{n-1,l}\| \le C t_{n-l+1}^{-\xi} + Ck \sum_{q=l}^{n-1} \|(\mathfrak{A}_n + \mu_0 I)^{\xi} \widehat{\mathfrak{U}}_{n-1,q+1}\| \, t_{q-l+1}^{-\gamma_0}.$$

Since $\xi < 1$ and $\gamma_0 < 1$, by Lemma 2.1 it follows from the last inequality that for $t_l \le t_n$,

$$\|(\mathfrak{A}_n + \mu_0 I)^{\xi} \widehat{\mathfrak{U}}_{n-1,l}\| \le C t_{n-l+1}^{-\xi},$$

and similarly it can be shown that

$$\sum_{j=1}^{2} \|(\mathfrak{A}_{j;n} + \mu_0 I)^{\xi} \widehat{\mathfrak{U}}_{n-1,l}\| \le C t_{n-l+1}^{-\xi}.$$

Finally, with the aid of these estimates, a trivial estimation on the basis of (3.39), with both sides multiplied from the left consequtively by $(\mathfrak{A}_n + \mu_0 I)^{\xi}$ and by $(\mathfrak{A}_{j;n} + \mu_0 I)^{\xi}$, $j = 1, 2$, shows that (3.40) holds for the solution of problem (3.37), (3.38) as well. □

3.3 A symmetric operator splitting method

In this section we consider a problem of the form (1.1) for which

$$\begin{aligned} \mathfrak{A}_n \;=\;& 2\mathfrak{A}_{1;n} + 2\mathfrak{A}_{2;n} - k(\mathfrak{A}_{1;n}^2 + 2\mathfrak{A}_{1;n}\mathfrak{A}_{2;n} + 2\mathfrak{A}_{2;n}\mathfrak{A}_{1;n} + \mathfrak{A}_{2;n}^2) \\ & + k^2(\mathfrak{A}_{1;n}\mathfrak{A}_{2;n}^2 + 2\mathfrak{A}_{1;n}\mathfrak{A}_{2;n}\mathfrak{A}_{1;n} + \mathfrak{A}_{2;n}^2\mathfrak{A}_{1;n}) \\ & - k^3 \mathfrak{A}_{1;n}\mathfrak{A}_{2;n}^2\mathfrak{A}_{1;n}. \end{aligned} \tag{3.49}$$

Clearly, using this representation leads to the following operator splitting method

$$Y_{n+1} = (I - k\mathfrak{A}_{1;n})\,(I - k\mathfrak{A}_{2;n})^2 (I - k\mathfrak{A}_{1;n}) Y_n + kF_n, \quad t_n \in \breve{\Omega}_k, \quad Y_0 = y^0. \tag{3.50}$$

It is seen that the transition operator possesses now a certain symmetry and this circumstance allows us to construct second order approximation schemes based on the principle of operator splitting. In the non-autonomous case it is expedient to examine problem (1.1), (3.49) (or, equivalently, problem (3.50)) with $\widehat{\mathfrak{A}}_{j;n}$ substituted for $\mathfrak{A}_{j;n}$, where $\widehat{\mathfrak{A}}_{j;n} = \mathfrak{A}_{j;n} + \mathfrak{B}_{j;n}$ with some $\mathfrak{B}_{j;n} \in \mathcal{B}(\mathfrak{X})$, $j = 1, 2$.

Our results here will be that under the same restrictions on the operators $\mathfrak{A}_{j;n}$, $j = 1, 2$, as above, it is possible to obtain stability estimates for problem

(1.1), (3.49), which are analogous to those already found for (1.1), (3.1). The first result concerns the case where the operators $\mathfrak{A}_{j;n}$, $j = 1, 2$, are independent of t_n and the second one is stated in the general case supposing that the dependence on t_n is allowed.

Theorem 3.5. *Let the operators $\mathfrak{A}_{j;n}$, $j = 1, 2$, be subject to the conditions of Theorem 3.3.*[13] *Then, with any fixed $\mu_0 \geq 0$ sufficiently large and with any fixed $\xi \in [0, 1]$, the solution Y_n of (1.1), (3.49) satisfies the stability estimate (3.36). Moreover, the first term on the left-hand side of (3.36) is estimated as stated for any fixed $\xi \geq 0$ even though* **COM2**$[\kappa]$ *is not assumed.*

Theorem 3.6. *Let the operators $\mathfrak{A}_{j;n}$, $j = 1, 2$, be subject to the conditions of Theorem 3.4. Then, for any fixed $\mu_0 \geq 0$ sufficiently large and for any fixed $\xi \in [0, 1)$, the solution Y_n of (1.1), (3.49) satisfies the estimate (3.36). Furthermore, if $\mathfrak{B}_{j;n}$, $j = 1, 2$, are small perturbations to $\mathfrak{A}_{j;n}$, respectively, in the sense that condition (3.41) is fulfilled, then (3.36) still holds if by Y_n we mean the solution of the perturbed problem (1.1), (3.49), that is, with $\widehat{\mathfrak{A}}_{j;n} = \mathfrak{A}_{j;n} + \mathfrak{B}_{j;n}$ substituted for $\mathfrak{A}_{j;n}$ in (3.49).*

Both theorems can be proved with the aid of the permutation techniques as above. Such an investigation is not essentially different from the one which has already been demonstrated. For this reason we do not give detailed proofs of the stated results but instead sketch how it might be done, as follows.

For convenience' sake we introduce

$$\ddot{\mathfrak{A}}_n = \ddot{\mathfrak{A}}_{1;n} + \ddot{\mathfrak{A}}_{2;n} - k\ddot{\mathfrak{A}}_{2;n}\ddot{\mathfrak{A}}_{1;n},$$

and

$$\ddot{\mathfrak{U}}_n = (I - k\ddot{\mathfrak{A}}_{2;n})\,(I - k\ddot{\mathfrak{A}}_{1;n}),$$

where

$$\ddot{\mathfrak{A}}_{j;n} = 2\mathfrak{A}_{j;n} - k\mathfrak{A}_{j;n}^2, \qquad j = 1, 2.$$

It is not hard to see that

$$\ddot{\mathfrak{U}}_n = (I - k\mathfrak{A}_{2;n})^2 (I - k\mathfrak{A}_{1;n})^2.$$

The fact that $\mathfrak{A}_{1;n}$ and $\mathfrak{A}_{2;n}$ both satisfy the Λ_1^k-condition implies that the same is true for $\ddot{\mathfrak{A}}_{j;n}$, $j = 1, 2$, which can be shown similarly to (3.12), and that for all $k \in (0, K]$ and $t_n \in \Omega_k$,

$$\|(2I - k\mathfrak{A}_{j;n})^{-1}\| \leq C, \qquad j = 1, 2.$$

[13] In particular, they are both independent of t_n.

The last estimate furthermore shows that with any fixed $\mu_0 > 0$ sufficiently large, for $j = 1, 2$, $u \in \mathfrak{X}$, and $\chi \in \mathfrak{X}^*$,

$$\|(\mathfrak{A}_{j;n} + \mu_0 I)\, u\| \le C\|(\ddot{\mathfrak{A}}_{j;n} + \mu_0 I)\, u\| \tag{3.51}$$

and

$$\|(\mathfrak{A}^*_{j;n} + \mu_0 I)\, \chi\|_* \le C\|(\ddot{\mathfrak{A}}^*_{j;n} + \mu_0 I)\, \chi\|_*. \tag{3.52}$$

Furthermore, with the aid of the following identity, where we use the notation $\mathfrak{C} = \mathrm{Com}\,(\mathfrak{A}_{1;n}|\mathfrak{A}_{2;n})$,

$$\begin{aligned}\mathrm{Com}\,(\ddot{\mathfrak{A}}_{1;n}|\ddot{\mathfrak{A}}_{2;n}) &= 4\mathfrak{C} - 2k(\mathfrak{A}_{1;n} + \mathfrak{A}_{2;n})\,\mathfrak{C} - 2k\mathfrak{C}\,(\mathfrak{A}_{1;n} + \mathfrak{A}_{2;n})\\ &\quad + k^2(\mathfrak{A}_{1;n}\mathfrak{C}\mathfrak{A}_{2;n} + \mathfrak{A}_{1;n}\mathfrak{A}_{2;n}\mathfrak{C} + \mathfrak{C}\mathfrak{A}_{2;n}\mathfrak{A}_{1;n} + \mathfrak{A}_{2;n}\mathfrak{C}\mathfrak{A}_{1;n}),\end{aligned}$$

using (3.13), (3.51), (3.52), and the above permutation techniques we can prove that the hypotheses **COM1**$[\kappa]$ and **COM2**$[\kappa]$ are still satisfied with $\ddot{\mathfrak{A}}_{j;n}$ substituted for $\mathfrak{A}_{j;n}$. We will show, for example, a possible estimation of the quantity $|\langle\chi, k\mathfrak{A}_{1;n}\mathfrak{C}u\rangle|$ when deriving (3.2) with $\ddot{\mathfrak{A}}_{j;n}$ in place of $\mathfrak{A}_{j;n}$. By **COM1**$[\kappa]$ and (3.13) we have, for $j = 1, 2$,

$$\begin{aligned}|\langle\chi, k\mathfrak{A}_{1;n}\mathfrak{C}u\rangle| &\le Ck\|(\mathfrak{A}^*_{j;n} + \mu_0 I)\,\mathfrak{A}^*_{1;n}\chi\|_*^{\kappa}\|\mathfrak{A}^*_{1;n}\chi\|_*^{1-\kappa}\|(\mathfrak{A}_{3-j;n} + \mu_0 I)\,u\|\\ &\quad + Ck\|(\mathfrak{A}^*_{j;n} + \mu_0 I)\,\mathfrak{A}^*_{1;n}\chi\|_*\|(\mathfrak{A}_{3-j;n} + \mu_0 I)\,u\|^{\kappa}\|u\|^{1-\kappa}\\ &\le C\|\mathfrak{A}^*_{1;n}\chi\|_*^{\kappa}\|\chi\|_*^{1-\kappa}\|(\mathfrak{A}_{3-j;n} + \mu_0 I)\,u\|\\ &\quad + C\|\mathfrak{A}^*_{1;n}\chi\|_*\|(\mathfrak{A}_{3-j;n} + \mu_0 I)\,u\|^{\kappa}\|u\|^{1-\kappa},\end{aligned}$$

where, as above, $\mathfrak{A}^*_{1;n}$ can be replaced by $(\mathfrak{A}^*_{1;n} + \mu_0 I)$,[14] and further $\mathfrak{A}_{j;n}$ can be replaced by $\ddot{\mathfrak{A}}_{j;n}$ for $j = 1, 2$, by (3.51) and (3.52).

Therefore, with the aid of the above reasonings we conclude that under the conditions of Theorems 3.1 and 3.2, respectively, the estimates (3.4) and (3.25) hold even with $\ddot{\mathfrak{A}}_n$, $\ddot{\mathfrak{A}}_{j;n}$, and $\ddot{\mathfrak{U}}_n$ substituted for $\mathfrak{A}_n$, $\mathfrak{A}_{j;n}$, and $\mathfrak{U}_n$, respectively. Now using the identity $(I - k\mathfrak{A}_{1;n})\ddot{\mathfrak{A}}_n^{\xi} = \mathfrak{A}_n^{\xi}(I - k\mathfrak{A}_{1;n})$ for any $\xi \in \mathbb{N}$, it is not hard to see that (3.4) will hold also with $\breve{\mathfrak{U}}_n = (I - k\mathfrak{A}_{1;n})\,(I - k\mathfrak{A}_{2;n})^2(I - k\mathfrak{A}_{1;n})$ substituted for $\mathfrak{U}_n$ for any fixed $\xi \in \mathbb{N} \cup \{0\}$ and by the convexity inequality (A.18) for any fixed $\xi \ge 0$. Also, since $\|2I - k\mathfrak{A}_{j;n}\| \le C$, we have for $j = 1, 2$ and $u \in \mathfrak{X}$,

$$\|(\ddot{\mathfrak{A}}_{j;n} + \mu_0 I)\, u\| \le C\|(\mathfrak{A}_{j;n} + \mu_0 I)\, u\|.$$

It thus follows that the estimate (3.25) holds even with $\breve{\mathfrak{U}}_n$ substituted for $\mathfrak{U}_n$, both for $\xi = 0$ and $\xi = 1$ and, applying (A.18), for any fixed $\xi \in [0, 1]$.

[14] Without loss of generality μ_0 can be assumed to be positive and sufficiently large.

Therefore using the above observations, Theorem 3.5 now straightforwardly follows. Also, employing analogues of the identities (2.28) and (3.47) in the case where $\breve{\mathfrak{U}}_n$ plays the role of $\mathfrak{U}_n$ and following the proof of Theorem 3.4 shows the assertion of Theorem 3.6 as well.

3.4 Comments and bibliographical remarks

Sections 3.1 and 3.2: The present considerations follow in essence our earlier work [13, 15, 34].
Section 3.3: Similar results for the unperturbed problem (1.1), (3.49) have been announced in [21].

Chapter 4

Equations with Memory

In this chapter we extend the field of our study by including equations with a memory term in our scope. Such equations arise, for example, from discretization of integro-differential equations (see Part IV).

More precisely, instead of problem (1.1) we now deal with the following discrete Cauchy problem

$$\partial Y_n + \mathfrak{A}_n Y_n = \sum_{l=0}^{n} \mathfrak{P}_{n,l} Y_l + F_n \quad \text{for } t_n \in \breve{\Omega}_k, \qquad Y_0 = y^0, \tag{4.1}$$

with $\mathfrak{P}_{n,l} \in \mathcal{B}(\mathfrak{X})$ given for $t_l \le t_n$ (depending generally on k) and all other objects retaining their former meaning (see Chapter 1). In view of the many important applications (particularly, in the case where $\mathfrak{A}_n$ depends on t_n, but not exclusively) it is expedient to consider together with problem (4.1) its perturbed version, with some $\mathfrak{B}_n \in \mathcal{B}(\mathfrak{X})$,

$$\partial Y_n + (\mathfrak{A}_n + \mathfrak{B}_n) Y_n = \sum_{l=0}^{n} \mathfrak{P}_{n,l} Y_l + F_n \quad \text{for } t_n \in \breve{\Omega}_k, \qquad Y_0 = y^0. \tag{4.2}$$

The problems (4.1) and (4.2) differ from (1.1) and (2.32), respectively, by the presence of the memory term $\sum_{l=0}^{n} \mathfrak{P}_{n,l} Y_l$ in the corresponding equation.[1] Obviously, both (4.1) and (4.2) are uniquely solvable.

To the problems (4.1) and (4.2) we associate the problems for the corresponding homogeneous equations with floating initial condition, for $t_n, t_j \in \breve{\Omega}_k$ with $t_j \le t_n$,

$$\partial \dot{Y}_n + \mathfrak{A}_n \dot{Y}_n = \sum_{l=j}^{n} \mathfrak{P}_{n,l} \dot{Y}_l, \quad \dot{Y}_j \text{ is given}, \tag{4.3}$$

[1] For the sake of simplicity throughout this chapter it is accepted that $\mathfrak{Q}_n = \mathfrak{R} = I$.

and

$$\partial \dot{Y}_n + (\mathfrak{A}_n + \mathfrak{B}_n)\dot{Y}_n = \sum_{l=j}^{n} \mathfrak{P}_{n,l}\dot{Y}_l, \quad \dot{Y}_j \text{ is given.} \tag{4.4}$$

The unique solvability of these problems does not give rise to doubts as well. Let $\mathfrak{W}_{n,j}$ and $\widehat{\mathfrak{W}}_{n,j}$ be the solution operators for the problems (4.3) and (4.4), respectively, that is, for $\dot{Y}_n$ the solution of (4.3),

$$\dot{Y}_n = \mathfrak{W}_{n-1,j}\dot{Y}_j, \tag{4.5}$$

and for $\dot{Y}_n$ the solution of (4.4),

$$\dot{Y}_n = \widehat{\mathfrak{W}}_{n-1,j}\dot{Y}_j. \tag{4.6}$$

Then, by linearity and causality it follows that the solution Y_n of problem (4.1) admits the representation

$$Y_n = \mathfrak{W}_{n-1,0}y^0 + k\sum_{j=0}^{n-1} \mathfrak{W}_{n-1,j+1}F_j, \tag{4.7}$$

while the solution Y_n of (4.2) can be represented by

$$Y_n = \widehat{\mathfrak{W}}_{n-1,0}y^0 + k\sum_{j=0}^{n-1} \widehat{\mathfrak{W}}_{n-1,j+1}F_j. \tag{4.8}$$

It is easily seen that (4.7) and (4.8) take the same form as (1.4) and (2.33), respectively. Therefore all we need for showing stability is to find a suitable estimate for the operators $\mathfrak{W}_{n-1,j}$ and $\widehat{\mathfrak{W}}_{n-1,j}$, which means that, in fact, for our aims it suffices to analyse the problems (4.3) and (4.4) only.

For further success we now derive suitable equations for $\mathfrak{W}_{n-1,j}$ and $\widehat{\mathfrak{W}}_{n-1,j}$. Since, by analogy with (1.4), for $\dot{Y}_n$ the solution of (4.3) we have

$$\dot{Y}_n = \mathfrak{U}_{n-1,j}\dot{Y}_j + k\sum_{l=j}^{n-1} \mathfrak{U}_{n-1,l+1}\sum_{q=j}^{l} \mathfrak{P}_{l,q}\dot{Y}_q \quad \text{for } t_j \le t_n,$$

where $\mathfrak{U}_{n,j}$ is the discrete evolution operator for problem (1.1), using (4.5) yields, for $t_j \le t_n$,

$$\mathfrak{W}_{n-1,j} = \mathfrak{U}_{n-1,j} + k\sum_{l=j}^{n-1} \mathfrak{U}_{n-1,l+1}\sum_{q=j}^{l} \mathfrak{P}_{l,q}\mathfrak{W}_{q-1,j},$$

or, equivalently, changing the order of summation,

$$\mathfrak{W}_{n-1,j} = \mathfrak{U}_{n-1,j} + k \sum_{l=j}^{n-1} \sum_{q=l}^{n-1} \mathfrak{U}_{n-1,q+1} \mathfrak{P}_{q,l} \mathfrak{W}_{l-1,j}. \tag{4.9}$$

Similarly, with the aid of (4.6) it follows that

$$\widehat{\mathfrak{W}}_{n-1,j} = \widehat{\mathfrak{U}}_{n-1,j} + k \sum_{l=j}^{n-1} \sum_{q=l}^{n-1} \widehat{\mathfrak{U}}_{n-1,q+1} \mathfrak{P}_{q,l} \widehat{\mathfrak{W}}_{l-1,l}, \quad t_j \leq t_n, \tag{4.10}$$

where $\widehat{\mathfrak{U}}_{n,j}$ is the discrete evolution operator for problem (2.32).

In the next two sections we will examine the stability of the problems (4.1) and (4.2) using the equations (4.9) and (4.10) and distinguishing between the cases where the operator $\mathfrak{A}_n$ depends and does not depend on t_n. The main idea employed is that the memory term $\sum_{l=j}^{n} \mathfrak{P}_{n,l} \dot{Y}_l$ is considered as a (nonlocal) perturbation to the principal term $\mathfrak{A}_n \dot{Y}_n$ or $(\mathfrak{A}_n + \mathfrak{B}_n)\dot{Y}_n$. Acting in such a way leads to an estimate which presents a discrete analogue of an integral inequality and can be solved by using a suitable discrete version of Gronwall's lemma. Note that these techniques are applicable in the case of finite intervals only. It is therefore assumed that $T < \infty$ throughout the whole chapter.

On the other hand, since Lemma 2.1 above does not meet our needs below, we present now a more general discrete lemma of Gronwall type which is workable in the following two sections.

Lemma 4.1. *Let $w_{n,j}$ be a nonnegative solution of the inequality, with some $\zeta \in [0,1)$, for all $t_n, t_j \in \Omega_k$ such that $t_j \leq t_n$,*

$$w_{n,j} \leq C t_{n-j+1}^{-\zeta} + \sum_{q=j}^{n-1} \phi_{n,q} w_{q,j}, \tag{4.11}$$

where $\phi_{n,q} \geq 0$, $0 \leq q \leq n-1$, are given[2] and satisfy the condition, with some $T_1, K \in (0,T]$ and $\eta \in [0,1)$, for all $k \in (0,K]$ and $t_j \leq t_m \leq t_s \leq t_n$,

$$t_{n-j+1}^{\zeta} \sum_{q=m}^{s-1} \phi_{n,q} t_{q-j+1}^{-\zeta} \leq \eta \text{ whenever } t_{s-m} \leq T_1. \tag{4.12}$$

Then, for all $k \in (0,K]$ and $t_j, t_n \in \Omega_k$ such that $t_j \leq t_n$,

$$w_{n,j} \leq C t_{n-j+1}^{-\zeta}.$$

[2] $\phi_{n,q}$ generally depend on k.

Proof. Denoting $\widetilde{w}_{n,j} = t_{n-j+1}^{\zeta} w_{n,j}$, it follows from (4.11) that for $t_j, t_n \in \Omega_k$ with $t_j \le t_n$,

$$\widetilde{w}_{n,j} \le C + t_{n-j+1}^{\zeta} \sum_{q=j}^{n-1} \phi_{n,q} t_{q-j+1}^{-\zeta} \widetilde{w}_{q,j}. \tag{4.13}$$

Further, given $t_j \in \Omega_k$, we put, for $l \ge 1$,

$$\tau_l = \big(t_j + (l-1)T_1, \min\{T, t_j + lT_1\}\big].$$

Clearly, by (4.12) and (4.13) we have

$$\max_{t_n \in \tau_l} \widetilde{w}_{n,j} \le C + \eta \sum_{q=1}^{l} \max_{t_n \in \tau_q} \widetilde{w}_{n,j},$$

which shows the claim since l does not really exceed $\lfloor T/T_1 \rfloor + 1$. □

4.1 The general case

In the general case where $\mathfrak{A}_n$ may vary depending on t_n, our analysis will be based on making use of some results of Section 2.3 where the stability of the discrete evolution operators $\mathfrak{U}_{n,j}$ and $\widehat{\mathfrak{U}}_{n,j}$ for the problems (1.1) and (2.32), respectively, is examined. In principle, each stability assertion of Section 2.3 can be employed to get an analogue for either problem (4.1) or (4.2). Here we restrict ourselves to considering some selected situations only.

Theorem 4.1. *Let seminorms* $]\![\cdot]\!]_n$, $\|\!| \cdot \|\!|_n$, $\|\!| \cdot \|\!|_{1;n}$ *and the operators* $\mathfrak{A}_n$, $\mathfrak{B}_n$ *satisfy the conditions posed in the former part of the statement of Theorem 2.10, with some* $\vartheta \in (0,1]$ *and* $\xi, \gamma_1, \gamma_0 \in [0,1)$. *Suppose further that seminorms* $\|\!| \cdot \|\!|_{3;n}$ *and* $\|\!| \cdot \|\!|_{*3;n}$ *form respectively* $\Lambda_1^k(\gamma_3)$*- and* $\Lambda_1^k(*;\gamma_{*3})$*-concordant pairs with the operator* $\mathfrak{A}_n$ *for some* $\gamma_3, \gamma_{*3} \ge 0$ *such that* $\gamma_{*3} < \min\{\vartheta, 1-\gamma_1, 1-\gamma_0, 1-\gamma_3\}$, *and that for all* $\chi \in \mathfrak{X}^*$, $u \in \mathfrak{X}$, *and* $t_l, t_n \in \breve{\Omega}_k$ *such that* $t_l \le t_n$,

$$|\langle \chi, \mathfrak{P}_{n,l} u \rangle| \le \pi_{n,l} \|\!|\chi\|\!|_{*3;n} \|\!|u\|\!|_{3;l}, \tag{4.14}$$

where $\pi_{n,l} \ge 0$ *satisfy the conditions, for all* $k \in (0,K]$ *and* $t_j \le t_m \le t_s \le t_n$,

for any $\eta > 0$ *there exists* $T_1 \in (0,T]$ *such that whenever* $t_{s-m} \le T_1$,

$$k t_{n-j+1}^{\gamma_3} \textstyle\sum_{l=m}^{s-1} \sum_{q=l}^{n-1} t_{n-q}^{-(\gamma_3+\gamma_{*3})} \pi_{q,l} t_{l-j+1}^{-\gamma_3} \le \eta, \tag{4.15}$$

and

$$kt_{n-j+1}^{\xi}\sum_{l=j}^{n-1}\sum_{q=l}^{n-1}t_{n-q}^{-(\xi+\gamma_{*3})}\pi_{q,l}t_{l-j+1}^{-\gamma_3}\leq C. \tag{4.16}$$

Then the solution Y_n of both (4.1) and (4.2) satisfies the stability estimate (2.81).

Proof. In view of the representations (4.7) and (4.8) it suffices to show the estimate, for all $u \in \mathfrak{X}$ and $t_j \leq t_n$,

$$]\![\mathfrak{W}_{n-1,j}u]\!]_n +]\![\widehat{\mathfrak{W}}_{n-1,j}u]\!]_n \leq Ct_{n-j+1}^{-\xi}\|u\|. \tag{4.17}$$

Since both terms on the left-hand side of (4.17) are similarly estimated, we will demonstrate our argument for showing the desired result for the first term only. To that end, we first note by Theorem 2.10 that with $p = \lfloor (n+q)/2 \rfloor$, for all $u \in \mathfrak{X}$ and $t_j \leq t_l \leq t_q \leq t_{n-1}$,

$$|\!|\!|\mathfrak{U}_{n-1,j}u|\!|\!|_{3;n} \leq Ct_{n-j+1}^{-\gamma_3}\|u\|, \tag{4.18}$$

and using as well (4.14) and Theorem 2.11

$$\begin{aligned} |\!|\!|\mathfrak{U}_{n-1,q+1}\mathfrak{P}_{q,l}\mathfrak{W}_{l-1,j}u|\!|\!|_{3;n} &\leq Ct_{n-q}^{-\gamma_3}\|\mathfrak{U}_{p,q+1}\mathfrak{P}_{q,l}\mathfrak{W}_{l-1,j}u\| \\ &\leq Ct_{n-q}^{-(\gamma_3+\gamma_{*3})}\pi_{q,l}|\!|\!|\mathfrak{W}_{l-1,j}u|\!|\!|_{3;j}. \end{aligned} \tag{4.19}$$

With the aid of (4.18) and (4.19), since $|\!|\!|\cdot|\!|\!|_{3;n}$ is a seminorm, a simple estimation on the basis of formula (4.9)[3] gives, taking (4.14) into account, for all $t_j \leq t_n$,

$$\sup_{\|u\|=1}|\!|\!|\mathfrak{W}_{n-1,j}u|\!|\!|_{3;n} \leq Ct_{n-j+1}^{-\gamma_3} + Ck\sum_{l=j}^{n-1}\sum_{q=l}^{n-1}t_{n-q}^{-(\gamma_3+\gamma_{*3})}\pi_{q,l}\sup_{\|u\|=1}|\!|\!|\mathfrak{W}_{l-1,j}u|\!|\!|_{3;n}. \tag{4.20}$$

Applying now Lemma 4.1 to this inequality and taking (4.15) into account, it follows that since $\gamma_3 + \gamma_{*3} < 1$,

$$\sup_{\|u\|=1}|\!|\!|\mathfrak{W}_{n-1,j}u|\!|\!|_{3;n} \leq Ct_{n-j+1}^{-\gamma_3}, \quad t_j \leq t_n. \tag{4.21}$$

With this fact in mind and using once again the above method, only this time with $]\![\cdot]\!]_n$ in place of $|\!|\!|\cdot|\!|\!|_{3;n}$, instead of (4.20) we obtain the inequality, for $t_j \leq t_n$,

$$\sup_{\|u\|=1}]\![\mathfrak{W}_{n-1,j}u]\!]_n \leq Ct_{n-j+1}^{-\xi} + Ck\sum_{l=j}^{n-1}\sum_{q=l}^{n-1}t_{n-q}^{-(\xi+\gamma_{*3})}\pi_{q,l}\sup_{\|u\|=1}|\!|\!|\mathfrak{W}_{l-1,j}u|\!|\!|_{3;n},$$

[3] When estimating $|\!|\!|\widehat{\mathfrak{W}}_{n-1,j}u|\!|\!|_n$, (4.10) is used instead.

thus the desired estimate for $]\!]\mathfrak{W}_{n-1,j}u]\!]_n$ follows after a trivial calculation using (4.21) and (4.16) as well. □

Theorem 4.2. *Let seminorms $]\!]\cdot]\!]_n$, $\|\!|\cdot|\!\|_n$, $\|\!|\cdot|\!\|_{1;n}$, $\|\!|\cdot|\!\|_{*;n}$, $\|\!|\cdot|\!\|_{0;n}$, and $\|\!|\cdot|\!\|_{*0;n}$ and the operators $\mathfrak{A}_n$, $\mathfrak{B}_n$ satisfy the conditions of Theorem 2.12, with some $J \in \mathbb{N}$, $\xi \geq 0$, $\gamma, \gamma_1, \gamma_*, \gamma_0, \gamma_{*0} \in [0,1)$, and $\vartheta, \vartheta_1, \vartheta_0 \in [0,1]$ such that $\gamma_* < \xi_0$, $\gamma_0 + \gamma_{*0} < 1 + \vartheta_0$, $\max\{\xi, \gamma, \gamma_1\} < \xi_1$, where $\xi_0 = 1 + \min\{\vartheta - \gamma, \vartheta_1 - \gamma_1\}$ and $\xi_1 = 1 - \gamma_* + \min\{\vartheta, \vartheta_1\}$. Let further seminorms $\|\!|\cdot|\!\|_{3;n}$ and $\|\!|\cdot|\!\|_{*3;n}$ form respectively $\Lambda_J^k(\gamma_3)$- and $\Lambda_J^k(*;\gamma_{*3})$-concordant pairs with the operator $\mathfrak{A}_n$, for some $\gamma_3, \gamma_{*3} \geq 0$ such that $\gamma_3 < \xi_1$, $\gamma_{*3} < \xi_0$, and let $\mathfrak{P}_{n,l}$ be under the conditions (4.14), (4.15), and (4.16). Then the solution Y_n of problem (4.1) satisfies (2.81). If in addition $\gamma_{*0} < \xi_0$, $\gamma_0 < \xi_1$, and $\max\{\xi, \gamma_3\} < 1 + \vartheta_0 - \gamma_{*0}$, then (2.81) remains valid for the solution of problem (4.2) as well.*

The proof is similar to that of Theorem 4.1, with reference to Theorem 2.12 for this once.

4.2 A particular case

Here we consider, as a particular case, the situation when the operator $\mathfrak{A}_n \equiv \mathfrak{A}$ does not depend on t_n. It will be seen below that under this requirement some further developments are possible. Using the notation formerly introduced in what follows, $\mathfrak{U}^n$ denotes the discrete semigroup generated by the operator $\mathfrak{A}$ and $\widehat{\mathfrak{U}}_{n,j}$ the discrete evolution operator generated by $\mathfrak{A} + \mathfrak{B}_n$.

In this section the following condition will be employed, with some $\mu_0 \geq 0$ and with $\widetilde{\mathfrak{P}}_{n,l} = (\mathfrak{A} + \mu_0 I)^{-1}\mathfrak{P}_{n,l}$, for all $k \in (0, K]$ and $t_j, t_m, t_s, t_n \in \Omega_k$ subject to $t_j \leq t_m \leq t_s \leq t_n$,

$$\text{for any } \eta > 0 \text{ there exists } T_1 \in (0,T] \text{ such that whenever } t_{s-m} \leq T_1,$$
$$\textstyle\sum_{l=m}^{s-1} \left(\|\widetilde{\mathfrak{P}}_{n-1,l}\| + k \sum_{q=l}^{n-1} t_{n-q}^{-1} \|\widetilde{\mathfrak{P}}_{q,l} - \widetilde{\mathfrak{P}}_{n-1,l}\| \right) \leq \eta. \tag{4.22}$$

We start by examining problem (4.1) in the considered case.

Theorem 4.3. *Let the operator $\mathfrak{A}$ satisfy the Λ_J^k-condition with some $J \in \mathbb{N}$ and let (4.22) hold. Then the solution Y_n of problem (4.1) satisfies the estimate, for all $k \in (0, K]$ and $t_n \in \Omega_k$,*

$$\|Y_n\| \leq C \left(\|y^0\| + k \sum_{j=0}^{n-1} \|F_j\| \right). \tag{4.23}$$

Proof. By (4.7) it suffices to show that

$$\|\mathfrak{W}_{n-1,j}\| \leq C \quad \text{for all } t_j \leq t_n. \tag{4.24}$$

Applying now (4.9) with $\mathfrak{A}_n \equiv \mathfrak{A}$ gives, after a simple calculation, with $\widetilde{\mathfrak{P}}_{n,l}$ defined above, for $t_j \leq t_n$,

$$\begin{aligned} \mathfrak{W}_{n-1,j} &= \mathfrak{U}^{n-j} + k \sum_{l=j}^{n-1} \sum_{q=l}^{n-1} \mathfrak{U}^{n-q-1} (\mathfrak{A} + \mu_0 I) \, (\widetilde{\mathfrak{P}}_{q,l} - \widetilde{\mathfrak{P}}_{n-1,l}) \mathfrak{W}_{l-1,j} \\ &+ \sum_{l=j}^{n-1} \sum_{q=l}^{n-1} \mathfrak{U}^{n-q-1} k \mathfrak{A} \widetilde{\mathfrak{P}}_{n-1,l} \mathfrak{W}_{l-1,j} \\ &+ \mu_0 k \sum_{l=j}^{n-1} \sum_{q=l}^{n-1} \mathfrak{U}^{n-q-1} \widetilde{\mathfrak{P}}_{n-1,l} \mathfrak{W}_{l-1,j}. \end{aligned} \tag{4.25}$$

By Theorem 2.3 we have, whenever $t_l \leq t_{n-1}$,

$$\|\mathfrak{U}^n\| + t_{n+1} \|(\mathfrak{A} + \mu_0 I) \, \mathfrak{U}^n\| \leq C,$$

$$\|\sum_{q=l}^{n-1} \mathfrak{U}^{n-q-1} k \mathfrak{A}\| = \|\sum_{q=l}^{n-1} \mathfrak{U}^{n-q-1} (I - \mathfrak{U})\| = \|I - U^{n-l}\| \leq C,$$

and

$$\|\mu_0 k \sum_{q=l}^{n-1} \mathfrak{U}^{n-q-1}\| \leq \mu_0 k \sum_{q=l}^{n-1} \|\mathfrak{U}^{n-q-1}\| \leq C t_{n-l} \leq C.$$

Using the last three estimates for an estimation on the basis of (4.25), it follows that for all $t_j \leq t_n$,

$$\|\mathfrak{W}_{n-1,j}\| \leq C + C \sum_{l=j}^{n-1} \left(\|\widetilde{\mathfrak{P}}_{n-1,l}\| + k \sum_{q=l}^{n-1} t_{n-q}^{-1} \|\widetilde{\mathfrak{P}}_{q,l} - \widetilde{\mathfrak{P}}_{n-1,l}\| \right) \|\mathfrak{W}_{l-1,j}\|.$$

Finally, applying Lemma 4.1 to this inequality and taking (4.22) into account gives us (4.24). □

For the considered class of equations it is worth studying the perturbed problem (4.2) even if $\mathfrak{A}_n \equiv \mathfrak{A}$ is independent of t_n. The following result is applicable just in this case.

Theorem 4.4. *Let the conditions of Theorem 4.3 be satisfied and let, with a seminorm $\|\!|\cdot|\!\|_{*0;n}$ that forms a $\Lambda_1^k(*;\gamma_{*0})$-concordant pair with $\mathfrak{A}$ for some $\gamma_{*0} \in [0,1)$ and $\vartheta_0 \in (0,1]$, the operator $\mathfrak{B}_n$ be under the condition, for all $k \in (0,K]$, $t_n \in \Omega_k$, and $\chi \in \mathfrak{X}^*$,*

$$\|\mathfrak{B}_n^* \chi\|_* \leq C k^{\vartheta_0} \|\!|\chi|\!\|_{*0;n}. \tag{4.26}$$

Then the solution Y_n of problem (4.2) satisfies the stability estimate (4.23).

Proof. The claim can be shown by using, in fact, the same argument as in the proof of the preceding theorem.

First of all, applying (4.26), letting formally $\gamma = \gamma_1 = \gamma_* = \gamma_0 = 0$ and $\vartheta = \vartheta_1 = 1$, and leaving γ_{*0} and ϑ_0 just the same as in the statement of the present theorem, with the aid of Theorem 2.12 it follows that for all $t_j \leq t_n$ and $t_q \leq t_{n-1}$,

$$\|\widehat{\mathfrak{U}}_{n-1,j}\| + t_{n-q}\|\widehat{\mathfrak{U}}_{n-1,q+1}\mathfrak{A}\| + t_{n-q}^{\gamma_{*0}}\|\widehat{\mathfrak{U}}_{n-1,q+1}\mathfrak{B}_q\| \leq C, \tag{4.27}$$

since the seminorm $\|\mathfrak{A}^* \cdot\|_*$ obviously forms a $\Lambda_1^k(*;1)$-concordant pair with $\mathfrak{A}$. Also, in view of the identity, for $t_l \leq t_n$,

$$k\sum_{q=l}^{n-1} \widehat{\mathfrak{U}}_{n-1,q+1}(\mathfrak{A} + \mathfrak{B}_q) = \sum_{q=l}^{n-1}(\widehat{\mathfrak{U}}_{n-1,q+1} - \widehat{\mathfrak{U}}_{n-1,q}) = I - \widehat{\mathfrak{U}}_{n-1,l},$$

it is not hard to see that equation (4.10) can be taken into the following one, for $t_j \leq t_n$,

$$\begin{aligned}
\widehat{\mathfrak{W}}_{n-1,j} &= \widehat{\mathfrak{U}}_{n-1,j} + k\sum_{l=j}^{n-1}\sum_{q=l}^{n-1} \widehat{\mathfrak{U}}_{n-1,q+1}(\mathfrak{A} + \mu_0 I)\,(\widetilde{\mathfrak{P}}_{q,l} - \widetilde{\mathfrak{P}}_{n-1,l})\,\widehat{\mathfrak{W}}_{l-1,j} \\
&\quad + \sum_{l=j}^{n-1}(I - \widehat{\mathfrak{U}}_{n-1,l})\widetilde{\mathfrak{P}}_{n-1,l}\widehat{\mathfrak{W}}_{l-1,j} \\
&\quad + k\sum_{l=j}^{n-1}\sum_{q=l}^{n-1} \widehat{\mathfrak{U}}_{n-1,q+1}(\mu_0 I - \mathfrak{B}_q)\,\widetilde{\mathfrak{P}}_{n-1,l}\widehat{\mathfrak{W}}_{l-1,j}.
\end{aligned} \tag{4.28}$$

Now the desired result follows, using (4.27) for an estimation on the basis of (4.28) and acting as in the proof of Theorem 4.3. □

4.3 Comments and bibliographical remarks

Actually, Lemma 4.1 is a simple generalization of Lemma 2.2 in Bakaev, Larsson and Thomée [40], and the proof on page 74 follows essentially the ideas from [40].

As concerns the assertions presented in Sections 4.1 and 4.2, they have never before appeared.

Part II

RUNGE-KUTTA METHODS FOR ABSTRACT PARABOLIC EQUATIONS

Chapter 5

Discretization by Runge-Kutta Methods

In the preceding part of the book we deal with general stability assertions stated for evolution equations in discrete time. The main purpose of the present part is to use those results in the study of Runge-Kutta methods applied for discretization in time of abstract linear parabolic problems.

5.1 Introductory light

Runge-Kutta methods are a widespread tool in the numerical treatment of ordinary differential equations. Their use in the abstract framework meets, however, some special features which mainly result from the fact that the operator in question is generally unbounded or is an approximation to an unbounded one. The last circumstance should therefore be taken into consideration.

Let there be given a complex Banach space $\mathfrak{X}$ and an interval $[0,T]$, $0 < T \leq \infty$. We consider the following abstract linear Cauchy problem

$$y_t + A(t)y = f(t) \quad \text{for } 0 < t \leq T, \qquad y(0) = y^0, \tag{5.1}$$

where $A(t)$, $t \in [0,T]$, is a closed linear (generally unbounded) operator in $\mathfrak{X}$, with domain $\operatorname{Dom} A(t) \equiv \mathcal{Dom} \subseteq \mathfrak{X}$ independent of t,[1] $f(t) : [0,T] \to \mathfrak{X}$ the forcing term, and $y^0 \in \mathfrak{X}$ the initial condition.

[1]This condition on $\operatorname{Dom} A(t)$ is accepted throughout Part II. Note that our theory allows us to study fully discrete approximations to problem (5.1) in the case where $\operatorname{Dom} A(t)$ may depend on t but the approximation to $A(t)$ has domain independent of t.

Our results can also be used in the case where the operator $A(t)$ depends on an additional parameter, h, say. In fact, under the assumption that all the starting restrictions are satisfied uniformly with respect to h, our below estimates are then uniformly valid with respect to h as well. This kind of extension is most helpful for applications of our theory.

In our considerations the main tool of analysis is reducing the starting discrete problem, which arises from discretization of (5.1), to the Cauchy problem for an evolution equation in discrete time. In order to use the general stability assertions of Part I, we therefore have to verify the restrictions accepted therein. We show that they can readily be verified if the operator $A(t)$ in (5.1) satisfies some reasonable conditions which are stated below.

Our investigation in this part of the book involves certain seminorms which, as a rule, depend on t; any such seminorm necessarily is furnished with t as a subscript whenever the dependence on t is generally available.[2] At the same time we keep the style of notation accepted in Part I for objects arising from discretization in time and depending in general on k and t_n.

In what follows one of the main assumptions will be that $A(t) \in \mathcal{S}(\varphi, \sigma)$ for some $\varphi \in (0, \pi/2)$ and $\sigma \in \mathbb{R}$, both independent of t. This assumption allows us to think of (5.1) as a parabolic type problem. Moreover, the discretizations of problem (5.1) that we deal with may be considered as discrete parabolic type problems in some sense. At the same time we will employ as well various Hölder-continuity type conditions on the operator $A(t)$. Now we pose certain necessary hypotheses, all of which are assumed to hold uniformly with respect to $t, s \in [0, T]$.

First we introduce the following idea, with some $\mu_1 \geq 0$, $\vartheta \in (0, 1]$ and with some seminorm $|\cdot|_t$ defined on $\mathcal{D}om$.

höl$^\oplus[\vartheta; |\cdot|_t]$: For all $u \in \mathcal{D}om$,

$$\|(A(t) - A(s))\,u\| \leq C|t-s|^\vartheta \|(A(t) + \mu_1 I)\,u\| + C|u|_t. \tag{5.2}$$

If the second term on the right is removed, we say that $A(t)$ satisfies hypothesis **höl**$[\vartheta]$.

If $A(t)$ is a multi-dimensional elliptic partial differential operator (or a suitable approximation to such an operator), hypothesis **höl**$^\oplus[\vartheta; |\cdot|_t]$ cannot be verified in some function spaces.[3] This problem is settled by using a weaker hypothesis which is thought of as a local version of **höl**$^\oplus[\vartheta; |\cdot|_t]$.

[2]This convention means that no dependence on t is allowed if t does not appear by the seminorm.

[3]For instance, this non-verification holds in L_1, L_∞, and in spaces with related norms.

More precisely, we say that $A(t)$ satisfies hypothesis $\mathbf{h\ddot{o}l}^{\oplus}_{(k)}[\vartheta; |\cdot|_t]$ if (5.2) holds for all $k \in (0, K]$ and $t, s \in [0, T]$ such that $|t - s| \le k$.[4]

Further, before stating other useful Hölder-continuity type conditions, we need to define a new notion.

Definition 5.1. *Let $\lambda_0 \in \mathbb{C}$ belong to the resolvent set of the operator $A(t)$ for any $t \in [0, T]$*[5] *and let $\mathcal{P}_*$ and $\mathcal{P}$ be linear manifolds (independent of t) such that* $\operatorname{Ran}\left((\lambda_0 I - A(t))^{-1}\right)^* \subseteq \mathcal{P}_* \subseteq \mathfrak{X}^*$ *and $\mathcal{D}om \subseteq \mathcal{P} \subseteq \mathfrak{X}$ for all $t \in [0, T]$. A bilinear functional $\mathcal{A}(t; \chi, u)$, $\chi \in \mathcal{P}_*$, $u \in \mathcal{P}$ is then called a* testing functional *for $A(t)$ if, for any $t \in [0, T]$,*

$$\langle \chi, A(t)u \rangle = \mathcal{A}(t; \chi, u) \quad \textit{whenever } \chi \in \mathcal{P}_* \textit{ and } u \in \mathcal{D}om\,.$$

Incidentally, the set $\mathcal{P}_* \times \mathcal{P} \subseteq \mathfrak{X}^* \times \mathfrak{X}$ is thought of as the domain of $\mathcal{A}(t; \cdot, \cdot)$ and we write $\mathcal{P}_* \times \mathcal{P} = \operatorname{Dom} \mathcal{A}(t; \cdot, \cdot)$.

Clearly, given linear manifolds $\mathcal{P}_* \subseteq \mathfrak{X}^*$ and $\mathcal{P} \subseteq \mathfrak{X}$, $A(t)$ has in general more than one testing functional $\mathcal{A}(t; \cdot, \cdot)$ for which $\operatorname{Dom} \mathcal{A}(t; \cdot, \cdot) = \mathcal{P}_* \times \mathcal{P}$. The question of uniqueness of such a functional is outside of the scope of this book.

Now we introduce more Hölder-continuity type hypotheses with some $\vartheta, \vartheta_1 \in [0, 1]$, $\mathcal{A}(t; \cdot, \cdot)$ some testing functional for $A(t)$, with $\operatorname{Dom} \mathcal{A}(t; \cdot, \cdot) = \mathcal{P}_* \times \mathcal{P}$, and $|\cdot|_{\emptyset;t}$, $|\cdot|_{1;t}$, $|\cdot|_{*\emptyset;t}$, and $|\cdot|_{*1;t}$ some seminorms of which the first two are defined on $\mathcal{P}$ and the other two on $\mathcal{P}_*$.

$\mathbf{h\ddot{o}l}^{(\rightarrow)}(\vartheta, \vartheta_1; \mathcal{A}(t; \cdot, \cdot); |\cdot|_{\mathfrak{m};t}, \mathfrak{m} = \emptyset, 1, *\emptyset)$: For all $\chi \in \mathcal{P}_*$ and $u \in \mathcal{P}$,

$$|\mathcal{A}(t; \chi, u) - \mathcal{A}(s; \chi, u)| \le C |\chi|_{*\emptyset;s} \left(|t - s|^{\vartheta} |u|_{\emptyset;t} + |t - s|^{\vartheta_1} |u|_{1;t} \right),$$

$\mathbf{h\ddot{o}l}^{(\leftarrow)}(\vartheta, \vartheta_1; \mathcal{A}(t; \cdot, \cdot); |\cdot|_{\mathfrak{m};t}, \mathfrak{m} = \emptyset, *\emptyset, *1)$: For all $\chi \in \mathcal{P}_*$ and $u \in \mathcal{P}$,

$$|\mathcal{A}(t; \chi, u) - \mathcal{A}(s; \chi, u)| \le C |u|_{\emptyset;t} \left(|t - s|^{\vartheta} |\chi|_{*\emptyset;s} + |t - s|^{\vartheta_1} |\chi|_{*1;t} \right),$$

---

[4]This hypothesis may look strange because $A(t)$ does not depend on k. However, if $A(t)$ depends on an additional parameter, h, say, and $A(t)$ is bounded but not uniformly bounded with respect to h, there are situations when ('...' substitutes for the second term on the right-hand side of (5.2).)

$$\|(A(t) - A(s))\, u\| \le C\psi(h)\, |t - s|^{\vartheta} \|(A(t) + \mu_1 I)\, u\| + \dots,$$

with $\psi(h)$ an unbounded factor. Then, if k and h are in conformity by the condition $k^{\vartheta_1} \psi(h) \le C$, with some $\vartheta_1 \in (0, \vartheta)$, which is sometimes the case in applications, we come to hypothesis $\mathbf{h\ddot{o}l}^{\oplus}_{(k)}[\vartheta - \vartheta_1; |\cdot|_t]$.

[5]We assume that λ_0 is independent of t. We might not accept this assumption but such a generalization would be worthless, taking into account our subsequent aims.

and

höl$(\vartheta; \mathcal{A}(t;\cdot,\cdot); |\cdot|_{\mathfrak{m};t}, \mathfrak{m} = \emptyset, *\emptyset)$: For all $\chi \in \mathcal{P}_*$ and $u \in \mathcal{P}$,

$$|\mathcal{A}(t;\chi,u) - \mathcal{A}(s;\chi,u)| \le C|t-s|^{\vartheta}|u|_{\emptyset;t}|\chi|_{*\emptyset;s}.$$

However, when deriving error estimates, if we are interested in getting orders of convergence ≥ 1, it is not natural to be based on Hölder-continuity type hypotheses and we employ instead some Lipschitz-continuity type conditions. As such we introduce the following.

lip: The operator $A(t)$ satisfies hypothesis **höl**$[\vartheta]$ for $\vartheta = 1$,

and

lip$(\mathcal{A}(t;\cdot,\cdot); |\cdot|_{\mathfrak{m};t}, \mathfrak{m} = \emptyset, *\emptyset)$: Hypothesis **höl**$(\vartheta; \mathcal{A}(t;\cdot,\cdot); |\cdot|_{\mathfrak{m};t}, \mathfrak{m} = \emptyset, *\emptyset)$ is satisfied with $\vartheta = 1$.

In what follows we often use seminorms with special properties. For the sake of convenience we therefore introduce certain new terminology.

Definition 5.2. *Let $A(t) \in \mathcal{S}(\varphi,\sigma)$ for some $\varphi \in (0,\pi/2)$, $\sigma \in \mathbb{R}$ and let $|\cdot|_t$, $t \in [0,T]$ be a seminorm with* Dom $|\cdot|_t \subseteq \mathfrak{X}$. *The pair $(|\cdot|_t; A(t))$ is called $(\xi|m|\varphi,\sigma)$*-concordant *for some $\xi \ge 0$ and $m \in \mathbb{N}$ such that $m \ge \xi$ if, for all $\lambda \notin \operatorname{Int}\Sigma_\varphi \cup \{0\}$, $t \in [0,T]$, and $u \in \mathfrak{X}$,*

$$\operatorname{Ran}((\lambda+\sigma)I - A(t))^{-m} \subseteq \operatorname{Dom}|\cdot|_t$$

and

$$\left|((\lambda+\sigma)I - A(t))^{-m}u\right|_t \le C|\lambda|^{\xi-m}\|u\|. \tag{5.3}$$

Similarly, if $|\cdot|_{*;t}$, $t \in [0,T]$, is a seminorm with Dom $|\cdot|_{*;t} \subseteq \mathfrak{X}^*$, *the pair $(|\cdot|_{*;t}; A(t))$ is called $(*;\xi|m|\varphi,\sigma)$*-concordant *for some $\xi \ge 0$ and $m \in \mathbb{N}$ such that $m \ge \xi$ if, for all $\lambda \notin \operatorname{Int}\Sigma_\varphi \cup \{0\}$, $t \in [0,T]$, and $\chi \in \mathfrak{X}^*$,*

$$\operatorname{Ran}\left(((\lambda+\sigma)I - A(t))^{-m}\right)^* \subseteq \operatorname{Dom}|\cdot|_{*;t}$$

and

$$\left|(((\lambda+\sigma)I - A(t))^{-m})^*\chi\right|_{*;t} \le C|\lambda|^{\xi-m}\|\chi\|_*. \tag{5.4}$$

If the size of m is not important in the context, the pairs $(|\cdot|_t; A(t))$ and $(|\cdot|_{*;t}; A(t))$ are then called $(\xi|\varphi,\sigma)$-concordant *and $(*;\xi|\varphi,\sigma)$*-concordant, *respectively. Apart from everything else, if the resolvent estimates (5.3) and (5.4) hold for all $t, s \in [0,T]$ with $|\cdot|_s$ in place of $|\cdot|_t$ and with $|\cdot|_{*;s}$ in place of $|\cdot|_{*;t}$, respectively, then the pairs $(|\cdot|_t; A(t))$ and $(|\cdot|_{*;t}; A(t))$ are called* strictly $(\ldots)$-concordant.

With regard to possible applications of the abstract theory developed below, we add a few words about classical parabolic PDE problems that can fit into our framework.

For example, let Ω be a convex, bounded domain in $\mathbb{R}^d$, $d \geq 1$, with smooth boundary, and let $A(t)$ be an elliptic partial differential operator of second order defined by

$$A(t)v = -\sum_{j=1}^{d}\sum_{l=1}^{d} \frac{\partial}{\partial x_j}\left(a_{jl}(t,x)\frac{\partial v}{\partial x_l}\right), \quad x = (x_1, \ldots, x_d) \in \Omega,$$

with Dirichlet boundary conditions, where the coefficients $a_{jl}(t,x) = a_{lj}(t,x)$ are smooth and real-valued on $[0,T] \times \mathrm{Cl}\,\Omega$. The operator $A(t)$ is assumed to be uniformly elliptic in the following standard sense, with some $a_0 > 0$,

$$\sum_{j=1}^{d}\sum_{l=1}^{d} a_{jl}(t,x)\mu_j\mu_l \geq a_0 \sum_{j=1}^{d} \mu_j^2 \quad \text{for all } \mu_j \in \mathbb{R},\ j = 1, \ldots, d.$$

Then, thinking of $A(t)$ as a closed linear operator in $L_\infty(\Omega)$ with suitably chosen domain, it can be shown that $A(t) \in \mathcal{S}(\varphi, 0)$ for any fixed $\varphi \in (0, \pi/2)$ (cf., e.g., [38, Theorem 1.1] or [43, Theorem 2.1]). By duality and interpolation the same is valid if $A(t)$ is considered as an operator in $L_p(\Omega)$, $p \in [1, \infty)$.

For applications of operator theory to the analysis of fully discrete approximations to parabolic PDE problems, it is, however, more interesting to have sectorial estimates for discrete analogues of elliptic operators. In particular, it is shown in [38, 43] that finite element approximations to the above operator $A(t)$ possess, in fact, the same sectorialness property as $A(t)$ itself.

Some further results on sectorialness for second and higher order elliptic operators in Lebesgue, Sobolev, and Hölder spaces can be found in Agmon [2], Bakaev [39], Campanato [56], Colombo & Vespri [58], Lunardi [108, Chapter 3], and Stewart [155]. Note that not all of the mentioned results can suitably be applied for the aims of Part II because there are times that the operator in question has a time-variable domain even if it is subject to a uniform resolvent estimate with respect to t, which is impermissible for Runge-Kutta discretization in time. This disadvantage is avoided when dealing with discrete versions of elliptic operators, but, unfortunately, there are very few known results on estimating resolvents in the discrete case. Apart from what is already cited above concerning finite element operators, we can mention the work of Schatz, Thomée & Wahlbin [141], where the case of L_∞ norm and Neumann boundary conditions is studied, and our own work [39] where in particular sectorial estimates in Hölder norms are presented.

Furthermore, as concerns the above **höl/lip** hypotheses, they easily are verified for elliptic differential and finite element operators, assuming a suitable Hölder/Lipschitz-continuity property for the coefficients.

Finally, in the matter of (...)-concordant pairs involving elliptic differential operators, such pairs often occur in interpolation theory (see, e.g., Komatsu [94, § 1,2] or Triebel [164, Chapter 1]). At the same time (...)-concordant pairs involving Hölder or Lebesgue norms and finite element operators are studied in our paper [39] which contains related results for differential operators as well.

In the remainder of the section we show some auxiliary results involving (...)-concordant pairs, for subsequent use.

Lemma 5.1. *Suppose that $A(t) \in \mathcal{S}(\varphi, \sigma)$ and seminorms $|\cdot|_t$ and $|\cdot|_{*;t}$ form respectively $(\xi|m|\varphi,\sigma)$- and $(*;\xi|m|\varphi,\sigma)$-concordant pairs with $A(t)$ for some $\varphi \in (0, \pi/2)$, $\sigma \in \mathbb{R}$, $m \in \mathbb{N}$, and $\xi \geq 0$ such that $m \geq \max\{\xi, 1\}$. Then, for any fixed integer $m_1 > \xi$, the pairs $\big(|\cdot|_t; A(t)\big)$ and $\big(|\cdot|_{*;t}; A(t)\big)$ are $(\xi|m_1|\varphi,\sigma)$- and $(*;\xi|m_1|\varphi,\sigma)$-concordant, respectively.*

Proof. We show only that the pair $\big(|\cdot|_t; A(t)\big)$ is $(\xi|m_1|\varphi,\sigma)$-concordant, noting that the $(*;\xi|m_1|\varphi,\sigma)$-concordance of the latter pair $\big(|\cdot|_{*;t}; A(t)\big)$ is proved in a similar manner.

Observe now that if $m_1 \geq m$, by the $(\xi|m|\varphi,\sigma)$-concordance of the pair $\big(|\cdot|_t; A(t)\big)$ and by the fact that $A(t) \in \mathcal{S}(\varphi, \sigma)$, it then follows that for all $\lambda \notin \operatorname{Int}\Sigma_\varphi$ and $u \in \mathfrak{X}$,

$$\begin{aligned}\big|((\lambda+\sigma)I-A(t))^{-m_1}u\big|_t &= \big|((\lambda+\sigma)I-A(t))^{-m}((\lambda+\sigma)I-A(t))^{-(m_1-m)}u\big|_t \\ &\leq C|\lambda|^{\xi-m}\big\|((\lambda+\sigma)I-A(t))^{-(m_1-m)}u\big\| \leq C|\lambda|^{\xi-m_1}\|u\|.\end{aligned}$$

It thus remains to consider the case $\xi < m_1 < m$. Clearly, it suffices to show the claim when $\xi < m_1 = m-1$. It is not hard to see that for $\lambda \notin \operatorname{Int}\Sigma_\varphi$ and $u \in \mathfrak{X}$,

$$((\lambda+\sigma)I-A(t))^{-(m-1)}u = (m-1)\int_\lambda^\infty ((\mu+\sigma)I-A(t))^{-m}u\,d\mu,$$

where the integral is taken along the ray $\mu = xe^{i\arg\lambda}$, $|\lambda| \leq x < \infty$. Using this equality, since $m-1 > \xi$ and $|\cdot|_t$ is a seminorm, a simple estimation now yields, for $\lambda \notin \operatorname{Int}\Sigma_\varphi$,

$$\begin{aligned}\big|((\lambda+\sigma)I - A(t))^{-(m-1)}u\big|_t &\le C\int_{|\lambda|}^{\infty} \big|((xe^{i\arg\lambda}+\sigma)I - A(t))^{-m}u\big|_t dx \\ &\le C\int_{|\lambda|}^{\infty} x^{\xi-m}dx\,\|u\| = C|\lambda|^{\xi-(m-1)}\|u\|,\end{aligned}$$

which shows again the $(\xi|m_1|\varphi,\sigma)$-concordance of the pair $\big(|\cdot|_t; A(t)\big)$. □

The next definition introduces a helpful notion when working with functions of sectorial operators.

Definition 5.3. *A function $\omega(z) : \mathbb{C} \to \mathbb{C}$ is called (φ)-regular for some $\varphi \in (0,\pi/2)$ if it is holomorphic in the open sector $-\Sigma_\varphi$ as well as at* 0.

Note that if $A(t) \in \mathcal{S}(\varphi,\sigma)$ for some $\varphi \in (0,\pi/2)$ and $\sigma \in \mathbb{R}$ and if $\omega(z)$ is a rational function which is (φ)-regular and bounded at ∞, then since it holds $kA(t) \in \mathcal{S}(\varphi,k\sigma)$ for any $k > 0$, using the operator calculus formula (A.10) with $kA(t)$ in place of $\mathfrak{L}$, one can define the operator $\omega(-kA(t)) \in \mathcal{B}(\mathfrak{X})$ for any $k \in (0,K]$ and $t \in [0,T]$, with some $K > 0$ sufficiently small. The same conclusion is in force if $\omega(z)$ is (φ)-regular (not necessarily rational) and if for some $\omega(\infty) \in \mathbb{C}$, it holds $\omega(-z) = \omega(\infty) + \mathcal{O}\big(|z|^{-1}\big)$ as $|z| \to \infty$ uniformly with respect to $z \in \Sigma_{\varphi_1}$, for any fixed $\varphi_1 \in (0,\varphi)$. Now we show some estimates involving such operators.

For further needs, for any rational function $\omega(z)$, we define its integer-valued characteristic $\deg[\omega]$ by

$$\deg[\omega] = \lim_{|z|\to\infty} \frac{\log|\omega(z)|}{\log|z|}.$$

Lemma 5.2. *Suppose that $A(t) \in \mathcal{S}(\varphi,\sigma)$ and seminorms $|\cdot|_t$ and $|\cdot|_{*;t}$ form respectively $(\xi|m|\varphi,\sigma)$- and $(*;\xi|m|\varphi,\sigma)$-concordant pairs with $A(t)$ for some $\varphi \in (0,\pi/2)$, $\sigma \in \mathbb{R}$, $\xi \ge 0$, and $m \in \mathbb{N}$ such that $m \ge \max\{\xi,1\}$. Let further $\omega(z)$ be a rational function which is (φ)-regular and satisfies the condition $\deg[\omega] \le -m$. Then there exists a fixed $K > 0$ such that for all $k \in (0,K]$, $t \in [0,T]$, $u \in \mathfrak{X}$, and $\chi \in \mathfrak{X}^*$,*

$$|\omega(-kA(t))\,u|_t \le Ck^{-\xi}\|u\| \tag{5.5}$$

and

$$\big|\big(\omega(-kA(t))\big)^*\chi|_{*;t} \le Ck^{-\xi}\|\chi\|_*.$$

If $\sigma \ge 0$ in addition, then the above is in force for $K = \infty$.

Proof. We will show only (5.5), noting that the latter estimate is proved in a similar way.[6]

It follows from Lemma A.2 that with some $\varphi_1 \in (0, \varphi)$ and $R \geq 0$, for all $t \in [0, T]$, $z \notin \operatorname{Int}(\Sigma_{\varphi_1} \cup \mathcal{D}(R))$, and $u \in \mathfrak{X}$,

$$\left|(zI - A(t))^{-m} u\right|_t \leq C|z|^{\xi - m}\|u\|. \tag{5.6}$$

Moreover, in the case $\sigma \geq 0$, (5.6) holds with $R = 0$. Using (5.6), we find as well for $z \notin \operatorname{Int}(\Sigma_{\varphi_1} \cup \mathcal{D}(kR))$, since $z/k \notin \operatorname{Int}(\Sigma_{\varphi_1} \cup \mathcal{D}(R))$,

$$\begin{aligned} \left|(zI - kA(t))^{-m} u\right|_t &= k^{-m}\left|((z/k)I - A(t))^{-m} u\right|_t \\ &\leq Ck^{-m}|z/k|^{\xi - m}\|u\| = Ck^{-\xi}|z|^{\xi - m}\|u\|. \end{aligned} \tag{5.7}$$

Next note that the function $\omega(z)$ has no poles at least in some disk $\mathcal{D}(d)$, $d > 0$. Further, we set $K > 0$ subject to $K \leq d/R$ if $\sigma < 0$, and we take $K = \infty$ if $\sigma \geq 0$ (in the last case we have $R = 0$). Let now Γ be the contour coinciding with the boundary of the set $\Sigma_{\varphi_1} \cup \mathcal{D}(d)$. Using now the operator calculus formula (A.10) with Γ in place of $\Xi_{R+\varepsilon}$ and with $kA(t)$ in place of $\mathfrak{L}$, we get, since $\omega(\infty) = 0$,

$$\omega(-kA(t)) = \frac{1}{2\pi i} \int_\Gamma \omega(-z)\,(zI - kA(t))^{-1} dz. \tag{5.8}$$

Integrating, if $m \geq 2$, by parts and multiplying further both sides from the right by $u \in \mathfrak{X}$, we obtain

$$\omega(-kA(t))\,u = \frac{1}{2\pi i} \int_\Gamma \Psi_m(z)\,(zI - kA(t))^{-m} u\,dz, \tag{5.9}$$

with

$$\Psi_m(z) = \begin{cases} \omega(-z) & \text{if } m = 1, \\ (m-1) \displaystyle\int_{\infty e^{-i\varphi_1}}^{z} (z - z_1)^{m-2} \omega(-z_1)\,dz_1 & \text{if } m \geq 2, \end{cases}$$

where the path of integration from $\infty e^{-i\varphi_1}$ to z lies on the contour Γ. For further success we need to estimate suitably the function $\Psi_m(z)$. In the case $m = 1$ it is evident that

$$|\Psi_m(z)| = |\omega(-z)| \leq C(1 + |z|)^{-1} \quad \text{for } z \in \Gamma. \tag{5.10}$$

[6] In fact, we then have to use the below representation (5.8) with both sides converted into the adjoint operators.

Turning next to the case $m \geq 2$, we find, since $\deg[\omega] \leq -m$,

$$\begin{aligned} |\Psi_m(z)| &\leq (m-1) \int_{\infty e^{-i\varphi_1}}^{z} |z - z_1|^{m-2} |\omega(-z_1)| \, |dz_1| \\ &\leq C \int_{\infty e^{-i\varphi_1}}^{z} |z - z_1|^{m-2} (1 + |z_1|)^{-m} \, |dz_1|, \end{aligned}$$

which implies, for all $z \in \Gamma$ with $\Im z \leq 0$,

$$|\Psi_m(z)| \leq C \int_{|z|}^{\infty} (1+x)^{-2} dx = C(1 + |z|)^{-1}. \tag{5.11}$$

Observe that by Cauchy's Theorem,

$$\int_{\infty e^{-i\varphi_1}}^{z} (z - z_1)^{m-2} \omega(-z_1) \, dz_1 = -\int_{z}^{\infty e^{i\varphi_1}} \dots,$$

and thus (5.11) remains valid for all $z \in \Gamma$. Using now (5.9) together with (5.10), (5.11), and (5.7) and since $|\cdot|_t$ is a seminorm, a simple estimation gives, for $k \in (0, K]$,

$$\begin{aligned} |\omega(-kA(t)) \, u|_t &\leq C \int_{\Gamma} (1 + |z|)^{-1} \big|(zI - kA(t))^{-m} u\big|_t |dz| \\ &\leq Ck^{-\xi} \int_{\Gamma} (1 + |z|)^{-1} |z|^{\xi - m} |dz|, \end{aligned}$$

thus (5.5) follows at least in the case $\xi < m$.

Now let $\xi = m$. It follows from Lemma 5.1 that the pair $\big(|\cdot|_t; A(t)\big)$ is, in fact, $(\xi|m+1|\varphi, \sigma)$-concordant. Therefore, if a rational function $\omega_1(z)$ with $\deg[\omega_1] \leq -(m+1)$ is (φ)-regular, applying the above argument shows that for all $k \in (0, K]$ and $u \in \mathfrak{X}$,

$$|\omega_1(-kA(t)) \, u|_t \leq Ck^{-\xi} \|u\|. \tag{5.12}$$

Next, let $\lambda_0 \in \mathbb{C}$ be subject to

$$\lambda_0^{-m} = \lim_{|z| \to \infty} \left(\omega(-z) z^m \right),$$

and furthermore let $\omega_1(z)$ be given by

$$\omega_1(z) = \omega(z) - \begin{cases} (1 - \lambda_0 z)^{-m} & \text{if } |\arg \lambda_0| \leq \pi/2, \\ (-1 - \lambda_0 z)^{-m} & \text{otherwise.} \end{cases} \tag{5.13}$$

It is easily seen that $\deg[\omega_1] \leq -(m+1)$ and $\omega_1(z)$ is (φ)-regular so that ω_1 satisfies (5.12). Since $A(t) \in \mathcal{S}(\varphi, \sigma)$ and the pair $\big(|\cdot|_t; A(t)\big)$ is $(\xi|m|\varphi, \sigma)$-concordant, we have for all $k \in (0, K]$ and $u \in \mathfrak{X}$,

$$|(I + \lambda_0 kA(t))^{-m} u|_t \leq Ck^{-\xi}\|u\| \quad \text{if } |\arg \lambda_0| \leq \pi/2, \tag{5.14}$$

and

$$|(I - \lambda_0 kA(t))^{-m} u|_t \leq Ck^{-\xi}\|u\| \quad \text{if } |\arg(-\lambda_0)| < \pi/2. \tag{5.15}$$

By (5.13) with $-kA(t)$ substituted for z, we get an operator equality

$$\omega(-kA(t)) = \omega_1(-kA(t)) + \begin{cases} (I + \lambda_0 kA(t))^{-m} & \text{if } |\arg \lambda_0| \leq \pi/2, \\ (-I + \lambda_0 kA(t))^{-m} & \text{otherwise.} \end{cases} \tag{5.16}$$

Therefore, using (5.16) together with (5.12), (5.14), and (5.15), the desired estimate (5.5) in the case $\xi = m$ follows as well, after a trivial estimation. □

Now we pose separately an important case of Lemma 5.2.

Lemma 5.3. *Let $A(t) \in \mathcal{S}(\varphi, \sigma)$ for some $\varphi \in (0, \pi/2)$ and $\sigma \in \mathbb{R}$. Let further $\omega(z)$ be a rational function which is (φ)-regular and bounded at ∞. Then, if $K > 0$ is sufficiently small, we have, for all $k \in (0, K]$ and $t \in [0, T]$,*

$$\|\omega(-kA(t))\| \leq C. \tag{5.17}$$

Furthermore, if $\deg[\omega] \leq -m$ for some $m \in \mathbb{N}$, then, for all $k \in (0, K]$ and $t \in [0, T]$, $\operatorname{Ran} \omega(-kA(t)) \subseteq \operatorname{Dom} A^m(t)$ and

$$\|(kA(t))^m \, \omega(-kA(t))\| \leq C.$$

In the case $\sigma \geq 0$, the above is still in force for $K = \infty$.

This fact follows directly from Lemma 5.2 taking $\xi = 0$, $m = 1$, and $|\cdot|_t = \|\cdot\|$.

Lemma 5.4. *Suppose that for some $\varphi \in (0, \pi/2)$, $\sigma \in \mathbb{R}$, and $\omega(\infty) \in \mathbb{C}$, $A(t) \in \mathcal{S}(\varphi, \sigma)$ and a (not necessarily rational) function $\omega(z)$ is (φ)-regular and obeys the condition $\omega(-z) = \omega(\infty) + \mathcal{O}\big(|z|^{-1}\big)$ as $|z| \to \infty$ uniformly with respect to $z \in \Sigma_{\varphi_1}$, for any fixed $\varphi_1 \in (0, \varphi)$. Then the operator $\omega(-kA(t))$ satisfies the estimate (5.17).*

This result can be shown by using the same techniques as those employed in the proof of Lemma 5.2.

Lemma 5.5. *Let $A(t) \in \mathcal{S}(\varphi, \sigma)$ and let a (not necessarily rational) function $\omega_n(z)$, $n \in \mathbb{N} \cup \{0\}$, be holomorphic (in z) in a neighbourhood of Σ_φ, for some $\varphi \in (0, \pi/2)$ and $\sigma \in \mathbb{R}$. Suppose further that for some $d > 0$, the function $\omega_n(z)$ obeys the conditions, for all $n \in \mathbb{N} \cup \{0\}$,*

$$|\omega_n(z)| \leq Ce^{n|z|} \quad \text{for all } z \in \mathcal{D}(d),$$

$$|\omega_n(-z)| \leq Ce^{-cn|z|} \quad \text{for all } z \in \Sigma_\varphi \cap \mathcal{D}(d),$$

and

$$|\omega_n(-z)| \leq C(1 + |z|)^{-1} \quad \text{for all } z \in \Sigma_\varphi.$$

Then it holds, with some $K > 0$ sufficiently small, for all $t \in [0, T]$, $k \in (0, K]$, and $n \in \mathbb{N} \cup \{0\}$ such that $nk \leq C$,

$$\|\omega_n(-kA)\| \leq C. \tag{5.18}$$

Furthermore, if $\sigma \geq 0$ in addition, then the restriction $nk \leq C$ is not needed.

Proof. Note that (see Appendix A.2) the operator $\omega_n(-kA(t))$ can be represented by formula (5.8) with $\omega_n(\cdot)$ substituted for $\omega(\cdot)$, where Γ is replaced by Γ_n the boundary of the set $\Sigma_\varphi \cap \mathcal{D}(d/(2n+2) + Rk)$, for some $R \geq 0$ sufficiently large.[7] Next we split Γ_n into three parts by $\Gamma_n = \Gamma_n^{(1)} \cup \Gamma_n^{(2)} \cup \Gamma^{(3)}$, where

$$\Gamma_n^{(1)} = \{z \in \mathbb{C} : |z| = d/(2n+2) + Rk, \ z \notin \operatorname{Int} \Sigma_\varphi\},$$

$$\Gamma_n^{(2)} = \{z \in \mathbb{C} : d/(2n+2) + Rk \leq |z| \leq d, \ z \in \Sigma_\varphi\},$$

and

$$\Gamma^{(3)} = \{z \in \mathbb{C} : |z| \geq d, \ z \in \Sigma_\varphi\}.$$

Now, estimating separately the integrals over each part of the contour Γ_n with the aid of the accepted conditions on $\omega_n(z)$ and taking into account the reasonings employed in the proof of Theorem 2.1, the claim follows as in the proof of Lemma 5.2 in the case $m = 1$. □

Lemma 5.6. *Let $A(t) \in \mathcal{S}(\varphi, \sigma)$ and let a rational function $\omega_n(z)$, $n \in \mathbb{N} \cup \{0\}$, be (φ)-regular for any fixed n, for some $\varphi \in (0, \pi/2)$ and $\sigma \in \mathbb{R}$. Suppose further that for some $d \in (0, 1)$, $d_1 > 1$, $q > 0$ and for any fixed $\varphi_1 \in (0, \varphi)$, the function $\omega_n(z)$ obeys the conditions, for all $n \in \mathbb{N} \cup \{0\}$,*

$$|\omega_n(z)| \leq Ce^{n|z|} \quad \text{for all } z \in \mathcal{D}(d),$$

[7] One can take $R = 0$ if $\sigma \geq 0$.

$$|\omega_n(-z)| \leq Ce^{-cn|z|} \quad \textit{for all } z \in \Sigma_{\varphi_1} \cap \mathcal{D}(1),$$

$$|\omega_n(-z)| \leq Ce^{-cn|z|^{-q}} \quad \textit{for all } z \in \Sigma_{\varphi_1} \textit{ such that } |z| \geq 1,$$

and

$$|\omega_n(z)| \leq Ce^{cn|z|^{-q}} \quad \textit{for all } z \notin \operatorname{Int} \mathcal{D}(d_1).$$

Then, with some $K > 0$ sufficiently small, the operator $\omega(-kA(t))$ satisfies the estimate (5.18) for all $t \in [0,T]$, $k \in (0,K]$, and $n \in \mathbb{N} \cup \{0\}$ such that $nk \leq C$. Furthermore, the condition $nk \leq C$ is not needed if $\sigma \geq 0$.

Proof. Taking φ_1 and R the same as in the proof of Lemma 5.2 and letting $\Gamma_n = \bigcup_{j=1}^{4} \Gamma_n^{(j)}$, where

$$\Gamma_n^{(1)} = \{z \in \mathbb{C} : |z| = d/(2n+2) + Rk,\ z \notin \operatorname{Int} \Sigma_{\varphi_1}\},$$

$$\Gamma_n^{(2)} = \{z \in \mathbb{C} : d/(2n+2) + Rk \leq |z| \leq 1,\ z \in \Sigma_{\varphi_1}\},$$

$$\Gamma_n^{(3)} = \{z \in \mathbb{C} : 1 \leq |z| \leq d_1(n+1)^{1/q},\ z \in \Sigma_{\varphi_1}\},$$

and

$$\Gamma_n^{(4)} = \{z \in \mathbb{C} : |z| = d_1(n+1)^{1/q},\ z \notin \operatorname{Int} \Sigma_{\varphi_1}\},$$

we use the Taylor operator calculus (see, e.g., Taylor [160]), so that we have the representation

$$\omega_n(-kA(t)) = \omega_n(\infty)I + \frac{1}{2\pi i}\int_{\Gamma_n} \omega_n(-z)(zI - kA(t))^{-1}dz.$$

Now, with this equality in mind, the claim easily follows in a manner similar to that used in the proof of Lemma 5.2 in the case $m = 1$, taking the accepted condtions on $\omega_n(z)$ into account and estimating separately the integrals over $\Gamma_n^{(j)}$, $j = 1, 2, 3, 4$. □

5.2 Discretization

In this section we consider the application of the Runge-Kutta discretization to the abstract linear Cauchy problem (5.1). For subsequent reference, we introduce the reader to a proper classification of the Runge-Kutta methods through a series of definitions. Apart from everything else, we give examples of families of Runge-Kutta methods which go into the theory.

Let $\Omega_k = \{t_n : t_n = nk \text{ for } n = 0, \dots, N\}$ be a uniform grid in the interval $[0, T]$, with stepsize k.[8] Furthermore, let problem (5.1) be discretized in time by means of a ν-stage Runge-Kutta method[9] with Butcher array

$$\left(\begin{array}{c|c} \mathbf{c} & \mathfrak{a} \\ \hline & \mathbf{b}^T \end{array} \right),$$

where[10] $\mathbf{b} = \big(b_1, \dots, b_\nu\big)^T \in \mathbb{C}^\nu$, $\mathbf{c} = \big(c_1, \dots, c_\nu\big)^T \in \mathbb{R}^\nu$, and $\mathfrak{a} = (a_{jl})_{\nu\times\nu} \in \mathbb{C}^{\nu\times\nu}$ are given. The application of the RK method to problem (5.1) leads to the following discrete problem,[11] with some given $\mathfrak{R} \in \mathcal{B}(\mathfrak{X})$ generally depending on k,[12]

$$Y_{n+1} = Y_n - k\sum_{j=1}^{\nu} b_j(A(t_{nj})Y_{nj} - f(t_{nj})) \quad \text{for } t_n \in \breve{\Omega}_k, \qquad Y_0 = \mathfrak{R}y^0, \tag{5.19}$$

where Y_{nj} are calculated from the system of equations

$$Y_{nj} = Y_n - k\sum_{l=1}^{\nu} a_{jl}(A(t_{nl})Y_{nl} - f(t_{nl})), \quad j = 1, \dots, \nu. \tag{5.20}$$

In (5.19), (5.20): $t_{nj} = t_n + c_j k$ and Y_n is the approximation to $y(t_n)$.

The stability function of the method is defined in the standard way by

$$r(z) = 1 + z\mathbf{b}^T(\mathfrak{i} - z\mathfrak{a})^{-1}\mathbf{e},$$

where $\mathfrak{i}$ stands for the identity $(\nu \times \nu)$-matrix and $\mathbf{e} = \big(\underbrace{1, \dots, 1}_{\nu}\big)^T$. Next, let $g_{jl}(z)$, $j, l = 1, \dots, \nu$, be the entries of the matrix $(\mathfrak{i} - z\mathfrak{a})^{-1}$ and let $\mathbf{w}(z) = \big(w_1(z), \dots, w_\nu(z)\big)^T$ and $\mathbf{v}(z) = \big(v_1(z), \dots, v_\nu(z)\big)^T$ be given by

$$\mathbf{w}^T(z) = \mathbf{b}^T(\mathfrak{i} - z\mathfrak{a})^{-1} \text{ and } \mathbf{v}(z) = (\mathfrak{i} - z\mathfrak{a})^{-1}\mathbf{e}. \tag{5.21}$$

---

[8] As above, N and k are in conformity through $Nk = T$, and the case $T = \infty$ can be trivially settled (cf. Section 1.1). Also, $\breve{\Omega}_k$ possesses the same meaning as in Section 1.1.

[9] In what follows we often use RK method instead of Runge-Kutta method, for brevity.

[10] For simplicity we assume that $0 \le c_j \le 1$, $j = 1, \dots, \nu$, throughout. This restriction is by no means crucial.

[11] These equations are a formal procedure only. Sufficient conditions for problem (5.19) and (5.20) to be uniquely solvable are given later in the chapter.

[12] It is not in general use to construct Runge-Kutta approximations with correction of the initial condition y^0. In the sequel it is shown, however, that sometimes involving $\mathfrak{R}$ is essential.

Thus for the function $\alpha(z) = 1 - r(z)$ we have the following representation

$$\alpha(z) = -z\mathbf{w}^T(z)\mathbf{e} = -z\mathbf{b}^T(\mathfrak{i} - z\mathfrak{a})^{-1}\mathbf{e}. \tag{5.22}$$

The function $\alpha(z)$ will play an important role in the theory below. Also, there are times when we employ the function

$$\rho(z) = -z^{-1}\alpha(z) = \mathbf{b}^T(\mathfrak{i} - z\mathfrak{a})^{-1}\mathbf{e}. \tag{5.23}$$

Clearly, all the functions $\alpha(z)$, $\rho(z)$, $v_j(z)$, $w_j(z)$, and $g_{jl}(z)$ are rational.

Throughout this part of the book we will consider only such RK methods that satisfy the additional condition[13]

$$r(0) = r'(0) = 1,$$

which is therefore always assumed in what follows without explicit mention.

Definition 5.4. *The RK method is said to be* $A(\varphi)$-stable *for some* $\varphi \in (0, \pi/2)$ *if*

$$|r(z)| \leq 1 \quad \text{for all } z \in -\Sigma_\varphi,$$

and A-stable *if the last inequality holds for* $\varphi = \pi/2$.

Definition 5.5. *We say that the RK method is of* type $C(\varphi)$ *for some* $\varphi \in (0, \pi/2)$ *if*
(i) it is $A(\varphi)$*-stable;*
(ii) each of the functions $w_j(-z)$, $j = 1, \ldots, \nu$ *is* (φ)*-regular and* $\deg[w_j] \leq 0$, $j = 1, \ldots, \nu$.

Assume in addition to (i) and (ii) that
(iii) The eigenvalues of the coefficient matrix $\mathfrak{a} = (a_{jl})_{\nu\times\nu}$ *are contained in the sector* $\Sigma_{\pi-\varphi}$*;*
and
(iv)

$$\deg[g_{jl}] \leq 0, \quad j, l = 1, \ldots, \nu.$$

Then we say that the method is of type $V(\varphi)$ *or of* type $V^*(\varphi)$ *if it satisfies as well*
(v)

$$\deg[v_j] \leq -1, \quad j = 1, \ldots, \nu,$$

[13] In principle, this condition is not needed for showing stability in itself. At the same time it is satisfied for all practical discretization methods. We therefore prefer to assume this condition throughout, which slightly facilitates our subsequent proofs.

or
(vi)

$$\deg[w_j] \leq -1, \quad j = 1, \ldots, \nu,$$

respectively.[14]

It is worth noting that unlike some authors (cf., e.g., González & Palencia [73, 74]), we do not suppose that the $A(\varphi)$-stability implies (iii). Also, observe that (iii) and (iv) together imply (ii), in view of (5.21). Furthermore, if the matrix $\mathfrak{a}$ is non-singular, we have

$$\deg[g_{jl}] \leq -1, \quad j, l = 1, \ldots, \nu, \tag{5.24}$$

thus it immediately follows that

$$\deg[v_j] \leq -1 \text{ and } \deg[w_j] \leq -1 \quad \text{for } j = 1, \ldots, \nu. \tag{5.25}$$

In particular, it is easily seen that (ii) is then a direct consequence of (iii).

Obviously, dealing with $A(\varphi)$-stable methods, we have only three possibilities:

(S_I)
$$|r(\infty)| < 1,$$

(S_{II})
$$r(\infty) = 1,$$

and

(S_{III})
$$|r(\infty)| = 1 \text{ and } r(\infty) \neq 1.$$

Definition 5.6. *We say that the RK method belongs to the* class S_J, $J = I, II, III$, *if it satisfies the respective condition* (S_J).[15] *The method is called* $A_J(\varphi)$-stable *(*A_J-stable*), $J = I, II, III$, if it is $A(\varphi)$-stable (A-stable) and belongs to the respective class S_J. Furthermore, the method is of* type $C_J(\varphi)$ *(of* type $V_J(\varphi)$, *of* type $V_J^*(\varphi)$*), $J = I, II, III$, if it is of type $C(\varphi)$ (of type $V(\varphi)$, of type $V^*(\varphi)$) and simultaneously belongs to the respective class S_J.*

[14] This terminology may be justified by noting that the use of methods of type $C(\varphi)$ and of type $V(\varphi)$ ($V^*(\varphi)$) is reasonable in the cases with a constant and a variable operator, respectively.

[15] This classification is relevant for single step methods (not only for the Runge-Kutta ones), which justifies the use of S in the abbreviations.

Therefore, we have placed at our disposal a certain classification of the Runge-Kutta methods. We take this classification as a basis for application of our techniques in this book. When establishing the below results, we nonetheless distinguish between the classes and types introduced by Definition 5.6.

Next we recall some widely practised concepts of method's order, which are needed for our investigation in Chapter 7. The first definition is rather general and may be employed for any single step discretization methods but not necessarily for Runge-Kutta ones.

Definition 5.7. *A single step discretization method is said to have* order p *if the local error of the method, after one step of length k of its application to an ordinary differential equation $dy/dt = f(t, y)$ with sufficiently smooth right-hand side, is $\mathcal{O}(k^{p+1})$ as $k \to 0$.*

Dealing with Runge-Kutta methods, the notion of stage order is useful in certain situations as well.

Definition 5.8. *An RK method is said to be of* stage order q *if the error of the method, when it is applied with stepsize k to an ordinary differential equation with sufficiently smooth right-hand side, is $\mathcal{O}(k^{q+1})$ as $k \to 0$ at each separate internal stage.*

In the case with a constant operator we employ one more concept of order (cf. Brenner, Crouzeix, and Thomée [50] or Akrivis & Crouzeix [4]).

Definition 5.9. *A discretization method is said to be of* polynomial order q *if, when it is applied to a linear ordinary differential equation $dy/dt + \lambda y = f(t)$, $\lambda \in \mathbb{C}$, the truncation error vanishes whenever the solution $y(t)$ is a polynomial in t of degree at most $q - 1$.*

It is easily checked that an RK method is of polynomial order $q + 1$ if it is of order p and of stage order $q \leq p$.

In conclusion, we point out some concrete families of the Runge-Kutta methods which are classified in due course according to the above definitions. We first recall, however, some of the so-called simplifying assumptions introduced in the theory of the Runge-Kutta methods by Butcher (see, e.g., [53]). In fact, for subsequent use we need the following only three.

$$\mathcal{B}(m) \qquad \sum_{s=1}^{\nu} b_s c_s^{j-1} = j^{-1} \quad \text{for } 1 \leq j \leq m;$$

$$\mathcal{C}(m) \qquad \sum_{s=1}^{\nu} a_{ls} c_s^{j-1} = j^{-1} c_l^j \quad \text{for } 1 \leq j \leq m \text{ and } 1 \leq l \leq \nu;$$

and

$$\mathcal{D}(m) \qquad \sum_{s=1}^{\nu} b_s c_s^{j-1} a_{sl} = j^{-1} b_l (1 - c_l^j) \quad \text{for } 1 \leq l \leq \nu \text{ and } 1 \leq j \leq m.$$

It is well known that the fact that the method is of order p implies $\mathcal{B}(p)$ and the fact that the method is of stage order q is equivalent to $\mathcal{C}(q)$ (see, e.g., Butcher [53, Chapter 3]).

First of all we discuss the family of Gauss-Legendre methods (see, e.g., Butcher [53], Dekker & Verwer [65] or Hairer & Wanner [81]). Given $\nu \geq 1$, for the ν-stage Gauss-Legendre method, the coefficients c_j, $j = 1, \ldots, \nu$, coincide with the zeros of $P_\nu(2c-1)$, where $P_\nu(x)$ is the Legendre polynomial of degree ν, while b_j, a_{jl}; $j, l = 1, \ldots, \nu$, are specified by the symplifying assumptions $\mathcal{B}(\nu)$ and $\mathcal{C}(\nu)$. It is known (see, e.g., [53, 65, 81]) that all methods of this family are A-stable and, for any fixed $\nu \geq 1$, the ν-stage Gauss-Legendre method is of order 2ν and of stage order ν.[16] Also, it follows from Saff & Varga [137] that the eigenvalues of the coefficient matrix $\mathfrak{a}$ are localized in the set $\operatorname{Int} \Sigma_{\pi/2}$. It is pointed out in Dekker & Verwer [65] that the matrix $\mathfrak{a}$ is non-singular so that (5.24) and (5.25) hold. Apart from everything else, it is not hard to see that $r(\infty) = 1$ if ν is even and $r(\infty) = -1$ if ν is odd. It is therefore clear from the above reasonings that if ν is even, the corresponding Gauss-Legendre method is of type $C_{II}(\varphi)$ for any fixed $\varphi \in (0, \pi/2)$.

Further, using the well-known representation of the stability function $r(z)$, we find, for $\alpha(z) = 1 - r(z)$,

$$\alpha(z) = 2 \sum_{\substack{0 \leq m \leq \nu, \\ m \text{ is odd}}} \frac{(2\nu - m)!}{m!\,(\nu - m)!} z^m \left(\sum_{m=0}^{\nu} \frac{(2\nu - m)!}{m!\,(\nu - m)!} z^m \right)^{-1}.$$

With the aid of this equality it follows that for odd ν,

$$1 - \alpha(\infty)^{-1} \alpha(z) = 2 \sum_{\substack{0 \leq m \leq \nu, \\ m \text{ is even}}} \frac{(2\nu - m)!}{m!\,(\nu - m)!} z^m \left(\sum_{m=0}^{\nu} \frac{(2\nu - m)!}{m!\,(\nu - m)!} z^m \right)^{-1},$$

which implies directly that $\deg[\alpha(.) - \alpha(\infty)] = -1$. Also, it is easily seen that $\alpha(\infty) = 2$ for odd ν and, hence, $r(\infty) = -1$. Therefore, if ν is odd, the

[16] The last fact is a direct consequence of the simplifying assumption $\mathcal{C}(\nu)$.

corresponding Gauss-Legendre method is both of type $V_{III}(\varphi)$ and of type $V_{III}^*(\varphi)$ for any fixed $\varphi \in (0, \pi/2)$, by (5.25).

In particular, the 1-stage Gauss-Legendre method has Butcher array

$$\left(\begin{array}{c|c} \frac{1}{2} & \frac{1}{2} \\ \hline & 1 \end{array} \right).$$

This manner of discretization is often referred to as the Crank-Nicolson method.

Next we consider the so-called Radau IA and IIA methods. Given $\nu \in \mathbb{N}$, for the ν-stage Radau IA method, c_j, $j = 1, \ldots, \nu$, are the roots of the equation

$$P_{\nu-1}(2c - 1) + P_\nu(2c - 1) = 0,$$

while b_j and a_{jl}; $j, l = 1, \ldots, \nu$, are specified by the simplifying assumptions $\mathcal{B}(\nu)$ and $\mathcal{D}(\nu)$. For the ν-stage Radau IIA method, c_j, $j = 1, \ldots, \nu$, are the roots of the equation

$$P_{\nu-1}(1 - 2c) + P_\nu(1 - 2c) = 0,$$

while the other coefficients are specified by the simplifying conditions $\mathcal{B}(\nu)$ and $\mathcal{C}(\nu)$. It is known (see, e.g., [53, 65, 81]) that the methods of both families are A-stable, all of which satisfy the condition

$$r(\infty) = 0 \text{ or, equivalently, } \alpha(\infty) = 1,$$

and, for any $\nu \geq 1$, the ν-stage Radau IA or IIA method is of order $2\nu - 1$. At the same time the stage order is $\nu - 1$ for Radau IA methods and ν for Radau IIA ones. Also, by Saff and Varga [137] the eigenvalues of the coefficient matrix $\mathfrak{a}$ are localized in $\operatorname{Int} \Sigma_{\pi/2}$. Moreover, the matrix $\mathfrak{a}$ itself is non-singular (see, e.g., [65]) so that (5.24) and (5.25) hold. Therefore any Radau IA or IIA method is both of type $V_I(\varphi)$ and of type $V_I^*(\varphi)$ for any fixed $\varphi \in (0, \pi/2)$.

In particular, the widely popular backward Euler method is nothing but the 1-stage Radau IIA method with the Butcher array

$$\left(\begin{array}{c|c} 1 & 1 \\ \hline & 1 \end{array} \right).$$

Further, we discuss the behaviour of the Lobatto IIIA, IIIB, and IIIC methods (see, e.g., [53, 65, 81]). For all these methods, $c_1 = 0$, $c_\nu = 1$ and c_j, $j = 1, \ldots, \nu$, are the roots of the equation

$$P'_{\nu-1}(2c - 1) = 0,$$

while the other coefficients are specified by the simplifying assumptions $\mathcal{B}(\nu)$ and $\mathcal{C}(\nu)$ for Lobatto IIIA methods, by $\mathcal{B}(\nu)$ and $\mathcal{D}(\nu)$ for Lobatto IIIB methods, and by $\mathcal{B}(\nu)$ and $\mathcal{C}(\nu-1)$ and the additional condition

$$a_{j1} = b_1, \quad j = 1, \dots, \nu,$$

for Lobatto IIIC methods. It is known (see, e.g., [53, 65, 81]) that all these methods are A-stable and the order of a ν-stage method is always $2\nu - 2$.

For the ν-stage Lobatto IIIA and IIIB methods the stage order is ν and $\nu - 2$, respectively, and the stability function $r(z) = 1 - \alpha(z)$ coincides with the $(\nu-1, \nu-1)$-Padé approximation to e^z but the coefficient matrix $\mathfrak{a}$ is singular. Nevertheless, it follows from [65] that

$$\deg[\det(\mathfrak{i} - \cdot\, \mathfrak{a})] = \nu - 1,$$

which shows that $\deg[g_{jl}] \le 0$ and, thus, $\deg[w_j] \le 0$. Therefore, for any fixed $\varphi \in (0, \pi/2)$, the ν-stage Lobatto IIIA or IIIB method is of type $C_{III}(\varphi)$ if ν is even and of type $C_{II}(\varphi)$ if ν is odd.

Next, for the ν-stage Lobatto IIIC method, the stage order is $\nu - 1$, and for all methods of this family, the matrix $\mathfrak{a}$ is non-singular (see, e.g., [65]) so that (5.24) and (5.25) hold. Also, since $\alpha(\infty) = 1$, it is easily seen that

$$\deg[\alpha_\infty - \alpha(\cdot)] = \deg[1 - \alpha(\cdot)] = -2.$$

The above reasonings show that any Lobatto IIIC method is both of type $V_I(\varphi)$ and of type $V_I^*(\varphi)$ for any fixed $\varphi \in (0, \pi/2)$.

Finally, we look at the family of 3-stage RK methods for which the Butcher array is given by

$$\left(\begin{array}{c|ccc} c_1 & c_1 & 0 & 0 \\ c_2 & c_2 - \gamma & \gamma & 0 \\ c_3 & c_3 - \zeta - \eta & \zeta & \eta \\ \hline & b_1 & b_2 & b_3 \end{array}\right), \tag{5.26}$$

where[17] $\tau_1 \ne \tau_2 \ne \tau_3 \ne \tau_1$, $3\tau_1\tau_2\tau_3 + \tau_1 + \tau_2 + \tau_3 = 0$, $c_j = (1 + \tau_j)/2$ for $j = 1, 2, 3$, and

$$b_1 = \tfrac{1+3\tau_2\tau_3}{3(\tau_1-\tau_2)(\tau_1-\tau_3)}, \quad b_2 = \tfrac{1+3\tau_3\tau_1}{3(\tau_2-\tau_3)(\tau_2-\tau_1)}, \quad b_3 = \tfrac{1+3\tau_1\tau_2}{3(\tau_3-\tau_1)(\tau_3-\tau_2)},$$
$$\eta = \tfrac{1-\tau_3}{2}, \quad \gamma = \tfrac{\tau_3-\tau_1+3\tau_1\tau_3}{2(1+3\tau_1\tau_3)}, \quad \zeta = \tfrac{(\tau_1-\tau_3)(1+2\tau_1)}{2(1+3\tau_1\tau_2)(\tau_2-\tau_1)}.$$

[17] Clearly, this family is actually described by two parameters.

It follows from [64] that the methods of this family are of order 4 and of stage order 1; and moreover, under the additional condition

$$\tau_1 > 0 \text{ and } 1 + 3\tau_1\tau_3 < 0, \tag{5.27}$$

they are A-stable, and it then holds that $|r(\infty)| < 1$. Also, it is easily seen that if (5.27) is fulfilled, the coefficient matrix is non-singular. Therefore all methods of this family that satisfy (5.27) are both of type $V_I(\varphi)$ and of type $V_I^*(\varphi)$ for any fixed $\varphi \in (0, \pi/2)$.

Many further examples of applicable methods can be found within other families of RK methods (see, e.g., [53, 65, 81]). The interested reader might carry out the corresponding analysis by himself/herself similarly to the above.

5.3 Connection with discrete evolution equations

Now we discuss the question in which manner the discrete problem (5.19) and (5.20) reduces to the discrete Cauchy problem (1.1) examined in Part I. In fact, we show that the reducibility is truly performed if the restrictions on the operator $A(t)$ and on the RK method suitably conform to each other.

Below in some intermediate estimations, we deal with matrices whose entries are members of $\mathfrak{X}$ or of $\mathcal{B}(\mathfrak{X})$.[18] If $\mathrm{u} = (u_{jl})_{p\times r}$ is such a matrix, we use the notation

$$\|\mathrm{u}\| = \max_{1\le j\le p, 1\le l\le r} \|u_{jl}\|,$$

where $\|u_{jl}\|$ is the vector norm of u_{jl} if $u_{jl} \in \mathfrak{X}$ and the operator norm of u_{jl} if $u_{jl} \in \mathcal{B}(\mathfrak{X})$.[19] In particular, working with the product space $\mathfrak{X}^{(\nu)} = \underbrace{\mathfrak{X} \times \cdots \times \mathfrak{X}}_{\nu}$ which consists of all ν-dimensional vectors $\mathbf{u} = \left(u_1, \ldots, u_\nu\right)^T$ with $u_j \in \mathfrak{X}$, $1 \le j \le \nu$, for the sake of being consistent, we assume that $\mathfrak{X}^{(\nu)}$ is endowed with the norm[20]

$$\|\mathbf{u}\| = \max_{1\le j\le \nu} \|u_j\|.$$

[18] In the sequel we also meet the situation when the entries of the matrix are unbounded operators, however such matrices are involved with no estimations.

[19] It is always clear from the context what u is and, thus, in which sense $\|\mathrm{u}\|$ is understood.

[20] Note that the space $\mathfrak{X}^{(\nu)}$ may be endowed with any suitable norm composed of the norms of the components in $\mathfrak{X}$. All such norms are equivalent to each other.

We are confident that using the same symbol $\|\cdot\|$ differently gives rise to no confusion.

Also, in any subsequent calculations involving operators, a matrix $\mathfrak{m} \in \mathbb{C}^{p\times r}$ will be identified with the operator[21] $\mathfrak{m} \otimes I : \mathfrak{X}^{(r)} \to \mathfrak{X}^{(p)}$.

We first consider the case with a constant operator, that is, where $A(t)$ is independent of t. (We therefore write A instead of $A(t)$.) Let the RK method be of type $C(\varphi)$ for some $\varphi \in (0, \pi/2)$. Then, for any fixed $\sigma \in \mathbb{R}$, if $K > 0$ is sufficiently small, the functions $\alpha(-k\lambda)$ and $w_j(-k\lambda)$, $j = 1, \dots, \nu$, are defined for any $k \in (0, K]$ and $\lambda \in \left(\operatorname{Int}\Sigma_\varphi \cup \{0\}\right) + \sigma$, and one can take $K = \infty$ if $\sigma \geq 0$. It is then not hard to see that in the scalar case $A(t) = \lambda I$, problem (5.19), (5.20) reduces to the following one

$$\partial Y_n + k^{-1}\alpha(-k\lambda)Y_n = \sum_{j=1}^{\nu} w_j(-k\lambda)f(t_{nj}) \quad \text{for } t_n \in \breve{\Omega}_k, \qquad Y_0 = \mathfrak{R}y^0.$$

In the operator case $A \in \mathcal{S}(\varphi, \sigma)$, dealing with (5.19), (5.20), one should eliminate Y_{nj}, $j = 1, \dots, \nu$, from (5.19) and (5.20), that is, solve (5.20) for Y_{nj} and insert them into (5.19). This procedure certainly is realized if the system of equations (5.20) is uniquely solvable. It may however happen that (5.20) is not uniquely solvable while the operators $r(-kA)$ and $w_j(-kA)$, $j = 1, \dots, \nu$, are nevertheless defined and bounded on $\mathfrak{X}$. In particular, it follows from Lemma 5.3 that the definedness and boundedness of $r(-kA)$ and of $w_j(-kA)$, $j = 1, \dots, \nu$, are observed for all $k \in (0, K]$, with $K > 0$ sufficiently small, if $A \in \mathcal{S}(\varphi, \sigma)$ and the RK method is of type $C(\varphi)$[22] for some $\varphi \in (0, \pi/2)$ and $\sigma \in \mathbb{R}$.[23] Thus, under these conditions, problem (1.1) with

$$\mathfrak{A}_n = \mathfrak{A} = k^{-1}\alpha(-kA), \ \mathfrak{Q}_n = I, \text{ and } F_n = \mathbf{w}^T(-kA)\Phi_n, \tag{5.28}$$

where

$$\Phi_n = \left(f(t_{n1}), \dots, f(t_{n\nu})\right)^T \in \mathfrak{X}^{(\nu)}, \tag{5.29}$$

may be thought of as a generalized form of problem (5.19), (5.20) in the case with constant operator. On the other hand, problem (1.1), (5.28), (5.29) is

---

[21]The symbol $\otimes$ denotes Kronecker's product.
[22]In the homogeneous case $f(t) \equiv 0$ it suffices to assume that the method is $A(\varphi)$-stable.
[23]One can take $K = \infty$ if $\sigma \geq 0$.

equivalently represented as follows

$$\begin{aligned} Y_{n+1} &= (I - \alpha(-kA))Y_n + k\sum_{j=1}^{\nu} w_j(-kA)f(t_{nj}) \\ &= r(-kA)Y_n + k\sum_{j=1}^{\nu} w_j(-kA)f(t_{nj}) \quad \text{for } t_n \in \breve{\Omega}_k, \\ Y_0 &= \mathfrak{R}y^0. \end{aligned} \tag{5.30}$$

In the literature methods of the form (5.30) are called rational. In principle rational methods for autonomous equations may be introduced and examined independently, without connection with the RK method (5.19) and (5.20). However in this book we deal with the non-autonomous case as well and we prefer to keep a common starting point for our consideration, which is problem (5.19), (5.20), whether the dependence of $A(t)$ on t is present or not. Moreover, we strive for creating a unified terminology within the book which means that we want related terms for different objects to be in conformity with each other. In view of these reasons we introduce the following.

Definition 5.10. *The discrete function $Y_n : \Omega_k \to \mathfrak{X}$ specified by the recurrence (5.30) is called the* extended solution *of the Runge-Kutta procedure (5.19) and (5.20) in the autonomous case.*

Further, we turn to the general case where the operator $A(t)$ may depend on t. For convenience we use

$$\mathfrak{D}_n = \text{diag}\,(A(t_{n1}), \ldots, A(t_{n\nu})), \quad \mathfrak{D}(t) = \text{diag}\,(A(t), \ldots, A(t)),$$

$$(\mathfrak{i} + k\mathfrak{a}\mathfrak{D}(t))^{-1} = \left(g_{jl}(-kA(t))\right)_{\nu\times\nu},$$

and

$$\mathfrak{A}(t) = k^{-1}\alpha(-kA(t)). \tag{5.31}$$

In what follows we consider operator equations of the form, given $\mathbf{u} = (u_1, \ldots, u_\nu)^T \in \mathfrak{X}^{(\nu)}$,

$$(\mathfrak{i} + k\mathfrak{a}\mathfrak{D}_n)\mathbf{z} = \mathfrak{a}\mathbf{u}, \tag{5.32}$$

or of the form, given $u \in \mathfrak{X}$,

$$(\mathfrak{i} + k\mathfrak{a}\mathfrak{D}_n)\mathbf{z} = \mathbf{e} \otimes u, \tag{5.33}$$

which are solved for $\mathbf{z} = (z_1, \dots, z_\nu)^T \in \underbrace{\mathcal{D}om \times \cdots \times \mathcal{D}om}_{\nu}$. If equation (5.32) (equation (5.33)) is uniquely solvable for any $\mathbf{u} \in \mathfrak{X}^{(\nu)}$ (for any $u \in \mathfrak{X}$), we denote the corresponding solution operator by $(\mathfrak{i} + k\mathfrak{a}\mathfrak{D}_n)^{-1}\mathfrak{a}$ (by $(\mathfrak{i} + k\mathfrak{a}\mathfrak{D}_n)^{-1}\mathbf{e}$) so that $\mathbf{z} = (\mathfrak{i} + k\mathfrak{a}\mathfrak{D}_n)^{-1}\mathfrak{a}\mathbf{u}$ ($\mathbf{z} = (\mathfrak{i} + k\mathfrak{a}\mathfrak{D}_n)^{-1}\mathbf{e} \otimes u$).

Now we present sufficient conditions for problem (5.19), (5.20) to be equivalently representable in the form (1.1). In what follows we use different forms of Hölder-continuity of $A(t)$ and this main idea separates the below assertions from each other. We start, however, by giving an auxiliary result which will be in common use for the corresponding proofs.

Lemma 5.7. *Let $A(t) \in \mathcal{S}(\varphi, \sigma)$ and let the RK method be of type $V(\varphi)$ or of type $V^*(\varphi)$, with some $\varphi \in (0, \pi/2)$ and $\sigma \in \mathbb{R}$. Then, with $K > 0$ sufficiently small, for all $k \in (0, K]$ and $t \in [0, T]$, the operators and matrix operators $\alpha(-kA(t))$, $\rho(-kA(t))$, $\mathbf{v}(-kA(t))$, $\mathbf{w}(-kA(t))$, $(\mathfrak{i} + k\mathfrak{a}\mathfrak{D}(t))^{-1}$, $k\mathfrak{D}(t)\,(\mathfrak{i} + k\mathfrak{a}\mathfrak{D}(t))^{-1}\mathfrak{a}$, $k\mathfrak{D}(t)\,(\mathfrak{i} + k\mathfrak{a}\mathfrak{D}(t))^{-1}\mathbf{e}$ are defined on $\mathfrak{X}$ or on $\mathfrak{X}^{(\nu)}$, with*

$$\|\alpha(-kA(t))\| + \|\rho(-kA(t))\| + \|\mathbf{v}(-kA(t))\| + \|\mathbf{w}(-kA(t))\| \le C \quad (5.34)$$

and

$$\|(\mathfrak{i} + k\mathfrak{a}\mathfrak{D}(t))^{-1}\| + \|k\mathfrak{D}(t)\,(\mathfrak{i} + k\mathfrak{a}\mathfrak{D}(t))^{-1}\mathfrak{a}\| \le C. \quad (5.35)$$

Furthermore, for all $k \in (0, K]$ and $t \in [0, T]$,

$$\|k\mathfrak{D}(t)\,\mathbf{v}(-kA(t))\| \le C \quad (5.36)$$

or

$$\|k\mathfrak{D}(t)\,\mathbf{w}(-kA(t))\| \le C, \quad (5.37)$$

if the method is of type $V(\varphi)$ or of type $V^*(\varphi)$, respectively.

Proof. Since the Runge-Kutta method is of type $V(\varphi)$ or of type $V^*(\varphi)$, it follows that the functions $g_{jl}(z)$, $j, l = 1, \dots, \nu$, are (φ)-regular. By (5.22), (5.23), and (5.21), the same is true for $\alpha(z)$, $\rho(z)$, $v_j(z)$, and $w_j(z)$, $j = 1, \dots, \nu$. Note further that $\deg[\alpha] \le 0$ since the method is $A(\varphi)$-stable, and, as a consequence, we have as well $\deg[\rho] \le -1$. Observe also that the condition $\deg[g_{jl}] \le 0$, $j, l = 1, \dots, \nu$, yields that $\deg[v_j] \le 0$, $\deg[w_j] \le 0$, and $\deg\left[\sum_{p=1}^{\nu} g_{jp}(\cdot)a_{pl}\right] \le -1$. Using therefore Lemma 5.3 we show (5.34) and (5.35).

In the case where the method is of type $V(\varphi)$, it follows that $\deg[v_j] \le -1$, which implies (5.36), again by Lemma 5.3. Similarly, if the method is of type $V^*(\varphi)$, we have $\deg[w_j] \le -1$, which clearly leads us to (5.37). $\square$

Now we are in a position to state our results on representability.

Theorem 5.1. *Suppose that $A(t) \in \mathcal{S}(\varphi, \sigma)$ and the RK method is of type $V(\varphi)$, for some $\varphi \in (0, \pi/2)$ and $\sigma \in \mathbb{R}$. Let further a seminorm $|\cdot|_t$ form a $(\gamma_1|\varphi, \sigma)$-concordant pair with $A(t)$ and let $A(t)$ itself satisfy hypothesis* $\mathbf{h\ddot{o}l}^{\oplus}_{(k)}[\vartheta; |\cdot|_t]$, *for some $\gamma_1 \in [0, 1)$ and $\vartheta \in (0, 1]$. Then, with $K > 0$ sufficiently small, for all $k \in (0, K]$, the entries of the operator matrices $(\mathfrak{D}_n - \mathfrak{D}(t_n))(\mathfrak{i} + k\mathfrak{a}\mathfrak{D}_n)^{-1}\mathbf{e}$ and $(\mathfrak{D}_n - \mathfrak{D}(t_n))(\mathfrak{i} + k\mathfrak{a}\mathfrak{D}_n)^{-1}\mathfrak{a}$ are defined and bounded on $\mathfrak{X}$ and problem (5.19), (5.20) can be equivalently represented in the form (1.1) with*

$$\mathfrak{A}_n = \mathfrak{A}(t_n) + \mathfrak{B}_n, \tag{5.38}$$

$$\begin{aligned} \mathfrak{B}_n &= \mathbf{w}^T(-kA(t_n))\left(\mathfrak{i} - k(\mathfrak{D}_n - \mathfrak{D}(t_n))(\mathfrak{i} + k\mathfrak{a}\mathfrak{D}_n)^{-1}\mathfrak{a}\right) \\ &\quad \times (\mathfrak{D}_n - \mathfrak{D}(t_n))\mathbf{v}(-kA(t_n)), \end{aligned} \tag{5.39}$$

$$\mathfrak{Q}_n = I, \tag{5.40}$$

and

$$F_n = \mathbf{w}^T(-kA(t_n))\left(\mathfrak{i} - k(\mathfrak{D}_n - \mathfrak{D}(t_n))(\mathfrak{i} + k\mathfrak{a}\mathfrak{D}_n)^{-1}\mathfrak{a}\right)\Phi_n, \tag{5.41}$$

where Φ_n is given by (5.29).

The proof of the theorem is based on the following

Lemma 5.8. *Under the conditions of Theorem 5.1, with $K > 0$ sufficiently small, for all $k \in (0, K]$ and $t_n \in \breve{\Omega}_k$, the entries of the operator matrices $k(\mathfrak{D}_n - \mathfrak{D}(t_n))(\mathfrak{i} + k\mathfrak{a}\mathfrak{D}(t_n))^{-1}\mathfrak{a}$, $k(\mathfrak{D}_n - \mathfrak{D}(t_n))(\mathfrak{i} + k\mathfrak{a}\mathfrak{D}(t_n))^{-1}\mathbf{e}$, $(\mathfrak{i} + k\mathfrak{a}\mathfrak{D}_n)^{-1}\mathfrak{a}$, $(\mathfrak{i} + k\mathfrak{a}\mathfrak{D}_n)^{-1}\mathbf{e}$, $k(\mathfrak{D}_n - \mathfrak{D}(t_n))(\mathfrak{i} + k\mathfrak{a}\mathfrak{D}_n)^{-1}\mathfrak{a}$, and $k(\mathfrak{D}_n - \mathfrak{D}(t_n))(\mathfrak{i} + k\mathfrak{a}\mathfrak{D}_n)^{-1}\mathbf{e}$ belong to $\mathcal{B}(\mathfrak{X})$, with*

$$\begin{aligned} \|k(\mathfrak{D}_n - \mathfrak{D}(t_n))(\mathfrak{i} + k\mathfrak{a}\mathfrak{D}(t_n))^{-1}\mathfrak{a}\| + \|k(\mathfrak{D}_n - \mathfrak{D}(t_n))(\mathfrak{i} + k\mathfrak{a}\mathfrak{D}(t_n))^{-1}\mathbf{e}\| \\ \le C(k^{\vartheta} + k^{1-\gamma_1}) \end{aligned} \tag{5.42}$$

and

$$\begin{aligned} \|k(\mathfrak{D}_n - \mathfrak{D}(t_n))(\mathfrak{i} + k\mathfrak{a}\mathfrak{D}_n)^{-1}\mathfrak{a}\| + \|k(\mathfrak{D}_n - \mathfrak{D}(t_n))(\mathfrak{i} + k\mathfrak{a}\mathfrak{D}_n)^{-1}\mathbf{e}\| \\ \le C(k^{\vartheta} + k^{1-\gamma_1}). \end{aligned} \tag{5.43}$$

Furthermore, for all $k \in (0, K]$ and $t_n \in \breve{\Omega}_k$, the entries of the operator matrices $k\mathfrak{D}_n(\mathfrak{i} + k\mathfrak{a}\mathfrak{D}_n)^{-1}\mathfrak{a}$ and $k\mathfrak{D}_n(\mathfrak{i} + k\mathfrak{a}\mathfrak{D}_n)^{-1}\mathbf{e}$ are defined and bounded on $\mathfrak{X}$, and the following representations hold

$$\begin{aligned} k\mathfrak{D}_n(\mathfrak{i} + k\mathfrak{a}\mathfrak{D}_n)^{-1}\mathfrak{a} &= k\mathfrak{D}(t_n)(\mathfrak{i} + k\mathfrak{a}\mathfrak{D}(t_n))^{-1}\mathfrak{a} + (\mathfrak{i} + k\mathfrak{a}\mathfrak{D}(t_n))^{-1} \\ &\quad \times k(\mathfrak{D}_n - \mathfrak{D}(t_n))(\mathfrak{i} + k\mathfrak{a}\mathfrak{D}_n)^{-1}\mathfrak{a} \end{aligned} \tag{5.44}$$

and

$$\begin{aligned} k\mathcal{D}_n(\mathfrak{i}+k\mathfrak{a}\mathcal{D}_n)^{-1}\mathbf{e} &= k\mathcal{D}(t_n)\,(\mathfrak{i}+k\mathfrak{a}\mathcal{D}(t_n))^{-1}\mathbf{e}+(\mathfrak{i}+k\mathfrak{a}\mathcal{D}(t_n))^{-1} \\ &\quad \times\left(\mathfrak{i}-k(\mathcal{D}_n-\mathcal{D}(t_n))\,(\mathfrak{i}+k\mathfrak{a}\mathcal{D}_n)^{-1}\mathfrak{a}\right) \\ &\quad \times k(\mathcal{D}_n-\mathcal{D}(t_n))\mathbf{v}(-kA(t_n)). \end{aligned} \tag{5.45}$$

Proof. First of all, since the seminorm $|\cdot|_t$ forms a $(\gamma_1|\varphi,\sigma)$-concordant pair with $A(t)$, similarly to (5.7) it holds, for all $k\in(0,K]$ and $u\in\mathfrak{X}$,

$$|(I+kA(t))^{-1}u|_t\le Ck^{-\gamma_1}\|u\|.$$

Using this inequality as well as the fact that the operator $A(t)$ satisfies hypothesis $\textbf{höl}^{\oplus}_{(k)}[\vartheta;|\cdot|_t]$, and applying further (5.35) and (5.36) we see that (5.42) holds, since $|t_{nj}-t_n|\le k$, $j=1,\dots,\nu$.

Next, recalling that $\operatorname{Dom}A(t)\equiv\mathcal{D}om$ is independent of t, we prove that equation (5.32) possesses a unique solution $\mathbf{z}=\left(z_1,\dots,z_\nu\right)^T$ such that $z_j\in\mathcal{D}om$, $j=1,\dots,\nu$, for any $\mathbf{u}\in\mathfrak{X}^{(\nu)}$, and further we show (5.43). More precisely, since $\mathbf{z}\in\mathcal{D}om$ componentwise, (5.32) is rewritten in the form

$$(\mathfrak{i}+k\mathfrak{a}\mathcal{D}(t_n))\mathbf{z}+k\mathfrak{a}(\mathcal{D}_n-\mathcal{D}(t_n))\mathbf{z}=\mathfrak{a}\mathbf{u},$$

which gives as well

$$\mathbf{z}=-(\mathfrak{i}+k\mathfrak{a}\mathcal{D}(t_n))^{-1}\mathfrak{a}k(\mathcal{D}_n-\mathcal{D}(t_n))\mathbf{z}+(\mathfrak{i}+k\mathfrak{a}\mathcal{D}(t_n))^{-1}\mathfrak{a}\mathbf{u}. \tag{5.46}$$

Now we put[24] $\mathcal{G}_n=k(\mathcal{D}_n-\mathcal{D}(t_n))\,(\mathfrak{i}+k\mathfrak{a}\mathcal{D}(t_n))^{-1}\mathfrak{a}$ and multiply both sides of (5.46) from the left by $k(\mathcal{D}_n-\mathcal{D}(t_n))$ to obtain

$$k(\mathcal{D}_n-\mathcal{D}(t_n))\mathbf{z}=-\mathcal{G}_nk(\mathcal{D}_n-\mathcal{D}(t_n))\mathbf{z}+\mathcal{G}_n\mathbf{u}.$$

Applying further (5.42), we therefore find, if $K>0$ is sufficiently small, that for all $k\in(0,K]$ and $t_n\in\breve{\Omega}_k$,

$$\|(\mathfrak{i}+\mathcal{G}_n)^{-1}\|\le C \tag{5.47}$$

and

$$k(\mathcal{D}_n-\mathcal{D}(t_n))\mathbf{z}=(\mathfrak{i}+\mathcal{G}_n)^{-1}\mathcal{G}_n\mathbf{u}. \tag{5.48}$$

Inserted into (5.46), this yields

$$\mathbf{z}=-(\mathfrak{i}+k\mathfrak{a}\mathcal{D}(t_n))^{-1}\mathfrak{a}(\mathfrak{i}+\mathcal{G}_n)^{-1}\mathcal{G}_n\mathbf{u}+(\mathfrak{i}+k\mathfrak{a}\mathcal{D}(t_n))^{-1}\mathfrak{a}\mathbf{u}.$$

[24] It follows from (5.35) that the entries of $\mathcal{G}_n$ are bounded on $\mathfrak{X}$.

Observe that, as already shown, the entries of the operator matrix $(\mathfrak{i} + k\mathfrak{a}\mathcal{D}(t_n))^{-1}\mathfrak{a}$ are mappings of $\mathfrak{X}$ into $\mathcal{D}om$, hence the above argument is correct. Further, by (5.35), (5.42), and (5.47) it follows that the entries of the operator matrix $(\mathfrak{i} + k\mathfrak{a}\mathcal{D}_n)^{-1}\mathfrak{a}$ are defined and bounded on $\mathfrak{X}$. Now a simple estimation of the expression $(\mathfrak{i} + \mathcal{G}_n)^{-1}\mathcal{G}_n\mathbf{u}$ leads, with the aid of (5.42), (5.47), and (5.48), to the desired estimate for the first term on the left-hand side of (5.43), and the same for the second term follows by using a similar argument. It is also seen from the above reasonings that the entries of the operator matrices $k\mathcal{D}_n(\mathfrak{i}+k\mathfrak{a}\mathcal{D}_n)^{-1}\mathfrak{a}$ and $k\mathcal{D}_n(\mathfrak{i}+k\mathfrak{a}\mathcal{D}_n)^{-1}\mathbf{e}$ are defined and bounded on $\mathfrak{X}$.

It remains to show (5.44) and (5.45). We denote $\widetilde{\mathcal{G}}_n = k(\mathcal{D}_n - \mathcal{D}(t_n))\,(\mathfrak{i}+k\mathfrak{a}\mathcal{D}_n)^{-1}\mathfrak{a}$. It follows from (5.46) that

$$(\mathfrak{i} + k\mathfrak{a}\mathcal{D}_n)^{-1}\mathfrak{a} = -(\mathfrak{i} + k\mathfrak{a}\mathcal{D}(t_n))^{-1}\mathfrak{a}\widetilde{\mathcal{G}}_n + (\mathfrak{i} + k\mathfrak{a}\mathcal{D}(t_n))^{-1}\mathfrak{a}. \tag{5.49}$$

Multiplying now both sides of this equality from the left by $k\mathcal{D}(t_n)$ and using the identity

$$k\mathcal{D}(t_n)\,(\mathfrak{i} + k\mathfrak{a}\mathcal{D}(t_n))^{-1}\mathfrak{a} = \mathfrak{i} - (\mathfrak{i} + k\mathfrak{a}\mathcal{D}(t_n))^{-1}\mathfrak{a},$$

which can easily be verified due to the fact that $\mathfrak{a}$ and $(\mathfrak{i} + k\mathfrak{a}\mathcal{D}(t_n))^{-1}$ are permutable, we find that

$$k\mathcal{D}(t_n)\,(\mathfrak{i}+k\mathfrak{a}\mathcal{D}_n)^{-1}\mathfrak{a} = k\mathcal{D}(t_n)\,(\mathfrak{i}+k\mathfrak{a}\mathcal{D}(t_n))^{-1}\mathfrak{a} - \widetilde{\mathcal{G}}_n + (\mathfrak{i}+k\mathfrak{a}\mathcal{D}(t_n))^{-1}\mathfrak{a}\widetilde{\mathcal{G}}_n.$$

Inserting this result into the right-hand side of the identity

$$k\mathcal{D}_n(\mathfrak{i} + k\mathfrak{a}\mathcal{D}_n)^{-1}\mathfrak{a} = k\mathcal{D}(t_n)\,(\mathfrak{i} + k\mathfrak{a}\mathcal{D}_n)^{-1}\mathfrak{a} + \widetilde{\mathcal{G}}_n,$$

leads immediately to (5.44).

Further, acting similarly to the manner leading to (5.50) and (5.44), we get

$$\begin{aligned} k\mathcal{D}_n(\mathfrak{i} + k\mathfrak{a}\mathcal{D}_n)^{-1}\mathbf{e} \;&=\; k\mathcal{D}(t_n)\,(\mathfrak{i} + k\mathfrak{a}\mathcal{D}(t_n))^{-1}\mathbf{e} + (\mathfrak{i} + k\mathfrak{a}\mathcal{D}(t_n))^{-1} \\ &\quad \times k(\mathcal{D}_n - \mathcal{D}(t_n))\,(\mathfrak{i} + k\mathfrak{a}\mathcal{D}_n)^{-1}\mathbf{e} \end{aligned} \tag{5.50}$$

and

$$k(\mathcal{D}_n - \mathcal{D}(t_n))(\mathfrak{i} + k\mathfrak{a}\mathcal{D}_n)^{-1}\mathbf{e} = (\mathfrak{i} + \mathcal{G}_n)^{-1}k(\mathcal{D}_n - \mathcal{D}(t_n))(\mathfrak{i} + k\mathfrak{a}\mathcal{D}(t_n))^{-1}\mathbf{e}. \tag{5.51}$$

Thus we see that (5.45) follows when inserting (5.51) into the right-hand side of (5.50) and using (5.48) and the trivial identity

$$(\mathfrak{i} + \mathcal{G}_n)^{-1} = \mathfrak{i} - (\mathfrak{i} + \mathcal{G}_n)^{-1}\mathcal{G}_n.$$

Now the proof of the lemma is complete. □

Proof of Theorem 5.1. Letting $\mathbf{Y}_n \in \mathfrak{X}^{(\nu)}$ be defined by

$$\mathbf{Y}_n = \left(Y_{n1}, \ldots, Y_{n\nu}\right)^T,$$

we rewrite problem (5.19), (5.20) as follows

$$Y_{n+1} = Y_n - k\mathbf{b}^T \mathfrak{D}_n \mathbf{Y}_n + k\mathbf{b}^T \Phi_n \quad \text{for } t_n \in \breve{\Omega}_k, \qquad Y_0 = \mathfrak{R}y^0 \tag{5.52}$$

and

$$\mathbf{Y}_n = \mathbf{e} \otimes Y_n - k\mathfrak{a}\mathfrak{D}_n \mathbf{Y}_n + k\mathfrak{a}\Phi_n. \tag{5.53}$$

Since, by Lemma 5.8, the entries of the operator matrices $(\mathfrak{i} + k\mathfrak{a}\mathfrak{D}_n)^{-1}\mathfrak{a}$ and $(\mathfrak{i} + k\mathfrak{a}\mathfrak{D}_n)^{-1}\mathbf{e}$ are defined on $\mathfrak{X}$ for $k \in (0, K]$ and $t_n \in \breve{\Omega}_k$, it follows from (5.53) that

$$\mathbf{Y}_n = (\mathfrak{i} + k\mathfrak{a}\mathfrak{D}_n)^{-1}\mathbf{e} \otimes Y_n + k(\mathfrak{i} + k\mathfrak{a}\mathfrak{D}_n)^{-1}\mathfrak{a}\Phi_n. \tag{5.54}$$

Inserting further (5.54) into (5.52), we find that problem (5.19), (5.20) can be rewritten in the equivalent form

$$\begin{aligned} Y_{n+1} &= \left(I - k\mathbf{b}^T \mathfrak{D}_n (\mathfrak{i} + k\mathfrak{a}\mathfrak{D}_n)^{-1}\mathbf{e}\right) Y_n \\ &\quad + k\mathbf{b}^T \left(\mathfrak{i} - k\mathfrak{D}_n (\mathfrak{i} + k\mathfrak{a}\mathfrak{D}_n)^{-1}\mathfrak{a}\right) \Phi_n, \quad t_n \in \breve{\Omega}_k, \\ Y_0 &= \mathfrak{R}y^0. \end{aligned} \tag{5.55}$$

Therefore, to show the claim, it suffices to combine (5.55) with the above representations (5.44) and (5.45), noting that $\mathbf{b}^T \mathfrak{D}(t)(\mathfrak{i} + k\mathfrak{a}\mathfrak{D}(t))^{-1}\mathbf{e} = \mathfrak{A}(t)$ and $\mathbf{b}^T(\mathfrak{i} + k\mathfrak{a}\mathfrak{D}(t))^{-1} = \mathbf{w}^T(-kA(t))$.[25] □

Remark 5.1. *In particular, the assertion of Theorem 5.1 remains valid if, instead of* $\mathbf{h\ddot{o}l}^{\oplus}_{(k)}[\vartheta; |\cdot|_t]$*, it is based on either of the more restrictive hypotheses* $\mathbf{h\ddot{o}l}^{\oplus}[\vartheta; |\cdot|_t]$ *or* $\mathbf{h\ddot{o}l}[\vartheta]$.

So we have shown that under some reasonable conditions problem (5.19), (5.20) can be equivalently represented in the form (1.1), that is, as the Cauchy problem for an evolution equation in discrete time. In particular, this result yields that the unique solvability of problem (5.19), (5.20) then follows and it remains to examine stability and convergence only. We concentrate on these questions in subsequent chapters of the book.

[25] Note that, as follows from Lemma 5.8, the operators $\mathfrak{A}(t_n)$, $\mathfrak{B}_n$, $\mathfrak{A}_n$ are bounded on $\mathfrak{X}$ for $k \in (0, K]$ and $t_n \in \breve{\Omega}_k$.

5.4 Comments and bibliographical remarks

Section 5.1: The proof of Lemma 5.2 follows that of Lemma 2.2 in our paper [37].
Section 5.2: For general books in the field of RK methods, we refer the reader to Butcher [53], Dekker & Verwer [65], Hairer, Nørsett, and Wanner [80], Hairer & Wanner [81], and Stetter [155]. Although written in the context of ordinary differential equations, these manuals present a wide range of ideas, notions, terminology. Moreover, in [53, 65, 81, 155], they pay attention to the analysis of discretizations of stiff problems. We also mention a recent book of Hundsdorfer & Verwer [86] which deals with discretizations of both ODEs and PDEs. The books cited above are therefore related to the present one which is intended, in particular, for applications in order to examine discrete versions of partial differential equations. Most of the modern approaches to the study of discretization methods deal with abstract differential equations, and the situation of partial differential equations can then be settled as a particular case. In the abstract case they meet, however, new features of which the central one is the problem of infinite dimension. As concerns different aspects of general RK methods considered in Hilbert and Banach space settings and applied to linear, quasilinear, and fully nonlinear problems, we mention our work [16, 19, 20, 22, 25, 26, 27, 30, 35, 36] and the work of Beyn & Garay [47], González & Ostermann [71], González *et al.* [72], González & Palencia [73, 74, 75], Gudovich & Terteryan [78], Karakashian [90], Keeling [92], Lubich & Ostermann [105, 106, 107], Nakaguchi & Yagi [115], Ostermann & Roche [123], and Ostermann & Thalhammer [125].
Section 5.3: Our presentation here is, in fact, an improved version of our earlier work [19, 35]. (These results are also announced in [16, 22, 26].)

Chapter 6

Analysis of Stability

In this chapter we further investigate the Runge-Kutta discretizations discussed previously in Chapter 5. We concentrate here on examining stability for discrete problems of the form (5.19), (5.20) by using seminorms $|\cdot|_t$ that form concordant pairs with $A(t)$ in the sense of Definition 5.2 above. It has already been pointed out before that our techniques are based on reducing the starting problem (5.19), (5.20) to the discrete Cauchy problem (1.1). In fact, under the assumption that $A(t)$ is (φ, σ)-sectorial with some $\varphi \in (0, \pi/2)$ and $\sigma \in \mathbb{R}$ and satisfies one of the Hölder-continuity type hypotheses posed in the preceding chapter, we verify the necessary conditions on the operator $\mathfrak{A}_n$ in (1.1), mainly, the $\Lambda_J^k(\ldots)$-concordance and one of the Hölder-continuity type hypotheses stated in Part I. It turns out that an important part of this work is to examine the properties of the operator $\mathfrak{A}(t_n)$ defined by (5.31). Further, we show that the operator $\mathfrak{A}_n$ itself is characterized after the same pattern, by certain perturbation techniques.[1] Finally, the stability of problem (5.19), (5.20) is established by applying the abstract results found in Part I.

6.1 Some resolvent estimates

Our main aim in this section is to show certain weighted resolvent estimates of the operator $k\mathfrak{A}(t) = \alpha(-kA(t))$ by using seminorms with properties regulated by Definition 5.2. In fact, such estimates are related to the $\Lambda_J^k(\ldots)$-concordance of certain new seminorms derived from the original ones.

[1] It is seen from our below investigation that the operator $\mathfrak{A}(t_n)$ can be thought of as the principal part of $\mathfrak{A}_n$ in some sense.

Perhaps, it will not be a good style if it is often used to state, one after the other, even two assertions which are formally different but the latter of which is obvious as soon as the former is stated and proved. We therefore accept the following.

Convention. *In what follows there often occur different assertions which involve seminorms and operators in the space $\mathfrak{X}$ and possess, at the same time, dual versions in $\mathfrak{X}^*$ in which all objects in $\mathfrak{X}$ are replaced by their doubles in $\mathfrak{X}^*$. For example, the former estimate in Definition 5.2 possesses a dual version which is just the latter estimate involving the corresponding objects in $\mathfrak{X}^*$. In such situations we will not state very often the dual version explicitly, marking by [*] the corresponding assertion.*[2] *Furthermore, if we need to refer to the dual version subsequently, we refer to the explicit result (in $\mathfrak{X}$) marked by *. For example, Theorem 6.1 below possesses a dual version which may be referred to in the sequel, although not stated explicitly, as Theorem 6.1*.*

Now we present a series of helpful resolvent estimates.

Theorem 6.1. [*] *Suppose that $A(t) \in \mathcal{S}(\varphi, \sigma)$ and a seminorm $|\cdot|_t$ forms a $(\xi|m|\varphi, \sigma)$-concordant pair with $A(t)$, for some $\varphi \in (0, \pi/2)$, $\sigma \in \mathbb{R}$, $\xi \geq 0$, and $m \in \mathbb{N}$ such that $m-1 \leq \xi < m$. Let also the RK method be $A_I(\varphi)$-stable and let a rational function $\omega(z)$ be (φ)-regular and subject to $\deg[\omega] \leq -m$. Then there exist a set Υ of Λ_1-configuration and numbers $\delta \geq 0$ and $K > 0$ such that for all $k \in (0, K]$, $t \in [0, T]$, $\lambda \notin \operatorname{Int}\big(\Upsilon \cup \mathcal{D}(\delta k)\big)$, and $u \in \mathfrak{X}$,*

$$\left|\omega(-kA(t))\,(\lambda I - k\mathfrak{A}(t))^{-m}u\right|_t \leq C|\lambda|^{\xi-m}\|u\|. \tag{6.1}$$

In the case $\xi = m-1$, under the extra requirement that

$$\begin{aligned}&\text{the pair } \big(|\cdot|_t; A(t)\big) \text{ is } (\xi|\xi|\varphi,\sigma)\text{-concordant if } \xi \geq 1\\ &\text{and } |u|_t \leq C\|u\| \text{ for all } t \in [0,T] \text{ and } u \in \mathfrak{X} \text{ if } \xi = 0,\end{aligned} \tag{6.2}$$

it is enough to suppose that $\deg[\omega] \leq -\xi$ instead of $\leq -m$. If $\sigma \geq 0$ in addition, then the above still holds for $\delta = 0$ and $K = \infty$.

Proof. Assuming first that (6.2) is not generally satisfied, we then act under the condition $\deg[\omega] \leq -m$. Let $\varphi_1 \in (0, \varphi)$ and $R \geq 0$ be the same as in the proof of Lemma 5.2. Observe that since the method is $A(\varphi)$-stable, the poles of the functions $\alpha(-z)$ and $\rho(-z)$ lie outside of Σ_{φ_1}.[3] Let further

[2] This symbol appears just before the statement.

[3] We recall that the function $\rho(z)$ is defined by (5.23).

$d > 0$ be chosen sufficiently small so that $\rho(z)$, $\alpha(z)$, and $\omega(z)$ have no poles in $\mathcal{D}(d)$. Also, we choose $K \leq d/R$ if $R > 0$ and $K = \infty$ if $R = 0$.

By the accepted conditions there exists a set Υ' of Λ_1-configuration such that the function $\alpha(-z)$ maps Σ_{φ_1} into Υ'. Let Υ be a set of Λ_1-configuration such that $\Upsilon' \setminus \{0\} \subset \operatorname{Int}\Upsilon$. We select a fixed $d_0 > 0$ sufficiently small to have

$$d_0|\rho(z)| \leq \frac{1}{2} \qquad \text{for all } z \in \mathcal{D}(d). \tag{6.3}$$

Let now Γ_λ be the contour coinciding with the boundary of the set $\mathcal{D}(\min\{d_0|\lambda|, d\}) \cup \Sigma_{\varphi_1}$. We then apply the operator calculus formula (A.10) with $kA(t)$ in place of $\mathfrak{L}$ and with Γ_λ in place of $\Xi_{R+\varepsilon}$ to get, with $\delta = d_0^{-1}R$, for all $k \in (0, K]$, $\lambda \notin \operatorname{Int}\big(\Upsilon \cup \mathcal{D}(\delta k)\big)$, and $u \in \mathfrak{X}$,

$$\begin{aligned} G &:= \omega(-kA(t))\left((\lambda I - \alpha(-kA(t)))^{-m} - (\lambda - \alpha(\infty))^{-m} I\right) u \\ &= \frac{1}{2\pi i}\int_{\Gamma_\lambda} \widetilde{\alpha}_m(\lambda; z)\,\omega(-z)\,(zI - kA(t))^{-1} u\, dz, \end{aligned} \tag{6.4}$$

where

$$\widetilde{\alpha}_m(\lambda; z) = (\lambda - \alpha(-z))^{-m} - (\lambda - \alpha(\infty))^{-m}.$$

Integrating by parts on the right-hand side of (6.4), we get instead

$$G = \frac{1}{2\pi i}\int_{\Gamma_\lambda} \Psi_m(\lambda; z)\,(zI - kA(t))^{-m} u\, dz, \tag{6.5}$$

where

$$\Psi_m(\lambda; z) = \begin{cases} \widetilde{\alpha}_m(\lambda; z)\omega(-z) & \text{if } m = 1, \\ (m-1) \displaystyle\int_{\infty e^{-i\varphi_1}}^{z} (z - z_1)^{m-2}\widetilde{\alpha}_m(\lambda; z_1)\omega(-z_1)\, dz_1 & \text{if } m \geq 2, \end{cases}$$

while the path of integration from $\infty e^{-i\varphi_1}$ to z lies on Γ_λ.

Assume for a moment that we have shown the estimate, for $\lambda \notin \operatorname{Int}\Upsilon$ and $z \in \Gamma_\lambda$,

$$|\Psi_m(\lambda; z)| \leq \begin{cases} C(|\lambda| + |z| + |\lambda|\,|z|)^{-1} & \text{if } m = 1, \\ C(1 + |\lambda|)^{1-m}(|\lambda| + |z|)^{-1} & \text{if } m \geq 2. \end{cases} \tag{6.6}$$

Since the pair $\big(|\cdot|_t; A(t)\big)$ is $(\xi|m|\varphi, \sigma)$-concordant, as in (5.7) we find that for $z \in \Gamma_\lambda$ and $u \in \mathfrak{X}$,

$$\left|(zI - kA(t))^{-m} u\right|_t \leq Ck^{-\xi}|z|^{\xi - m}\|u\|. \tag{6.7}$$

If $\lambda \notin \operatorname{Int}\Upsilon$, $\delta k \leq |\lambda| \leq dd_0^{-1}$, noting that $\xi < m$, it follows from (6.5), (6.6), and (6.7) that

$$|G|_t \leq Ck^{-\xi} \int_{\Gamma_\lambda} |z|^{\xi-m-1} |dz| \, \|u\| \leq Ck^{-\xi} |\lambda|^{\xi-m} \|u\|. \tag{6.8}$$

In the case where $m \geq 2$ and $\lambda \notin \operatorname{Int}\Upsilon$, $|\lambda| > dd_0^{-1}$, we derive, using (6.5), (6.6), (6.7), and the fact that $1 \leq \xi < m$,

$$\begin{aligned} |G|_t &\leq Ck^{-\xi}(1+|\lambda|)^{1-m} \int_{\Gamma_\lambda} |z|^{\xi-m}(|\lambda|+|z|)^{-1} |dz| \, \|u\| \\ &\leq Ck^{-\xi}|\lambda|^{1-m}\|u\| \int_{\Gamma_\lambda} |z|^{\xi-m-1}|dz| \leq Ck^{-\xi}|\lambda|^{\xi-m}\|u\|. \end{aligned} \tag{6.9}$$

Finally, if $m = 1$, since $\xi \in [0, 1)$, by (6.5), (6.6), and (6.7) we have, for all $\lambda \notin \operatorname{Int}\Upsilon$, $|\lambda| > dd_0^{-1}$,

$$|G|_t \leq Ck^{-\xi}|\lambda|^{-1} \int_{\Gamma_\lambda} |z|^{\xi-2}|dz| \, \|u\| \leq Ck^{-\xi}|\lambda|^{\xi-1}\|u\|. \tag{6.10}$$

Observe furthermore that $\alpha(\infty) \neq 0$, since the RK method is of class S_I. Also, it necessarily holds $\alpha(\infty) \in \Upsilon'$. Therefore,

$$|\lambda - \alpha(\infty)|^{-1} \leq C(1+|\lambda|)^{-1} \qquad \text{for all } \lambda \notin \operatorname{Int}\Upsilon. \tag{6.11}$$

By Lemma 5.2 and (6.11) we then obtain, for $\lambda \notin \operatorname{Int}\Upsilon$ and $u \in \mathfrak{X}$,

$$\begin{aligned} |(\lambda - \alpha(\infty))^{-m}\omega(-kA(t))\, u|_{\xi;t} &\leq Ck^{-\xi}(1+|\lambda|)^{-m}\|u\| \\ &\leq Ck^{-\xi}|\lambda|^{\xi-m}\|u\|. \end{aligned} \tag{6.12}$$

The claim thus follows by combining (6.4), (6.8), (6.9), (6.10), and (6.12).

It remains to show (6.6). Since $\alpha(-z) \in \Upsilon'$ whenever $z \in \Sigma_{\varphi_1}$, we have, for all $\lambda \notin \operatorname{Int}\Upsilon$ and $z \in \Sigma_{\varphi_1}$,

$$|\lambda - \alpha(-z)|^{-1} \leq C(|\lambda| + |\alpha(-z)|)^{-1}. \tag{6.13}$$

Next, by the $A(\varphi)$-stability of the method, the zeros of $\alpha(-z)$ lie outside of the set $\Sigma_{\varphi_1} \setminus \{0\}$. Also, 0 is a simple zero of $\alpha(-z)$.[4] These observations yield that $\alpha(-z)$ admits the two-sided estimate

$$c|z|\,(1+|z|)^{-1} \leq |\alpha(-z)| \leq C|z|\,(1+|z|)^{-1} \quad \text{for all } z \in \Sigma_{\varphi_1}. \tag{6.14}$$

[4] We recall that $r(0) = r'(0) = 1$ for all RK methods under consideration.

Thus, since $\deg[\alpha(\infty)-\alpha(\cdot)] \leq -1$, with the aid of (6.11), (6.13), and (6.14) a simple estimation gives, for all $\lambda \notin \operatorname{Int}\Upsilon$ and $z \in \Sigma_{\varphi_1}$,

$$\begin{aligned}
&|(\lambda-\alpha(\infty))^{-1}-(\lambda-\alpha(-z))^{-1}| \\
&\qquad = |\alpha(\infty)-\alpha(-z)|\,|\lambda-\alpha(\infty)|^{-1}|\lambda-\alpha(-z)|^{-1} \\
&\qquad \leq C(1+|z|)^{-1}\,(|\lambda|+|\alpha(-z)|)^{-1} \\
&\qquad \leq C(|\lambda|+|z|+|\lambda z|)^{-1}.
\end{aligned} \tag{6.15}$$

Furthermore, it follows from (6.3) that

$$|\lambda-\alpha(-z)|^{-1} \leq 2|\lambda|^{-1} \qquad \text{for all } z \in \mathcal{D}(\min\{d_0|\lambda|, d\}). \tag{6.16}$$

It is now seen that if $m=1$, (6.6) immediately follows by (6.15), (6.16), and (6.11), since $|\omega(-z)| \leq C$ for $z \in \Gamma_\lambda$.

Let further $m \geq 2$. Assume first that $|\lambda| \leq dd_0^{-1}$. In view of the identity

$$\int_{\infty e^{-i\varphi_1}}^{z} (z-z_1)^{m-2}\widetilde{\alpha}_m(\lambda;z_1)\omega(-z_1)\,dz_1 = -\int_{z}^{\infty e^{i\varphi_1}} \cdots,$$

it suffices to examine the case $\Im z \leq 0$ only. Assuming therefore the last restriction and using next the identity

$$\widetilde{\alpha}_m(z) = \widetilde{\alpha}_1(z)\big((\lambda-\alpha(-z))^{-(m-1)}+\ldots+(\lambda-\alpha(\infty))^{-(m-1)}\big)$$

and the above estimates (6.11), (6.13), (6.15), and (6.16), we find, for $z \in \Gamma_\lambda \cap \partial\Sigma_{\varphi_1}$ since $\deg[\omega] \leq -(m-1)$ and $(1+|\lambda|)\,(1+|z_1|) \geq |\lambda|+|z_1|$,

$$\begin{aligned}
|\Psi_m(\lambda;z)| \;&\leq\; C\int_{\infty e^{-i\varphi_1}}^{z} |z-z_1|^{m-2}(|\lambda|+|z_1|)^{-1} \\
&\qquad \times\big((|\lambda|+|z_1|)^{1-m}+(1+|\lambda|)^{1-m}\big)(1+|z_1|)^{1-m}|dz_1| \\
&\leq\; C\int_{\infty e^{-i\varphi_1}}^{z} (|z_1|-|z|)^{m-2}(|\lambda|+|z_1|)^{-m}|dz_1| \\
&\leq\; C(|\lambda|+|z|)^{-1}.
\end{aligned} \tag{6.17}$$

Clearly, if z lies on $\partial\mathcal{D}(d_0|\lambda|)$, we also have to take into account the contribution coming from the integration over $\Gamma_\lambda \cap \partial\mathcal{D}(d_0|\lambda|)$. It is easily seen then that $|\Psi_m(\lambda;z)| \leq C|\lambda|^{-1}$ so that (6.17) is still valid for $z \in \partial\mathcal{D}(d_0|\lambda|)$ since $|z| = d_0|\lambda|$. These reasonings show (6.6) for $|\lambda| \leq dd_0^{-1}$.

In the case $|\lambda| > dd_0^{-1}$, a similar argument leads to the following estimate. For $z \in \Gamma_\lambda$,[5] since $|z_1| \geq |z| \geq d$,

$$
\begin{aligned}
|\Psi_m(\lambda; z)| &\leq C \int_{\infty e^{-i\varphi_1}}^{z} |z_1|^{m-2}(|z_1| + |z_1\lambda|)^{-1}(1+|\lambda|)^{1-m}|z_1|^{1-m}|dz_1| \\
&\leq C(1+|\lambda|)^{-m}(1+|z|)^{-1}.
\end{aligned}
$$

This result again implies (6.6).

Finally, we discuss separately the case $\xi = m-1$. Note that all the above conclusions, except for the estimate (6.12), are in force even if $\deg[\omega] \leq -(m-1) = -\xi$. However, using the last restriction on ω and assuming in addition condition (6.2), by Lemma 5.2 it follows that (6.12) holds as well, which shows the auxiliary assertion in the considered case.

This completes the proof. □

The following result is a consequence of Theorem 6.1.

Theorem 6.2. [*] *Let $A(t) \in \mathcal{S}(\varphi, \sigma)$ and let the RK method be $A_I(\varphi)$-stable, with some $\varphi \in (0, \pi/2)$ and $\sigma \in \mathbb{R}$. Then there exist a set Υ of Λ_1-configuration and numbers $\delta \geq 0$ and $K > 0$ such that for all $k \in (0, K]$, $t \in [0, T]$, and $\lambda \notin \operatorname{Int}\big(\Upsilon \cup \mathcal{D}(\delta k)\big)$,*

$$\|(\lambda I - k\mathfrak{A}(t))^{-1}\| \leq C|\lambda|^{-1}. \tag{6.18}$$

If $\sigma \geq 0$ in addition, then (6.18) still holds for $\delta = 0$ and $K = \infty$.

The result immediately follows if one takes $\xi = 0$, $m = 1$, and $|\cdot|_t = \|\cdot\|$ in the conditions of Theorem 6.1.

Further, we consider methods of class S_{II}.

Theorem 6.3. [*] *Let $A(t) \in \mathcal{S}(\varphi, \sigma)$, let the RK method be $A_{II}(\varphi)$-stable, and let a seminorm $|\cdot|_t$ form a $(\xi|m|\varphi, \sigma)$-concordant pair with $A(t)$, for some $\varphi \in (0, \pi/2)$, $\sigma \in \mathbb{R}$, $\xi \geq 0$, and $m \in \mathbb{N}$ such that $m - 1 \leq \xi < m$. Suppose also that a rational function $\omega(z)$ is (φ)-regular and satisfies the condition $\deg[\omega] \leq -\varsigma$ with some $\varsigma \in \mathbb{N}$ such that*

$$
\begin{array}{ll}
\varsigma > \xi(\varpi + 1) & \text{if } \xi = m-1 \text{ or } m \geq 2 \text{ and } \xi = m - 1/(\varpi+1), \\
\varsigma \geq \xi(\varpi + 1) & \text{otherwise,}
\end{array} \tag{6.19}
$$

[5] As above, without loss of generality it can be assumed that $\Im z \leq 0$.

where $\varpi = -\deg[\alpha]$. *Then there exist a set* Υ *of* Λ_1*-configuration and numbers* $\delta \geq 0$ *and* $K > 0$ *such that for all* $k \in (0, K]$, $t \in [0, T]$, $\lambda \notin \mathrm{Int}\,(\Upsilon \cup \mathcal{D}(\delta k))$, *and* $u \in \mathfrak{X}$,

$$|\omega(-kA(t))\,(\lambda I - k\mathfrak{A}(t))^{-m}u|_t \leq Ck^{-\xi}|\lambda|^{\xi-m}\|u\|. \tag{6.20}$$

In the particular case $\xi = 0$ *and* $m = 1$, *(6.20) still holds even if* $\deg[\omega] \leq 0$, *instead of* ≤ -1, *provided that*

$$|u|_t \leq C\|u\| \quad \text{for all } t \in [0, T] \text{ and } u \in \mathfrak{X}. \tag{6.21}$$

If $\sigma \geq 0$ *in addition, then the above is in force for* $\delta = 0$ *and* $K = \infty$.

Proof. We first prove the main assertion of the theorem without assuming (6.21). The particular case where $\xi = 0$, $m = 1$, and (6.21) holds is discussed separately in the final part of the proof.

Since the RK method is of class S_{II}, we have $\alpha(\infty) = 0$, which is equivalent to the inequality $\varpi \geq 1$. This circumstance leads to extra difficulties when dealing with methods of class S_{II} for the reason that $\alpha(z)$ vanishes both at 0 and at ∞.

Let Υ, Γ_λ, d, d_0, δ, and φ_1 be defined just the same as in the proof of Theorem 6.1. Note that the argument employed in the proof of Theorem 6.1 allows us to show the claim at least in the case $|\lambda| \geq dd_0^{-1}$. Therefore, we do no more than to consider the situation when $|\lambda| \leq dd_0^{-1}$. Also, we assume without loss of generality that ς is the smallest integer satisfying (6.19).

Now, noting that $\varsigma \geq 1$ and using the operator calculus formula (A.10), we have, for all $k \in (0, K]$, $\lambda \notin \mathrm{Int}\,(\Upsilon \cup \mathcal{D}(\delta k))$, and $u \in \mathfrak{X}$,

$$\begin{aligned}\widehat{G} &:= \omega(-kA(t))\,\big(\lambda I - \alpha(-kA(t))\big)^{-m}u \\ &= \frac{1}{2\pi i}\int_{\Gamma_\lambda} \omega(-z)\,(\lambda - \alpha(-z))^{-m}(zI - kA(t))^{-1}u\,dz.\end{aligned} \tag{6.22}$$

Integrating by parts, we get as well

$$\widehat{G} = \frac{1}{2\pi i}\int_{\Gamma_\lambda} \Theta_m(\lambda; z)\,(zI - kA(t))^{-m}u\,dz,$$

where

$$\Theta_m(\lambda; z) = \begin{cases} \omega_m(z) & \text{if } m = 1, \\ (m-1)\displaystyle\int_{\infty e^{-i\varphi_1}}^{z} (z - z_1)^{m-2}\omega_m(z_1)\,dz_1 & \text{if } m \geq 2, \end{cases}$$

with $\omega_m(z) = \omega(-z)\,(\lambda - \alpha(-z))^{-m}$, while the path of integration from $\infty e^{-i\varphi_1}$ to z lies on Γ_λ.

Let $m = 1$. Then noting that for $z \in \Gamma_\lambda$,

$$|\alpha(-z)| \geq c|z|\,(1+|z|)^{-(\varpi+1)} \tag{6.23}$$

and

$$|\omega(-z)| \leq C(1+|z|)^{-\varsigma}, \tag{6.24}$$

by (6.7), (6.13), and (6.16)[6] we have, since $0 \leq \xi < 1$ and $\varsigma \geq 1$, for all $\lambda \notin \operatorname{Int}\big(\Upsilon \cup \mathcal{D}(\delta k)\big)$,

$$\begin{aligned} |\widehat{G}|_t &\leq C\int_{\Gamma_\lambda} |\lambda - \alpha(-z)|^{-1}(1+|z|)^{-\varsigma}|z|^{\xi-1}|dz| \\ &\leq C|\lambda|^{\xi-1} + C\int_{d_0|\lambda|}^{d} \frac{x^{\xi-1}}{|\lambda|+x}\,dx + C\int_d^\infty \frac{x^{\xi-1-\varsigma}}{|\lambda|+x^{-\varpi}}\,dx \\ &\leq C\left(|\lambda|^{\xi-1} + |\lambda|^{(\varsigma-\xi)/\varpi-1} + \log(2+1/|\lambda|)\right), \end{aligned}$$

which shows (6.20) for $m = 1$, due to the fact that $\varsigma \geq \xi(\varpi+1)$ and, thus, the last two terms of the expression in parentheses are, in fact, dominated by the first one.

Now we study the case $m \geq 2$, $m - 1 \leq \xi < m - 1/(\varpi+1)$. For short we put

$$\ell(h) = \log(2+1/h) \qquad \text{for } h > 0. \tag{6.25}$$

We first show a suitable estimate for $|\Theta_m(\lambda; z)|$. By (6.13) and (6.23) it follows that for $z \in \Gamma_\lambda \cap \partial\Sigma_{\varphi_1}$,

$$\begin{aligned} |\lambda - \alpha(-z)|^{-m} &\leq C(|\lambda| + |\alpha(-z)|)^{-m} \\ &\leq \begin{cases} C(|\lambda|+|z|)^{-m} & \text{if} \quad |z| \leq 1, \\ C|z|^{-\varpi m} & \text{if} \quad 1 \leq |z| \leq |\lambda|^{-1/\varpi}, \\ C|\lambda|^{-m} & \text{if} \quad |z| \geq |\lambda|^{-1/\varpi}. \end{cases} \end{aligned}$$

Using this estimate and (6.24), since $m(\varpi+1) - 1 \geq \varsigma \geq \xi(\varpi+1) \geq m$, we

[6] Clearly, (6.13) and (6.16) hold for methods of class S_{II} as well.

have,[7] for $z \in \Gamma_\lambda \cap \partial\Sigma_{\varphi_1}$, $|z| \le \min\{1, |\lambda|^{-1/\varpi}\}$, $\Im z \le 0$,

$$\begin{aligned}
|\Theta_m(\lambda; z)| &\le C \int_{\infty e^{-i\arg\varphi_1}}^{z} \frac{(|z_1| - |z|)^{m-2}}{(|\lambda| + |\alpha(-z_1)|)^m (1 + |z_1|)^{\varsigma}} |dz_1| \\
&\le C \int_{|z|}^{1} \frac{(x - |z|)^{m-2}}{(|\lambda| + x)^m}\, dx + C \int_1^{|\lambda|^{-1/\varpi}} x^{\varpi m + m - 2 - \varsigma}\, dx \\
&\quad + C|\lambda|^{-m} \int_{|\lambda|^{-1/\varpi}}^{\infty} x^{m-2-\varsigma}\, dx \\
&\le C|z|^{-1} + C\ell(|\lambda|)^{\delta_{\varsigma,(\varpi+1)m-1}} |\lambda|^{(\varsigma+1-m)/\varpi - m}, \qquad (6.26)
\end{aligned}$$

where $\delta_{j,l}$ is the Kronecker delta. Clearly, the last estimate remains valid as well for $z \in \Gamma_\lambda$ such that $z \in \partial\mathcal{D}(d_0|\lambda|)$ or $z \in \partial\Sigma_{\varphi_1}$ and $1 \le |z| \le |\lambda|^{-1/\varpi}$. (The first integral in the second line does not enter if $|z| \ge 1$.) Besides, as in the proof of Theorem 6.1, the restriction $\Im z \le 0$ is not needed. A similar argument also yields since $m < \varsigma + 1$, for $z \in \Gamma_\lambda$ such that $|z| \ge |\lambda|^{-1/\varpi}$,

$$\begin{aligned}
|\Theta_m(\lambda; z)| &\le C|\lambda|^{-m} \int_{|z|}^{\infty} \frac{(x - |z|)^{m-2}}{x^{\varsigma}}\, dx \\
&\le C|\lambda|^{-m} \int_0^{\infty} \frac{x^{m-2}}{(x + |z|)^{\varsigma}}\, dx \le C|\lambda|^{-m} |z|^{m-1-\varsigma}. \qquad (6.27)
\end{aligned}$$

With the aid of (6.26) and (6.27) we then obtain, for all $\lambda \notin \mathrm{Int}\,(\Upsilon \cup \mathcal{D}(\delta k))$, $|\lambda| \le d_0^{-\varpi/(\varpi+1)}$,

$$\begin{aligned}
|\widehat{G}|_t &\le C\|u\| \int_{\Gamma_\lambda \cap \mathcal{D}(|\lambda|^{-1/\varpi})} |z|^{\xi - m} \\
&\quad \times \left(|z|^{-1} + \ell(|\lambda|)^{\delta_{\varsigma,(\varpi+1)m-1}} |\lambda|^{(\varsigma+1-m)/\varpi - m}\right) |dz| \\
&\quad + C|\lambda|^{-m} \|u\| \int_{|\lambda|^{-1/\varpi}}^{\infty} x^{\xi - \varsigma - 1}\, dx \\
&\le C\left(|\lambda|^{\xi - m} + \ell(|\lambda|)^{\delta_{\varsigma,(\varpi+1)m-1} + \delta_{\xi,m-1}} |\lambda|^{(\varsigma - \xi)/\varpi - m}\right) \|u\|,
\end{aligned}$$

which leads to (6.20), since the second term of the sum on the right-hand side is dominated by the first one, in view of (6.19).[8]

Let now $m \ge 2$ and $m - 1/(\varpi + 1) \le \xi < m$ so that $\varsigma = (\varpi + 1)m$. It is then not hard to see that for $z \in \Gamma_\lambda$, $|z| \le \min\{1, |\lambda|^{-1/\varpi}\}$, the second term

---

[7] The second integral in the second line does not enter if $|\lambda| \ge 1$.

[8] It is easily seen that $|\widehat{G}|_t$ is bounded as stated for $|\lambda| \ge d_0^{-\varpi/(\varpi+1)}$ as well.

in the second line and the term in the third line of (6.26) are bounded by C and $C|\lambda|^{1/\varpi}$, respectively, and, thus, both admit a common upper bound $C|z|^{-1}$. Therefore, instead of (6.26) we have $|\Theta_m(\lambda; z)| \le C|z|^{-1}$ for all $z \in \Gamma_\lambda$ with $|z| \le \min\{1, |\lambda|^{-1/\varpi}\}$ since the first integral in the second line of (6.26) is under the same bound as above. Clearly, the quantity $|\Theta_m(\lambda; z)|$ is estimated in the same way for all $z \in \Gamma_\lambda$ with $1 \le |z| \le |\lambda|^{-1/\varpi}$ too. Noting further that (6.27) remains unchanged as well, we see that $|\widehat{G}|_t \le C|\lambda|^{\xi - m}$ for $\lambda \notin \operatorname{Int}\left(\Upsilon \cup \mathcal{D}(\delta k)\right)$, which implies again (6.20).

It remains to show that if $\xi = 0$, $m = 1$, and (6.21) hold, the estimate (6.20) is still valid under the weaker restriction $\deg[\omega] \le 0$. Since the case $\deg[\omega] \le -1$ is actually settled in the preceding part of the proof, without loss of generality it is assumed that $\omega(z) \equiv 1$. Also, from now on we accept that (6.21) holds. Let further $\alpha_\star \in \mathbb{C}$ be defined by

$$\alpha_\star = \lim_{|z| \to \infty} \left(z^p \alpha(z)\right),$$

and let Γ_λ^1 and Γ_λ^2 be given by

$$\Gamma_\lambda^1 = \mathfrak{c}_\lambda \cup \mathfrak{l}_\lambda^{(+)} \cup \mathfrak{l}_\lambda^{(-)} \text{ and } \Gamma_\lambda^2 = \mathfrak{c}_\lambda \cup \mathfrak{l}_\lambda^{[+]} \cup \mathfrak{l}_\lambda^{[-]} \cup \widehat{\mathfrak{c}}_\lambda,$$

where

$$\begin{aligned} \mathfrak{c}_\lambda &= \left\{ z \in \mathbb{C} : \, |z| = \frac{1}{2}|\lambda/\alpha_\star|^{1/p}, \, |\arg(-z)| \le \pi - \varphi_1 \right\}, \\ \mathfrak{l}_\lambda^{(\pm)} &= \left\{ z \in \mathbb{C} : \, |z| \ge \frac{1}{2}|\lambda/\alpha_\star|^{1/p}, \, \arg z = \pm\varphi_1 \right\}, \\ \widehat{\mathfrak{c}}_\lambda &= \left\{ z \in \mathbb{C} : \, |z| \le d_0|\lambda|, \, |\arg(-z)| \le \pi - \varphi_1 \right\}, \end{aligned}$$

and

$$\mathfrak{l}_\lambda^{[\pm]} = \left\{ z \in \mathbb{C} : \, d_0|\lambda| \le |z| \le \frac{1}{2}|\lambda/\alpha_\star|^{1/p}, \, \arg z = \pm\varphi_1 \right\}.$$

We suppose that $\mathfrak{l}_\lambda^{(\pm)}$ and $\mathfrak{l}_\lambda^{[\pm]}$ are oriented downwards, which further specifies the orientation of the contours Γ_λ^1 and Γ_λ^2. Using next the operator calculus formula (6.4) with $m = 1$ and noting that $\alpha(\infty) = 0$ now, we get, for all $\lambda \notin \operatorname{Int}\left(\Upsilon \cup \mathcal{D}(\delta k)\right)$ and $u \in \mathfrak{X}$,

$$\begin{aligned} \widetilde{G} &:= (\lambda I - \mathfrak{A}(t))^{-1} \\ &= \lambda^{-1} u + \frac{1}{2\pi i} \int_{\Gamma_\lambda} \left((\lambda - \alpha(-z))^{-1} - \lambda^{-1}\right) (zI - kA(t))^{-1} u \, dz. \end{aligned}$$

Since the norm of the integrand is $\mathcal{O}(|z|^{-2})$, it follows from Cauchy's Theorem that after a suitable deformation of the contour Γ_λ, the last formula is in force with $\int_{\Gamma_\lambda^1} \cdots + \int_{\Gamma_\lambda^2} \cdots$ in place of $\int_{\Gamma_\lambda} \cdots$. Observe also that the resolvent $(zI - kA(t))^{-1}$ is holomorphic in a neighbourhood of the set surrounded by Γ_λ^2. Hence, by Cauchy's Theorem,

$$\int_{\Gamma_\lambda^2} \lambda^{-1}(zI - kA(t))^{-1}u\, dz = 0.$$

We therefore obtain

$$\begin{aligned}
\widetilde{G} &= (2\pi i)^{-1} \int_{\Gamma_\lambda^1} \left((\lambda - \alpha(-z))^{-1} - \lambda^{-1}\right) (zI - kA(t))^{-1}u\, dz \\
&\quad + (2\pi i)^{-1} \int_{\Gamma_\lambda^2} (\lambda - \alpha(-z))^{-1} (zI - kA(t))^{-1}u\, dz + \lambda^{-1}u \\
&=: \widetilde{G}_1 + \widetilde{G}_2 + \lambda^{-1}u. \qquad (6.28)
\end{aligned}$$

Since the pair $\big(|\cdot|_t; A(t)\big)$ is $(0|1|\varphi, \sigma)$-concordant, similarly to (5.7) we have, for all $z \in \Gamma_\lambda^1 \cup \Gamma_\lambda^2$ and $u \in \mathfrak{X}$,

$$\left|(zI - kA(t))^{-1}u\right|_t \le C|z|^{-1}\|u\|. \qquad (6.29)$$

Also, it is easily seen that for $\lambda \notin \operatorname{Int}\Upsilon$ and $z \in \Gamma_\lambda^1$,

$$\left|\lambda^{-1} - (\lambda - \alpha(-z))^{-1}\right| \le C|\alpha(-z)|\,|\lambda|^{-1}(|\lambda| + |\alpha(-z)|)^{-1} \le C|\lambda|^{-2}\,|z|^{-\varpi},$$

which with the aid of (6.29) gives, for $\lambda \notin \operatorname{Int}\big(\Upsilon \cup \mathcal{D}(\delta k)\big)$ and $u \in \mathfrak{X}$,

$$\|\widetilde{G}_1\| \le C|\lambda|^{-1}\|u\|. \qquad (6.30)$$

Next, to estimate $\widetilde{G}_2$ on the right-hand side of (6.28), observe that inside of the set (of z points) of which Γ_λ^2 is the boundary, the operator function $(zI - kA(t))^{-1}$ is holomorphic while the function $(\lambda - \alpha(-z))^{-1}$ has finitely many poles. One pole is localized in an annulus

$$\{z \in \mathbb{C} : d_1|\lambda| \le |z| \le d_2|\lambda|\}, \quad d_1, d_2 = \text{ const } > 0,$$

and the other ones lie outside of a disk $\mathcal{D}(d_3)$, $d_3 =$ const > 0. The norm of the contribution of the first pole to the total sum of residues of the integrand does not exceed $C|\lambda|^{-1}\|u\|$ while the other poles give a contribution whose norm does not exceed $C\|u\|$. It thus follows that for $\lambda \notin \operatorname{Int}\big(\Upsilon \cup \mathcal{D}(\delta k)\big)$ and $u \in \mathfrak{X}$,

$$\|\widetilde{G}_2\| \le C|\lambda|^{-1}\|u\|. \qquad (6.31)$$

Finally, using (6.30) and (6.31) as well as (6.21) for an estimation of the right-hand side of (6.28) readily leads to (6.20) in the case $\xi = 0$ and $m = 1$.

The proof is now complete. □

The following is, in fact, a consequence of Theorem 6.3.

Theorem 6.4. [*] *Let $A(t) \in \mathcal{S}(\varphi,\sigma)$ and let the RK method be $A_{II}(\varphi)$-stable, with some $\varphi \in (0,\pi/2)$ and $\sigma \in \mathbb{R}$. Then there exist a set Υ of Λ_1-configuration and numbers $\delta \geq 0$ and $K > 0$ such that for all $k \in (0,K]$, $t \in [0,T]$, and $\lambda \notin \operatorname{Int}\left(\Upsilon \cup \mathcal{D}(\delta k)\right)$,*

$$\|(\lambda I - k\mathfrak{A}(t))^{-1}\| \leq C|\lambda|^{-1}.$$

If $\sigma \geq 0$ in addition, then the last estimate still holds for $\delta = 0$ and $K = \infty$.

This result is easily seen if one takes $\xi = 0$, $m = 1$, $\omega(z) \equiv 1$, and $|\cdot|_t = \|\cdot\|$ in the conditions of the preceding theorem.

Remark 6.1. [*] *The results of Theorems 6.1 and 6.3 actually show that under the accepted conditions*[9] *the seminorm $|\!|\!|\cdot|\!|\!|_n = |\omega(-kA(t_n))\cdot|_{t_n}$ forms a $\Lambda_1^k(\xi|m)$-concordant pair with $\mathfrak{A}(t_n)$. As concerns Theorems 6.2 and 6.4, their assertions actually yield that for $A_I(\varphi)$- and $A_{II}(\varphi)$-stable methods, the operator $\mathfrak{A}(t_n)$ satisfies the Λ_1^k-condition, in particular, the Λ_1-condition if $\sigma \geq 0$.*

Remark 6.2. [*] *Under the conditions of Theorem 6.3, in the case $\xi = m - 1/(\varpi+1)$ and $\varsigma = \xi(\varpi+1)$, the resolvent estimate (6.20) holds with $m+1$ in place of m.*

This fact can be shown by using the same reasonings as those leading to (6.20).

In conclusion, we discuss methods of class S_{III}.

Theorem 6.5. [*] *Let $A(t) \in \mathcal{S}(\varphi,\sigma)$, let the RK method be $A_{III}(\varphi)$-stable, and let a seminorm $|\cdot|_t$ form a $(\xi|m|\varphi,\sigma)$-concordant pair with $A(t)$, for some $\varphi \in (0,\pi/2)$, $\sigma \in \mathbb{R}$, $\xi \geq 0$, and $m \in \mathbb{N}$ such that $m-1 \leq \xi < m$. Also, suppose that a rational function $\omega(z)$ is (φ)-regular and satisfies the condition $\deg[\omega] \leq -\varsigma$ for some $\varsigma \geq m$. Then there exist a set Υ of Λ_2-configuration, with singular points $\zeta_1 = 0$ and $\zeta_2 = \alpha(\infty)$, and numbers $\delta \geq 0$ and $K > 0$ such that for all $k \in (0,K]$, $t \in [0,T]$, $\lambda \notin \operatorname{Int}\left(\Upsilon \cup \mathcal{D}(\delta k)\right)$, and $u \in \mathfrak{X}$,*

$$|\omega(-kA(t))\,(\lambda I - k\mathfrak{A}(t))^{-m}u|_t \leq Ck^{-\xi}|\lambda|^{\xi-m}\|u\| \\ + Ck^{-\xi}\ell(|\lambda-\alpha(\infty)|)^{\kappa}|\lambda-\alpha(\infty)|^{\min\{0,(\varsigma-\xi)/\varpi-m\}}\|u\|, \quad (6.32)$$

[9]The main restriction is that the RK method is of class S_I or of class S_{II}.

where $\varpi = -\deg[\alpha(\cdot)-\alpha(\infty)]$, $\kappa = (\delta_{\varsigma,(\varpi+1)m-1}+\delta_{\xi,m-1})(1-\delta_{m,1})+\delta_{\xi,0}\delta_{\varsigma,\varpi}$, *and* $\ell(h)$ *is defined by (6.25). In the case* $\xi = m - 1/(\varpi+1)$ *and* $\varsigma = m(\varpi+1)-1$, *the last estimate holds with* $\kappa = 0$ *and with* $m+1$ *substituted for* m. *Also, for* $\xi = 0$ *and* $m = 1$, *(6.32) is still valid even when* $\varsigma = 0$,[10] *provided that the dominatedness property (6.21) is satisfied. If* $\sigma \geq 0$ *in addition, the above holds for* $\delta = 0$ *and* $K = \infty$.

Proof. For λ's bounded away from $\alpha(\infty)$, the proof is similar to that of Theorem 6.1. For λ's bounded away from 0, the claim follows in a manner similar to that used in the proof of Theorem 6.3. □

Now we state a result of particular character which is intended for subsequent use.

Theorem 6.6. [*] *Let* $A(t) \in \mathcal{S}(\varphi,\sigma)$ *and let the RK method be* $A_{III}(\varphi)$-*stable, for some* $\varphi \in (0,\pi/2)$ *and* $\sigma \in \mathbb{R}$. *Suppose also that a rational function* $\omega(z)$ *is* (φ)-*regular and satisfies the condition* $\deg[\omega] \leq -\varsigma$ *for some* $\varsigma \in \mathbb{N}\cup\{0\}$. *Then there exist a set* Υ *of* Λ_2-*configuration, with singular points* $\zeta_1 = 0$ *and* $\zeta_2 = \alpha(\infty)$, *and numbers* $\delta \geq 0$ *and* $K > 0$ *such that with any fixed* $\xi \geq 0$ *and* $m \in \mathbb{N}$ *such that* $\xi \leq m$ *and any fixed* $\mu_0 \geq 0$ *sufficiently large, for all* $k \in (0,K]$, $t \in [0,T]$, *and* $\lambda \notin \operatorname{Int}(\Upsilon \cup \mathcal{D}(\delta k))$,

$$\|(\mathfrak{A}(t)+\mu_0 I)^{\xi}(\lambda I - k\mathfrak{A}(t))^{-m}\omega(-kA(t))\| \leq Ck^{-\xi}\left(|\lambda|^{\xi-m} + |\lambda-\alpha(\infty)|^{\min\{0,\varsigma/\varpi-m\}}\right), \tag{6.33}$$

where $\varpi = -\deg[\alpha(\cdot)-\alpha(\infty)]$. *In particular,*[11] *for all* $k \in (0,K]$, $t \in [0,T]$, *and* $\lambda \notin \operatorname{Int}(\Upsilon \cup \mathcal{D}(\delta k))$,

$$\|(\lambda I - \mathfrak{A}(t))^{-1}\| \leq C\left(|\lambda|^{-1} + |\lambda-\alpha(\infty)|^{-1}\right). \tag{6.34}$$

If $\sigma \geq 0$ *in addition, the above holds for* $\delta = 0$ *and* $K = \infty$.

Proof. First of all note that the latter stated estimate (6.34) is a consequence of Theorem 6.5 in the particular case where $\xi = 0$, $m = 1$, $|\cdot|_t = \|\cdot\|$, and $\omega(z) \equiv 1$.

Further, for λ's bounded away from $\alpha(\infty)$, (6.33) follows by (1.2) and the convexity inequality (A.18) with $\alpha = \xi$ and $\beta = m$ since the operator $\omega(-kA(t))$ is uniformly bounded with respect to $k \in (0,K]$ and $t \in [0,T]$,

[10] It is therefore allowed that it would hold $\deg[\omega] = 0$.

[11] When $\omega(z) \equiv 1$, $\xi = 0$, $m = 1$

by Lemma 5.3. For λ's bounded away from 0, (6.33) is a direct consequence of the estimates

$$\|(\mathfrak{A}(t) + \mu_0 I)^{\xi}\| \leq Ck^{-\xi}$$

and

$$\|(\lambda I - k\mathfrak{A}(t))^{-m}\omega(-kA(t))\| \leq C|\lambda - \alpha(\infty)|^{\min\{0,\varsigma/\varpi - m\}}, \tag{6.35}$$

of which the former one follows from (1.43) with $\mathfrak{A}(t)$ substituted for $\mathfrak{L}$.[12] As concerns the latter inequality (6.35), for $\varsigma < m\varpi$ it is obtained by applying directly the representation (6.22) and next using the argument employed in the proof of Theorem 6.3. At the same time, if $\varsigma \geq m\varpi$, it is seen that as in the proof of Lemma 1.12,

$$\|(\lambda I - k\mathfrak{A}(t))^{-m}(\alpha(-kA(t)) - \alpha(\infty)I)^m\| \leq C,$$

while using Lemma 5.3 gives, with $\psi(z) = (\alpha(z) - \alpha(\infty))^{-m}\omega(z)$,

$$\|\psi(-kA(t))\| \leq C,$$

since $\deg[\psi] \leq 0$ and $\psi(z)$ is (φ)-regular. Combining the last two estimates clearly leads to (6.35) in the case $\varsigma \geq m\varpi$. □

Remark 6.3. [*] *Let $|\!|\!| \cdot |\!|\!|_n = |\omega(-kA(t_n)) \cdot |_{t_n}$. It is then seen that under the conditions of Theorem 6.5, the pair $\big(|\!|\!| \cdot |\!|\!|_n; \mathfrak{A}(t_n)\big)$ is $\Lambda_2^k(\xi|m)$-concordant for $\varsigma \geq \xi(\varpi + 1)$ unless $\xi = m - 1$ or $m \geq 2$ and $\xi = m - 1/(\varpi + 1)$ in which case the strict inequality $\varsigma > \xi(\varpi + 1)$ is required. In the particular case $\xi = 0$ and $m = 1$ it is however allowed that $\varsigma = 0$ provided that (6.21) is satisfied. At the same time the pair $\big(|\!|\!| \cdot |\!|\!|_n; \mathfrak{A}(t_n)\big)$ is $\Lambda_2^k(\xi|m+1)$-concordant for $\xi = m - 1/(\varpi + 1)$ and $\varsigma = \xi(\varpi + 1)$. As concerns Theorem 6.6, under its conditions the former stated estimate (6.33) actually shows that the pair $\big(\|(\mathfrak{A}(t_n) + \mu_0 I)^{\xi}\omega(-kA(t_n)) \cdot \|; \mathfrak{A}(t_n)\big)$ is $\Lambda_2^k(\xi)$-concordant for $\varsigma \geq \varpi\xi$ while the latter estimate (6.34) yields that the operator $\mathfrak{A}(t_n)$ satisfies the Λ_2^k-condition.*

Remark 6.4. *In some applications $(M; \varphi, \sigma)$-sectorial operators are of interest. More precisely, letting $\mathfrak{L} : \operatorname{Dom} \mathfrak{L} \subseteq \mathfrak{X} \to \mathfrak{X}$ be a closed linear operator, we write $\mathfrak{L} \in \mathcal{S}(M; \varphi, \sigma)$ and call the operator $\mathfrak{L}$ $(M; \varphi, \sigma)$*-sectorial *for some $\varphi \in (0, \pi/2)$ and $\sigma \in \mathbb{R}$ if the sectorial estimate (1) holds with some $M \geq 1$ substituted for C.*[13] *Assuming, instead of the requirement*

[12] The estimate (1.43) can be applied due to the fact that $\mathfrak{A}(t_n)$ satisfies the Λ_2^k-condition.

[13] As in Section 2.2, we denote by M a constant whose size, unlike that of C, is essential for our analysis and should be under control.

$A(t) \in \mathcal{S}(\varphi,\sigma)$, that for some $\varphi_1 \in (0,\varphi)$ it holds $A(t) \in \mathcal{S}(M;\varphi_1,\sigma)$, $M \geq 1$, the assertions of the above Theorems 6.2, 6.4, and 6.6 remain valid with CM substituted for C in the corresponding stability estimates.

In fact, this can be shown by using the same reasonings as above.

6.2 Hölder-continuity type results

Here we show that if the operator $A(t)$ satisfies one of the Hölder-continuity type conditions introduced in the preceding chapter, then the operator $\mathfrak{A}(t_n) = k^{-1}\alpha(-kA(t_n))$ also satisfies a certain Hölder-continuity condition, among those posed in Part I. The below assertions deal with the classes S_I and S_{III}.

Theorem 6.7. *Let $A(t) \in \mathcal{S}(\varphi,\sigma)$ and let the RK method be $A_I(\varphi)$-stable, for some $\varphi \in (0,\pi/2)$ and $\sigma \in \mathbb{R}$. Also, suppose that $\mathcal{A}(t;\cdot,\cdot)$ is a testing functional for $A(t)$ with $\operatorname{Dom}\mathcal{A}(t;\cdot,\cdot) = \mathcal{P}_* \times \mathcal{P}$. Furthermore let there be given seminorms $|\cdot|_{\mathfrak{m};t}$, $\mathfrak{m} = \emptyset, 1, *\emptyset, *1$, of which the first two are defined on $\mathcal{P}$ and the other two on $\mathcal{P}_*$ so that the pairs $\big(|\cdot|_{\mathfrak{m};t}; A(t)\big)$, $\mathfrak{m} = \emptyset, 1, *\emptyset, *1$, are $(\gamma|\varphi,\sigma)$-, $(\gamma_1|\varphi,\sigma)$-, $(*;\gamma_*|\varphi,\sigma)$-, and $(*;\gamma_{*1}|\varphi,\sigma)$-concordant, respectively, for some $\gamma,\gamma_1,\gamma_*,\gamma_{*1} \in [0,1)$. Then there exist seminorms $\|\cdot\|_{\mathfrak{m};n}$, $\mathfrak{m} = \emptyset, I, *\emptyset, *I$, such that the pairs $\big(\|\cdot\|_{\mathfrak{m};n}; \mathfrak{A}(t_n)\big)$, $\mathfrak{m} = \emptyset, I, *\emptyset, *I$, are respectively $\Lambda_1^k(\gamma|1)$-, $\Lambda_1^k(\gamma_1|1)$-, $\Lambda_1^k(*;\gamma_*|1)$-, and $\Lambda_1^k(*;\gamma_{*1}|1)$-concordant and there hold the following implications involving* **höl** *hypotheses for the operator $A(t)$ and* **HÖL** *hypotheses for the operator $\mathfrak{A}(t_n)$ substituted for $\mathfrak{A}_n$, with any fixed $\vartheta,\vartheta_1 \in [0,1]$,*

$$\mathbf{h\ddot{o}l}^{(\to)}(\vartheta,\vartheta_1;\mathcal{A}(t;\cdot,\cdot);|\cdot|_{\mathfrak{m};t},\mathfrak{m}=\emptyset,1,*\emptyset) \Longrightarrow \mathbf{H\ddot{O}L}^{(\to)}(\vartheta,\vartheta_1;\|\cdot\|_{\mathfrak{m};n},\mathfrak{m}=\emptyset,I,*\emptyset), \quad (6.36)$$

$$\mathbf{h\ddot{o}l}^{(\leftarrow)}(\vartheta,\vartheta_1;\mathcal{A}(t;\cdot,\cdot);|\cdot|_{\mathfrak{m};t},\mathfrak{m}=\emptyset,*\emptyset,*1) \Longrightarrow \mathbf{H\ddot{O}L}^{(\leftarrow)}(\vartheta,\vartheta_1;\|\cdot\|_{\mathfrak{m};n},\mathfrak{m}=\emptyset,*\emptyset,*I), \quad (6.37)$$

and

$$\mathbf{h\ddot{o}l}(\vartheta;\mathcal{A}(t;\cdot,\cdot);|\cdot|_{\mathfrak{m};t},\mathfrak{m}=\emptyset,*\emptyset) \Longrightarrow \mathbf{H\ddot{O}L}(\vartheta;\|\cdot\|_{\mathfrak{m};n},\mathfrak{m}=\emptyset,*\emptyset). \quad (6.38)$$

Proof. We will show the implication (6.36) only, noting that (6.37) and (6.38) can be proved with the aid of the same reasoning.

First of all observe that $\deg[\rho] = -1$ since the RK method is of class S_I. We therefore have, with some $b_1, b_0, b_{lj} \in \mathbb{C}$,

$$\rho(-z)^{-1} = b_1 z + b_0 + \sum_{l=1}^{l_0} \sum_{j=1}^{m_l} b_{lj}(z - z_l)^{-j} =: b_1 z + b_0 + P(z), \tag{6.39}$$

where $z_l, l = 1, \ldots, l_0$, are the poles of $\rho(-z)^{-1}$ and m_l is the multiplicity of a pole z_l.

Next, using the fact that $A(t) \in \mathcal{S}(\varphi, \sigma)$, it follows from Lemma A.1 that $\|(\lambda I - A(t))^{-1}\| \le C|\lambda|^{-1}$ for all $\lambda \notin \operatorname{Int}\left(\Sigma_{\varphi_1} \cup \mathcal{D}(R)\right)$, with some $R \ge 0$ and $\varphi_1 \in (0, \varphi)$. Also, since the RK method is $A(\varphi)$-stable and $r'(0) = 1$, all poles of $\rho(-z)^{-1}$ lie outside of Σ_{φ_1}. These observations yield that with $K > 0$ sufficiently small, the operators $(kA(t) - z_l I)^{-1}$, $l = 1, \ldots, l_0$, are defined on $\mathfrak{X}$ and uniformly bounded with respect to $k \in (0, K]$ and $t \in [0, T]$, and, clearly, the same is true for the operator $P(kA(t))$.

Below we need to get a suitable representation for the difference $\mathfrak{A}(t) - \mathfrak{A}(s)$. To that end, we put $\varrho(t) = \rho(-kA(t))$ and $\mathcal{P}(t) = P(kA(t))$ and note that using (6.39) leads to the identity

$$I = b_1 k \mathfrak{A}(t) + b_0 \varrho(t) + \mathcal{P}(t)\varrho(t),$$

which gives

$$\begin{aligned} \mathfrak{A}(t) - \mathfrak{A}(s) &= \left(b_1 k \mathfrak{A}(s) + b_0 \varrho(s) + \mathcal{P}(s)\varrho(s)\right)\mathfrak{A}(t) \\ &\quad - \mathfrak{A}(s)\left(b_1 k \mathfrak{A}(t) + b_0 \varrho(t) + \mathcal{P}(t)\varrho(t)\right) \\ &= b_0\left(\varrho(s)\mathfrak{A}(t) - \mathfrak{A}(s)\varrho(t)\right) + \left(\varrho(s)\mathcal{P}(s)\mathfrak{A}(t) - \mathfrak{A}(s)\mathcal{P}(t)\varrho(t)\right) \\ &=: Q_1 + Q_2. \end{aligned} \tag{6.40}$$

Further, for convenience' sake we put $\phi_l(t) = (kA(t) - z_l I)^{-1}$, $\varrho_{l,j}(t) = \varrho(t)\,(\phi_l(t))^j$, $\alpha_{l,j}(t) = (\phi_l(t))^j \alpha(-kA(t))$, and $\Delta\mathcal{A}(t, s; \chi, u) = \mathcal{A}(t; \chi, u) - \mathcal{A}(s; \chi, u)$. Recalling that $\operatorname{Dom} A(t) \equiv \mathcal{D}om$ is independent of t, we then have $Q_1 = b_0 \varrho(s)\,(A(t) - A(s))\,\varrho(t)$. Also, it is easily seen that $\operatorname{Ran} \varrho(t) \subseteq \mathcal{D}om \subseteq \mathcal{P}$ and $\operatorname{Ran} \varrho^*(t) \subseteq \mathcal{P}_*$ since $\deg[\rho] = -1$. Using the fact that $A(t)$ satisfies hypothesis $\mathbf{h\ddot{o}l}^{(\rightarrow)}(\vartheta, \vartheta_1; \mathcal{A}(t; \cdot, \cdot); |\cdot|_{\mathfrak{m};t}, \mathfrak{m} = \emptyset, 1, *\emptyset)$, we therefore obtain, for all $\chi \in \mathfrak{X}^*$ and $u \in \mathfrak{X}$,

$$\begin{aligned} |\langle \chi, Q_1 u\rangle| &= |b_0|\, \left|\Delta\mathcal{A}(t, s; \varrho^*(s)\,\chi, \varrho(t)\,u)\right| \\ &\le C|\varrho^*(s)\chi|_{*\emptyset;s}\left(|t-s|^{\vartheta}|\varrho(t)\,u|_{\emptyset;t} + |t-s|^{\vartheta_1}|\varrho(t)\,u|_{1;t}\right). \end{aligned} \tag{6.41}$$

Arguing as above and applying the identity, for all $\chi \in \mathfrak{X}^*$ and $u \in \mathfrak{X}$,

$$\begin{aligned}\langle \chi, Q_2 u\rangle &= \sum_{l=1}^{l_0}\sum_{j=1}^{m_l} b_{lj} \sum_{j_1=1}^{j} \Delta\mathcal{A}\big(t, s; \alpha^*_{l,j_1}(s)\,\chi, \varrho_{l,j+1-j_1}(t)\,u\big)\\ &\quad + \sum_{l=1}^{l_0}\sum_{j=1}^{m_l} b_{lj}\Delta\mathcal{A}\big(t, s; \varrho^*_{l,j}(s)\,\chi, \varrho(t)\,u\big),\end{aligned}$$

we have as well

$$\begin{aligned}|\langle \chi, Q_2 u\rangle| \le C &\sum_{l=1}^{l_0}\sum_{j=1}^{m_l} \big(|\alpha^*_{l,j}(s)\,\chi|_{*\emptyset;s} + |\varrho^*_{l,j}(s)\,\chi|_{*\emptyset;s}\big)\\ &\times \sum_{l=1}^{l_0}\sum_{j=0}^{m_l} \big(|t-s|^{\vartheta}|\varrho_{l,j}(t)\,u|_{\emptyset;t} + |t-s|^{\vartheta_1}|\varrho_{l,j}(t)\,u|_{1;t}\big). \qquad (6.42)\end{aligned}$$

Now, putting

$$|\!|\!|\cdot|\!|\!|_{*\emptyset;n} = \sum_{l=1}^{l_0}\sum_{j=1}^{m_l} \big(|\alpha^*_{l,j}(t_n)\ \cdot|_{*\emptyset;t_n} + |\varrho^*_{l,j}(t_n)\ \cdot|_{*\emptyset;t_n}\big) + |\varrho^*(t_n)\ \cdot|_{*\emptyset;t_n},$$

$$|\!|\!|\cdot|\!|\!|_{\emptyset;n} = \sum_{l=1}^{l_0}\sum_{j=0}^{m_l} |\varrho_{l,j}(t_n)\ \cdot|_{\emptyset;t_n},$$

and

$$|\!|\!|\cdot|\!|\!|_{I;n} = \sum_{l=1}^{l_0}\sum_{j=0}^{m_l} |\varrho_{l,j}(t_n)\ \cdot|_{1;t_n},$$

it follows from (6.40), (6.41), and (6.42) that the implication (6.36) holds.

Note that since $\gamma_* < 1$, the pair $\big(|\cdot|_{*\emptyset;t}; A(t)\big)$ is $(*;\gamma_*|1|\varphi,\sigma)$-concordant, thanks to Lemma 5.1. Using then Theorem 6.1 and taking into account that $\deg[\rho] = -1$, it is readily seen that the pair $\big(|\varrho^*(t_n)\ \cdot|_{*\emptyset;t_n}; \mathfrak{A}(t_n)\big)$ is $\Lambda^k_1(*;\gamma_*|1)$-concordant. Acting after the same pattern, it can be shown that the pairs $\big(|\alpha^*_{l,j}(t_n)\ \cdot|_{*\emptyset;t_n}; \mathfrak{A}(t_n)\big)$ and $\big(|\varrho^*_{l,j}(t_n)\ \cdot|_{*\emptyset;t_n}; \mathfrak{A}(t_n)\big)$, $l = 1,\dots,l_0$, $j = 1,\dots,m_l$, are $\Lambda^k_1(*;\gamma_*|1)$-concordant. Therefore the pair $\big(|\!|\!|\cdot|\!|\!|_{*\emptyset;n}; \mathfrak{A}(t_n)\big)$ is $\Lambda^k_1(*;\gamma_*|1)$-concordant since $|\!|\!|\cdot|\!|\!|_{*\emptyset;n}$ is the sum of seminorms that form $\Lambda^k_1(*;\gamma_*|1)$-concordant pairs with $\mathfrak{A}(t_n)$.

Similarly, we find that the pairs $\big(|\!|\!|\cdot|\!|\!|_{\mathfrak{m};n}; \mathfrak{A}(t_n)\big)$, $\mathfrak{m} = \emptyset, I$, are $\Lambda^k_1(\gamma|1)$- and $\Lambda^k_1(\gamma_1|1)$-concordant, respectively.

This completes the proof. □

The following result is helpful when dealing with methods of class S_{III}.

Theorem 6.8. *Suppose that* $A(t) \in \mathcal{S}(\varphi, \sigma)$ *and the RK method is* $A_{III}(\varphi)$*-stable, for some* $\varphi \in (0, \pi/2)$ *and* $\sigma \in \mathbb{R}$. *Let further* $\mathcal{A}(t;\cdot,\cdot)$ *be a testing functional for* $A(t)$ *with* $\operatorname{Dom} \mathcal{A}(t;\cdot,\cdot) = \mathcal{P}_* \times \mathcal{P}$. *Suppose also that there are given seminorms* $|\cdot|_{\mathfrak{m};t}$, $\mathfrak{m} = \emptyset, 1, *\emptyset, *1$, *of which the first two are defined on* $\mathcal{P}$ *and the other two on* $\mathcal{P}_*$ *so that the pairs* $\big(|\cdot|_{\mathfrak{m};t}; A(t)\big)$, $\mathfrak{m} = \emptyset, 1, *\emptyset, *1$, *are* $(1/2|\varphi,\sigma)$-, $(\gamma_1|\varphi,\sigma)$-, $(*;1/2|\varphi,\sigma)$-, *and* $(*;\gamma_{*1}|\varphi,\sigma)$*-concordant, respectively, with some* $\gamma_1, \gamma_{*1} \in [0, 1/2]$. *Then, under the condition* $\deg[\alpha(\cdot) - \alpha(\infty)] = -1$, *there exist seminorms* $\|\cdot\|_{\mathfrak{m};n}$, $\mathfrak{m} = \emptyset, I, *\emptyset, *I$, *that form respectively* $\Lambda_2^k(1/2)$-, $\Lambda_2^k(\gamma_1)$-, $\Lambda_2^k(*;1/2)$-, *and* $\Lambda_2^k(*;\gamma_{*1})$*-concordant pairs with* $\mathfrak{A}(t_n)$, *and there hold the following implications involving* **höl** *hypotheses for the operator* $A(t)$ *and* **HÖL** *hypotheses for the operator* $\mathfrak{A}(t_n)$ *substituted for* $\mathfrak{A}_n$, *with any fixed* $\vartheta, \vartheta_1 \in [0,1]$,

$$\mathbf{höl}^{(\rightarrow)}(\vartheta, \vartheta_1; \mathcal{A}(t;\cdot,\cdot); |\cdot|_{\mathfrak{m};t}, \mathfrak{m} = \emptyset, 1, *\emptyset) \\ \Longrightarrow \mathbf{HÖL}^{(\rightarrow)}(\vartheta, \vartheta_1; \|\cdot\|_{\mathfrak{m};n}, \mathfrak{m} = \emptyset, I, *\emptyset),$$

$$\mathbf{höl}^{(\leftarrow)}(\vartheta, \vartheta_1; \mathcal{A}(t;\cdot,\cdot); |\cdot|_{\mathfrak{m};t}, \mathfrak{m} = \emptyset, *\emptyset, *1) \\ \Longrightarrow \mathbf{HÖL}^{(\leftarrow)}(\vartheta, \vartheta_1; \|\cdot\|_{\mathfrak{m};n}, \mathfrak{m} = \emptyset, *\emptyset, *I),$$

and

$$\mathbf{höl}(\vartheta; \mathcal{A}(t;\cdot,\cdot); |\cdot|_{\mathfrak{m};t}, \mathfrak{m} = \emptyset, *\emptyset) \Longrightarrow \mathbf{HÖL}(\vartheta; \|\cdot\|_{\mathfrak{m};n}, \mathfrak{m} = \emptyset, *\emptyset).$$

The proof is similar to that of Theorem 6.7, using as well Theorems 6.5 and 6.5*, Remarks 6.3 and 6.3*, and the fact that $\deg[\rho] = -1$.

We state separately a more concrete assertion which will be needed in the sequel.

Theorem 6.9. *Let* $A(t) \in \mathcal{S}(\varphi, \sigma)$ *and let the RK method be* $A_I(\varphi)$*- or* $A_{III}(\varphi)$*-stable, for some* $\varphi \in (0, \pi/2)$ *and* $\sigma \in \mathbb{R}$. *Let further a seminorm* $|\cdot|_t$ *form a* $(\gamma_1|\varphi,\sigma)$*-concordant pair with* $A(t)$ *and let* $A(t)$ *itself satisfy hypothesis* $\mathbf{höl}^{\oplus}[\vartheta; |\cdot|_t]$, *for some* $\gamma_1 \in [0,1)$ *and* $\vartheta \in (0,1]$. *Then, with any fixed* $\mu_0 \geq 0$ *sufficiently large and with any fixed* $K > 0$ *sufficiently small, we have, for all* $k \in (0, K]$, $t, s \in [0, T]$, $\chi \in \mathfrak{X}^*$, *and* $u \in \mathfrak{X}$, *that*

$$|\langle \chi, (\mathfrak{A}(t) - \mathfrak{A}(s))\, u\rangle| \leq C\|(\rho(-kA(s)))^* \chi\|_* \\ \times \Big(|t-s|^{\vartheta}\, \|(\mathfrak{A}(t) + \mu_0 I)\, u\| + |(I + kA(t))^{-1} u|_t\Big). \quad (6.43)$$

If in addition $A(t)$ satisfies hypothesis **höl**$[\vartheta]$, *then the term* $|(I+kA(t))^{-1}u|_t$ *does not enter.*

Proof. The result can be shown following the proof Theorem 6.7 and noting, with the aid of Lemmas 5.3 and 1.5, that with $\mu_0 \geq 0$ sufficiently large, for all $t \in [0,T]$ and $u \in \mathfrak{X}$,

$$\|A(t)\varrho(t)\,u\| + \|\varrho(t)\,u\| \leq C(\|\mathfrak{A}(t)\,u\| + \|u\|) \leq C\|(\mathfrak{A}(t) + \mu_0 I)\,u\|$$

and

$$\|(I + kA(t))\,\omega(-kA(t))\| \leq C,$$

for any rational function $\omega(z)$ which is (φ)-regular and satisfies the condition $\deg[\omega] \leq -1$. □

In the remainder of the section we give two auxiliary results which show the Hölder-continuity of a bounded rational function of the operator $A(t)$ under the assumption that $A(t)$ itself satisfies a Hölder-continuity type condition among those stated before.

Lemma 6.1. *Suppose that $A(t) \in \mathcal{S}(\varphi, \sigma)$ and that $A(t)$ satisfies hypothesis* **höl**$[\vartheta]$, *for some $\varphi \in (0, \pi/2)$, $\sigma \in \mathbb{R}$, and $\vartheta \in (0,1]$. Let further $\omega(z)$ be a rational function which is (φ)-regular and satisfies the condition* $\deg[\omega] \leq 0$. *Then, with $K > 0$ sufficiently small, for all $k \in (0,K]$ and $t, s \in [0,T]$,*

$$\|\omega(-kA(t)) - \omega(-kA(s))\| \leq C|t-s|^{\vartheta}.$$

Proof. We can assume without loss of generality that $\deg[\omega] \leq -1$, otherwise we might deal with the function $\omega(z) - \omega(\infty)$. Having therefore accepted the last inequality, we let $\varphi_1 \in (0, \varphi)$ and $R \geq 0$ be just the same as in the statement of Lemma A.1 with $A(t)$ substituted for $\mathfrak{L}$. Let also $d > 0$ be chosen such that $\omega(z)$ has no poles in $\mathcal{D}(d)$. Further, we set $K > 0$ subject to $K \leq \min\{T, d/R\}$. Let next Γ be the contour coinciding with $\partial\big(\Sigma_{\varphi_1} \cup \mathcal{D}(d)\big)$. Since $\omega(\infty) = 0$ now, we can be based, as in the proof of Lemma 5.2, on formula (5.8) which clearly yields that

$$\begin{aligned}&\omega(-kA(t)) - \omega(-kA(s))\\ &\qquad = \frac{1}{2\pi i}\int_\Gamma \omega(-z)\left((zI - kA(t))^{-1} - (zI - kA(s))^{-1}\right)dz. \qquad (6.44)\end{aligned}$$

Using the fact that the domain of $A(t)$ is independent of t, we have as well

$$\begin{aligned}&(zI - kA(t))^{-1} - (zI - kA(s))^{-1}\\ &\qquad = (zI - kA(t))^{-1}k(A(t) - A(s))\,(zI - kA(s))^{-1}. \qquad (6.45)\end{aligned}$$

Since $A(t) \in \mathcal{S}(\varphi, \sigma)$, similarly to (5.7) it follows that

$$\|(zI - kA(t))^{-1}\| \le C|z|^{-1} \quad \text{for all } z \in \Gamma. \tag{6.46}$$

Also, using this estimate and the fact that $A(t)$ satisfies **höl**$[\vartheta]$, it is not hard to see that with some $\mu_1 \ge 0$, for all $z \in \Gamma$,

$$\begin{aligned}\|k(A(t) - A(s))\,(zI - kA(s))^{-1}\| & \\ &\le Ck|t-s|^{\vartheta}\|(A(s) + \mu_1 I)\,(zI - kA(s))^{-1}\| \\ &\le C|t-s|^{\vartheta}\left(1 + \|(z + k\mu_1)\,(zI - kA(s))^{-1}\|\right) \\ &\le C|t-s|^{\vartheta}. \end{aligned} \tag{6.47}$$

Combining therefore (6.44), (6.46), and (6.47) gives, after a simple estimation,

$$\|\omega(-kA(t)) - \omega(-kA(s))\| \le C|t-s|^{\vartheta} \int_{\Gamma} |\omega(-z)|\,|z|^{-1}|dz|,$$

which shows the claim in view of the evident estimate

$$|\omega(-z)| \le C(1 + |z|)^{-1} \quad \text{for } z \in \Gamma.$$

□

Lemma 6.2. *Let $A(t) \in \mathcal{S}(\varphi, \sigma)$ for some $\varphi \in (0, \pi/2)$ and $\sigma \in \mathbb{R}$ and let $\mathcal{A}(t;\cdot,\cdot)$ be a testing functional for $A(t)$ with* $\operatorname{Dom}\mathcal{A}(t;\cdot,\cdot) = \mathcal{P}_* \times \mathcal{P}$. *Suppose further that with some seminorms $|\cdot|_{\mathfrak{m};t}$, $\mathfrak{m} = \emptyset, *\emptyset$, of which the first is defined on $\mathcal{P}$ and the second on $\mathcal{P}_*$, the pairs $\big(|\cdot|_{\mathfrak{m};t}; A(t)\big)$, $\mathfrak{m} = \emptyset, *\emptyset$, are $(\gamma|\varphi,\sigma)$- and $(*;\gamma_*|\varphi,\sigma)$-concordant, respectively, and $A(t)$ satisfies hypothesis* **höl**$(\vartheta; \mathcal{A}(t;\cdot,\cdot); |\cdot|_{\mathfrak{m};t}, \mathfrak{m} = \emptyset, *\emptyset)$, *for some $\gamma, \gamma_* \in [0,1)$ and $\vartheta \in (0,1]$. Finally, let $\omega(z)$ be a rational function which is (φ)-regular and satisfies the condition* $\deg[\omega] \le 0$. *Then, with $K > 0$ sufficiently small, for all $k \in (0, K]$ and $t, s \in [0,T]$,*

$$\|\omega(-kA(t)) - \omega(-kA(s))\| \le C|t-s|^{\vartheta}.$$

Proof. The result is proved in a manner similar to that used in the proof of Lemma 6.1, with the difference that instead of (6.45) we now apply the identity, for all $\chi \in \mathfrak{X}^*$ and $u \in \mathfrak{X}$,

$$\begin{aligned}\langle \chi, \big((zI - kA(t))^{-1} - (zI - kA(s))^{-1}\big)\,u\rangle & \\ &= k\Delta\mathcal{A}\big(t; (z^*I - kA^*(t))^{-1}\chi, (zI - kA(s))^{-1}u\big).\end{aligned}$$

Next, estimating the expression on the right by analogy with the proof of Theorem 6.7, we find that for all $z \in \Gamma$,

$$\|(zI - kA(t))^{-1} - (zI - kA(s))^{-1}\| \leq C|t-s|^{\vartheta}|z|^{\gamma+\gamma_*-2}.$$

Using next the last inequality for an estimation on the basis of (6.44) leads to the desired result as in the proof of the preceding lemma. □

6.3 Stability estimates

In this section, which is the central part of the whole chapter, we examine the stability of the discrete problem (5.19), (5.20). To prove the problem, we use here the theory developed in Part I as well as the preparatory results found in the preceding two sections.

We begin with a discussion in the case with a time-independent operator, for which the question is settled trivially, after the above preparations.

Theorem 6.10. *Let the operator $A(t)$ be independent of t*[14] *and let*

$$\|\mathfrak{R}\| \leq C \qquad \textit{for all } k \in (0, K]. \tag{6.48}$$

Suppose further that the RK method is of type $C_I(\varphi)$ or of type $C_{II}(\varphi)$ ($A_I(\varphi)$- or $A_{II}(\varphi)$-stable if $f(t) \equiv 0$) and that $A \in \mathcal{S}(\varphi, \sigma)$, for some $\varphi \in (0, \pi/2)$ and $\sigma \in \mathbb{R}$. Then, if $T < \infty$, for Y_n the extended solution of problem (5.19), (5.20), we have, with any fixed $\mu_0 \geq 0$ sufficiently large and with any fixed $\xi \geq 0$, for all $k \in (0, K]$ and $t_n \in \Omega_k$,

$$\|(\mathfrak{A} + \mu_0 I)^{\xi} Y_n\| \leq C\left(t_{n+1}^{-\xi}\|y^0\| + k\sum_{l=0}^{n-1} t_{n-l+1}^{-\xi} \max_{j=1,\dots,\nu}\|f(t_{lj})\|\right). \tag{6.49}$$

Moreover, if in addition $\sigma \geq 0$ and $\mu_0 = 0$, then the last estimate is valid for $T = \infty$[15] *and $K = \infty$ as well.*

Proof. Since $\deg[w_j] \leq 0$, $j = 1, \dots, \nu$, it follows from Lemma 5.3 that $\|w_j(-kA)\| \leq C$, which further gives, by (5.28),

$$\|\mathbf{w}^T(-kA)\,\Phi_n\| \leq C \max_{j=1,\dots,\nu}\|f(t_{nj})\| \quad \text{for } t_n \in \breve{\Omega}_k. \tag{6.50}$$

Clearly, the last estimate remains valid for $A_I(\varphi)$- or $A_{II}(\varphi)$-stable methods if $f(t) \equiv 0$. Also, by Theorems 6.2 and 6.4 it follows that the operator $\mathfrak{A}$

[14] We can therefore write A and $\mathfrak{A}$ instead of $A(t)$ and $\mathfrak{A}(t)$, respectively.

[15] That is, for all $n = 0, 1, \dots$

satisfies the Λ_1^k-condition, in particular, the Λ_1-condition in the case $\sigma \geq 0$. To conclude the proof, it remains to apply Theorem 2.17, noting that the pair $(\|(\mathfrak{A}+\mu_0 I)^\xi \cdot \|; \mathfrak{A})$ is $\Lambda_1^k(\xi)$-concordant, in particular, $\Lambda_1(\xi)$-concordant if $\sigma \geq 0$ and $\mu_0 = 0$, by Lemma 1.12. □

Theorem 6.11. *Let the operator $A(t) \equiv A$ satisfy the conditions of Theorem 6.10 and let the RK method be of type $C_{III}(\varphi)$ ($A_{III}(\varphi)$-stable if $f(t) \equiv 0$), with some $\varphi \in (0, \pi/2)$ and $\sigma \in \mathbb{R}$. Also, suppose that (6.48) holds and $T < \infty$. Then the extended solution Y_n of (5.19), (5.20) satisfies (6.49) for $\xi = 0$. If, in addition to above, the operator $\mathfrak{R}$ is given by*

$$\mathfrak{R} = \omega(-kA), \tag{6.51}$$

where $\omega(z)$ is a rational function which is (φ)-regular and under the requirement, with some $\varsigma \in \mathbb{N}$,

$$\deg[\omega] \leq -\varsigma, \tag{6.52}$$

then, assuming that the RK method itself satisfies the condition[16]

$$\deg[w_j] \leq -\varsigma, \qquad j = 1, \dots, \nu, \tag{6.53}$$

(6.49) holds for any fixed $\xi \in [0, \varsigma/\varpi]$, where $\varpi = -\deg[\alpha(\cdot) - \alpha(\infty)]$. Moreover, if $\sigma \geq 0$ and $\mu_0 = 0$, the above is valid for $T = K = \infty$ as well.

Proof. It follows from Theorem 6.6 that the operator $\mathfrak{A}$ satisfies the Λ_2^k-condition, in particular, the Λ_2-condition if $\sigma \geq 0$. Note also that (6.50) holds for methods of type $C_{III}(\varphi)$ as well. Therefore, even if (6.51)−−(6.53) are not necessarily satisfied, (6.49) with $\xi = 0$ follows by Theorem 2.17 since the pair $(\|\cdot\|; \mathfrak{A})$ is then, obviously, $\Lambda_2^k(0)$-concordant ($\Lambda_2(0)$-concordant if $\sigma \geq 0$ and $\mu_0 = 0$).

Now we turn to the case where (6.51) and (6.53) hold, under the above conditions on $\omega(z)$, including (6.52). By (6.53) it then follows from Lemma 5.3 that the operators $(I + kA)^\varsigma w_j(-kA)$, $j = 1, \dots, \nu$, are defined on $\mathfrak{X}$ for $k \in (0, K]$, with

$$\|(I + kA)^\varsigma w_j(-kA)\| \leq C. \tag{6.54}$$

Note next that we can rewrite (5.30) equivalently in the form

$$\partial Y_n + \mathfrak{A} Y_n = \widetilde{\mathfrak{Q}} \widetilde{F}_n, \quad t_n \in \breve{\Omega}_k, \qquad Y_0 = \mathfrak{R} y^0,$$

with $\widetilde{\mathfrak{Q}} = (I + kA)^{-\varsigma}$ and $\widetilde{F}_n = (I + kA)^\varsigma \mathbf{w}^T(-kA)\Phi_n$, where, by (5.28) and (6.54),

$$\|\widetilde{F}_n\| \leq C \max_{j=1,\dots,\nu} \|f(t_{nj})\|. \tag{6.55}$$

[16]This condition is not needed if $f(t) \equiv 0$.

Now, by Theorem 6.6 (see also Remark 6.3) we conclude that the pairs $(\|(\mathfrak{A}+\mu_0 I)^{\xi}\,\omega(-kA)\cdot\|;\mathfrak{A})$ and $(\|(\mathfrak{A}+\mu_0 I)^{\xi}\,(I+kA)^{-\varsigma}\cdot\|;\mathfrak{A})$ are $\Lambda_2^k(\xi)$-concordant if $\xi \le \varsigma/\varpi$. Using this fact and (6.55), the result in the considered case thus follows by Theorem 2.17 since the operators $\omega(-kA)$ and $(I+kA)^{-\varsigma}$ are permutable with $r(-kA)$. Clearly, if $f(t)\equiv 0$, we have $\Phi_n = 0$ and condition (6.53) is not needed.

The more precise wording in the case $\sigma \ge 0$ and $\mu_0 = 0$ is obvious. □

Remark 6.5. *If, instead of the condition $A \in \mathcal{S}(\varphi,\sigma)$, one requires that $A \in \mathcal{S}(M;\varphi_1,\sigma)$, $M \ge 1$, for some fixed $\varphi_1 \in (0,\varphi)$ and $\sigma \in \mathbb{R}$,[17] the proofs of the preceding two theorems can be slightly modified in order to get the same results, with the only difference that the stability constant C is replaced by CM (cf. also Remark 6.4).*

Next we turn to the general case where the operator $A(t)$ depends on t. We now consider a variety of different situations. Since our techniques are applicable for finite intervals only, we suppose in the remainder of the section that $T < \infty$. In parallel, we give dual estimates for discrete evolution operators, which are often of interest in applications.

We inform the reader beforehand that the main circumstance employed below is that the starting problem (5.19), (5.20) can be equivalently represented in the form (1.1), that is, as the Cauchy problem for an evolution equation in discrete time. The main theorem on representability (Theorem 5.1) ensures this fact under certain reasonable restrictions of which assuming the hypotheses $\mathbf{höl}^{\oplus}(\ldots)$ or $\mathbf{höl}^{\oplus}_{(k)}(\ldots)$ is crucial. In particular, hypothesis $\mathbf{höl}^{\oplus}_{(k)}(\ldots)$ is included in the statements of Theorems 6.13, 6.14, and 6.16 to be in a position to apply Theorem 5.1 and Remark 5.1. Since the last results are applicable under the conditions of any assertion in the remainder of the section, each of the below proofs starts from the point when the representability of problem (5.19), (5.20) in the form (1.1) is considered to have already been verified so that we never make explicit mention of this fact in the sequel. Also, as a matter of course, we have interpreted $\mathfrak{A}_n$, $\mathfrak{Q}_n$, and F_n in (1.1) as given by (5.38), (5.39), (5.40), and (5.41), not taking care of their existence on each occasion, and use the notation $\mathfrak{U}_{n,j}$ introduced by (1.5) and (1.3) without explicit explanation.

Theorem 6.12. *Suppose that the RK method is of type $V_I(\varphi)$ and $A(t) \in \mathcal{S}(\varphi,\sigma)$, for some $\varphi \in (0,\pi/2)$ and $\sigma \in \mathbb{R}$. Let also a seminorm $|\cdot|_t$ form a $(\gamma_1|\varphi,\sigma)$-concordant pair with $A(t)$ and let $A(t)$ itself satisfy hypothesis*

[17] For the definition of $\mathcal{S}(M;\varphi,\sigma)$, see Remark 6.4.

höl$^{\oplus}[\vartheta; |\cdot|_t]$, *for some* $\gamma_1 \in [0,1)$ *and* $\vartheta \in (0,1]$. *Apart from everything else, let (6.48) hold. Then, with any fixed* $\xi \in [0,1)$ *and with any fixed* $\mu_0 > 0$ *sufficiently large, we have, for all* $k \in (0,K]$ *and* $t_n, t_l \in \Omega_k$ *with* $t_l \leq t_n$,

$$\|(\mathfrak{A}(t_n) + \mu_0 I)^{\xi} \mathfrak{U}_{n-1,l}\| \leq C t_{n-l+1}^{-\xi}, \tag{6.56}$$

and for Y_n *the solution of problem (5.19), (5.20),*

$$\|(\mathfrak{A}(t_n) + \mu_0 I)^{\xi} Y_n\| \leq C \left(t_{n+1}^{-\xi} \|y^0\| + k \sum_{l=0}^{n-1} t_{n-l}^{-\xi} \max_{j=1,\dots,\nu} \|f(t_{lj})\| \right). \tag{6.57}$$

Furthermore, with any fixed $0 \leq \xi_* < \min\{\vartheta, 1-\gamma_1\}$ *and with any fixed* $\mu_0 \geq 0$ *sufficiently large, for all* $k \in (0,K]$ *and* $t_n, t_l \in \Omega_k$ *such that* $t_l < t_n$,

$$\|(\mathfrak{A}^*(t_l) + \mu_0 I)^{\xi_*} \mathfrak{U}^*_{n-1,l+1}\|_* \leq C t_{n-l}^{-\xi_*}. \tag{6.58}$$

In the case $\xi_* = \min\{\vartheta, 1-\gamma_1\}$ *the last estimate remains valid if one adds* $\log(t_{n-l+1}/k)$ *to the factor* $t_{n-l}^{-\xi_*}$. *If in addition the operator* $A(t)$ *satisfies hypothesis* **höl**$[\vartheta]$, *then (6.56) and (6.57) are valid for any fixed* $\xi \in [0, 1+\vartheta)$. *Moreover, they retain their shapes even for* $\xi = 1+\vartheta$ *provided that* $t_j^{-\xi}$ *is replaced by* $t_j^{-\xi} + t_j^{-1}\log(t_{j+1}/k)$.

Proof. First of all it follows from (5.41), (5.34), and (5.43) that

$$\|F_n\| \leq C \max_{j=1,\dots,\nu} \|f(t_{nj})\|. \tag{6.59}$$

Also, observe that the operator $\mathfrak{A}(t_n)$ satisfies the Λ_1^k-condition, by Theorem 6.2.

Further, consider the general case where the operator $A(t)$ satisfies hypothesis **höl**$^{\oplus}[\vartheta; |\cdot|_t]$. Observing then that for methods of type $V_I(\varphi)$,

$$\deg[v_j] \leq -1 = \deg[\rho], \quad j = 1, \dots, \nu,$$

and noting that the function $\rho(z)^{-1}$ is (φ)-regular, it is seen that the operator $(\rho(-kA(t)))^{-1}$ is defined on $\operatorname{Ran} v_j(-kA(t))$. Thus by Lemmas 5.3 and 1.5 we have, with μ_1 coming from the statement of hypothesis **höl**$^{\oplus}[\vartheta; |\cdot|_t]$ and with some $\mu_0 > 0$ sufficiently large, for $j = 1, \dots, \nu$ and for all $u \in \mathfrak{X}$,

$$\begin{aligned} &\|(A(t) + \mu_1 I)\, v_j(-kA(t))\, u\| \\ &\qquad \leq C\|(\rho(-kA(t)))^{-1} v_j(-kA(t))\|\, \|A(t)\, \rho(-kA(t))\, u\| + C\|u\| \\ &\qquad \leq C\,(\|\mathfrak{A}(t)\, u\| + \|u\|) \leq C\|(\mathfrak{A}(t) + \mu_0 I)\, u\|. \end{aligned} \tag{6.60}$$

Using now (5.39) and taking into account the estimates (5.34), (5.43), and the accepted Hölder-continuity type condition, with the aid of (6.60) we obtain, for all $u \in \mathfrak{X}$,

$$
\begin{aligned}
\|\mathfrak{B}_n u\| &\le C\sum_{j=1}^{\nu} \|(\mathfrak{D}_n - \mathfrak{D}(t_n))\, v_j(-kA(t_n))\, u\| \\
&\le C\sum_{j=1}^{\nu} \Big(k^{\vartheta}\|(A(t_n) + \mu_1 I)\, v_j(-kA(t_n))\, u\| + |v_j(-kA(t_n))\, u|_{t_n}\Big) \\
&\le Ck^{\vartheta}\|(\mathfrak{A}(t_n) + \mu_0 I)\, u\| + C\sum_{j=1}^{\nu} |v_j(-kA(t_n))\, u|_{t_n}. \qquad (6.61)
\end{aligned}
$$

Since $\mathfrak{A}(t_n)$ satisfies the Λ_1^k-condition, by Lemma 1.1 we have $\|k\mathfrak{A}(t_n)\| \le C$, which further implies

$$k^{\vartheta}\|(\mathfrak{A}(t_n) + \mu_0 I)\, u\| \le C\|(\mathfrak{A}(t_n) + \mu_1 I)\, u\|^{1-\vartheta}\|u\|^{\vartheta}.$$

Using the last estimate and applying Lemma 1.7, it easily follows that the seminorm $k^{\vartheta}\|(\mathfrak{A}(t_n) + \mu_0 I)\,\cdot\,\|$ forms a $\Lambda_1^k(1-\vartheta|1)$-concordant pair with $\mathfrak{A}(t_n)$. Also, applying Lemma 5.1 shows that the seminorm $|\cdot|_t$ is, in fact, $(\gamma_1|1|\varphi,\sigma)$-concordant with $A(t)$ since $\gamma_1 < 1$, which therefore yields, with the aid of Theorem 6.1, that the seminorm $\sum_{j=1}^{\nu} |v_j(-kA(t_n))\,\cdot\,|_{t_n}$ forms a $\Lambda_1^k(\gamma_1|1)$-concordant pair with $\mathfrak{A}(t_n)$ since $\deg[v_j] \le -1$, $j = 1,\dots,\nu$. Further, by Lemma 1.8 we conclude that the seminorms $k^{\vartheta}\|(\mathfrak{A}(t_n) + \mu_0 I)\,\cdot\,\|$ and $\sum_{j=1}^{\nu} |v_j(-kA(t_n))\,\cdot|_{t_n}$ both form $\Lambda_1^k(\gamma_0|1)$-concordant pairs with $\mathfrak{A}(t_n)$, where $\gamma_0 = \max\{1-\vartheta, \gamma_1\}$. These observations clearly yield that the pair $(\|\mathfrak{B}_n\cdot\|; \mathfrak{A}(t_n))$ is $\Lambda_1^k(\gamma_0|1)$-concordant. Also, using Theorem 6.9 gives that[18] $\mathfrak{A}(t_n)$ satisfies hypothesis $\mathbf{H\ddot{O}L}^{(\to)}(\vartheta, 0; \|\!|\cdot|\!\|_{\mathfrak{m};n}, \mathfrak{m} = \emptyset, I, *\emptyset)$ with $\|\!|\cdot|\!\|_{\emptyset;n} = \|(\mathfrak{A}(t_n) + \mu_0 I)\,\cdot\,\|$ and $\|\!|\cdot|\!\|_{*\emptyset;n} = \|\cdot\|_*$ while the seminorm $\|\!|\cdot|\!\|_{I;n} = |(I + kA(t_n))^{-1}\cdot|_{t_n}$ forms a $\Lambda_1^k(\gamma_1|1)$-concordant pair with $\mathfrak{A}(t_n)$, by Theorem 6.1. By the above reasonings, applying Theorems 2.10, 2.16, and 2.11, all of them with $\mathfrak{A}(t_n)$ substituted for $\mathfrak{A}_n$, and taking (6.59) into account, we see that the estimates (6.56) and (6.57) hold for $\xi \in [0,1)$ and the estimate (6.58) holds for $\xi_* \in [0, \min\{\vartheta, 1-\gamma_1\}]$ since the seminorms $\|(\mathfrak{A}(t_n) + \mu_0 I)^{\xi}\,\cdot\,\|$ and $\|(\mathfrak{A}^*(t_n) + \mu_0 I)^{\xi_*}\,\cdot\,\|_*$ are respectively $\Lambda_1^k(\xi|1)$- and $\Lambda_1^k(*;\xi_*|1)$-concordant with $\mathfrak{A}(t_n)$ for any fixed $\xi \in [0,1]$, by Lemma 1.12.

[18] One has to take into account that $\|(\rho(-kA(t)))^*\|_* \le C$ for all $k \in (0,K]$ and $t \in [0,T]$, by Lemma 5.3 and duality.

In the case where $A(t)$ satisfies hypothesis $\mathbf{höl}^{\oplus}[\vartheta]$, instead of (6.61) we have, for all $u \in \mathfrak{X}$,

$$\|\mathfrak{B}_n u\| \leq Ck^{\vartheta}\|(\mathfrak{A}(t_n) + \mu_0 I)\, u\|$$

so that (6.56) and (6.57) hold for $\xi \in [1, 1+\vartheta]$ as well, by Theorems 2.10, 2.16, and 6.9, more precisely, by the latter part of the statement of each one. □

Theorem 6.13. *Suppose that $A(t) \in \mathcal{S}(\varphi, \sigma)$ and the RK method is both of type $V_I(\varphi)$ and of type $V_I^*(\varphi)$, for some $\varphi \in (0, \pi/2)$ and $\sigma \in \mathbb{R}$. Also, let seminorms $|\cdot|_{\oplus;t}$ and $|\cdot|_{\mathfrak{m};t}$, $\mathfrak{m} = \emptyset, 1, *\emptyset$, form respectively $(\gamma_{\oplus}|\varphi,\sigma)$-, $(\gamma|\varphi,\sigma)$-, $(\gamma_1|\varphi,\sigma)$-, and strictly $(*;\gamma_*|\varphi,\sigma)$-concordant pairs with $A(t)$, for some $\gamma_{\oplus}, \gamma, \gamma_1, \gamma_* \in [0,1)$ such that $\gamma + \gamma_* = 1$ and $\gamma_1 + \gamma_* < 1$, and let, with some $\mu_1 \geq 0$ and $\vartheta_{\bullet}, \vartheta_{*\bullet} \in [0,1)$, for all $t \in [0,T]$, $u \in \mathfrak{X}$, and $\chi \in \mathfrak{X}^*$,*[19]

$$k^{\vartheta_{\bullet}}|(A(t) + \mu_1 I)^{-\gamma} u|_{\emptyset;t} \leq C\|u\| \tag{6.62}$$

and

$$k^{\vartheta_{*\bullet}}|((A(t) + \mu_1 I)^{-\gamma_*})^* \chi|_{*\emptyset;t} \leq C\|\chi\|_*. \tag{6.63}$$

Next, letting $\mathcal{A}(t;\cdot,\cdot)$ be a testing functional for $A(t)$, suppose that $A(t)$ satisfies the hypotheses $\mathbf{höl}^{\oplus}_{(k)}[\vartheta_{\oplus}; |\cdot|_{\oplus;t}]$ and $\mathbf{höl}^{(\rightarrow)}(\vartheta, \vartheta_1; \mathcal{A}(t;\cdot,\cdot); |\cdot|_{\mathfrak{m};t}, \mathfrak{m} = \emptyset, 1, *\emptyset)$ for some $\vartheta_{\oplus}, \vartheta \in (0,1]$ and $\vartheta_1 \in [0,1]$, where $\vartheta > \vartheta_{\bullet} + \vartheta_{*\bullet}$. Apart from everything else, suppose that (6.48) holds. Then the stability estimate (6.58) holds for any fixed $\mu_0 > 0$ sufficiently large and for any fixed $\xi_* \in [0, \xi_0)$, where $\xi_0 = 1 + \min\{\vartheta - \gamma, \vartheta_1 - \gamma_1\}$, and it remains valid in the case $\xi_* = \xi_0$ if one adds $t_{n-l}^{-\gamma_*} \log(t_{n-l+1}/k)$ to the factor $t_{n-l}^{-\xi_*}$. Furthermore, if $\vartheta_1 > 0$ in addition, the stability estimates (6.56) and (6.57) hold for any fixed $\mu_0 > 0$ sufficiently large and for any fixed $\xi \in [0, \xi_1)$, where $\xi_1 = 1 - \gamma_* + \min\{\vartheta, \vartheta_1\}$. In the case $\xi = \xi_1$, both estimates retain their shapes with $t_q^{-\xi}$ replaced by $t_q^{-\xi} + t_q^{-\gamma_2} \log(t_{q+1}/k)$, where γ_2 is given by (2.57).

Proof. First of all, to make our below argument correct, we assume that $A(t)$ is extrapolated by $A(t) = A(T)$ for $t > T$.

Further, note that the estimate (6.59) is unchanged in the considered case. At the same time hypothesis $\mathbf{höl}^{(\rightarrow)}(\vartheta, \vartheta_1; \mathcal{A}(t;\cdot,\cdot); |\cdot|_{\mathfrak{m};t}, \mathfrak{m} = \emptyset, 1, *\emptyset)$ for $A(t)$ implies, by Theorem 6.7, that hypothesis $\mathbf{HÖL}^{(\rightarrow)}(\vartheta, \vartheta_1; |\!|\!|\cdot|\!|\!|_{\mathfrak{m};n}, \mathfrak{m} = \emptyset, I, *\emptyset)$ is fulfilled for $\mathfrak{A}(t_n)$ substituted for $\mathfrak{A}_n$, where the seminorms $|\!|\!|\cdot|\!|\!|_{\mathfrak{m};n}$,

[19]Since $A(t) \in \mathcal{S}(\varphi, \sigma)$, the negative fractional powers $(A(t) + \mu_1 I)^{-\tau}$, $\tau \geq 0$, are defined and bounded on $\mathfrak{X}$ if $\mu_1 \geq 0$ is sufficiently large (cf. Appendix A.3).

$\mathfrak{m} = \emptyset, I, *\emptyset$, form respectively $\Lambda_1^k(\gamma)$-, $\Lambda_1^k(\gamma_1)$-, and $\Lambda_1^k(*;\gamma_*)$-concordant pairs with $\mathfrak{A}(t_n)$, due to the conditions imposed on $|\cdot|_{\mathfrak{m};t}$, $\mathfrak{m} = \emptyset, 1, *\emptyset$.

Next, since $\deg[v_j] \leq -1$ and $\deg[w_j] \leq -1$ for $j = 1, \ldots, \nu$, applying Theorems 6.1 and 6.1* yields that the seminorms

$$\|\!|\cdot|\!\|_{\bullet\emptyset;n} = \sum_{j=1}^{\nu} |v_j(-kA(t_n)) \cdot|_{\emptyset;t_n}$$

and

$$\|\!|\cdot|\!\|_{\bullet I;n} = \sum_{j=1}^{\nu} |v_j(-kA(t_n)) \cdot|_{1;t_n}$$

form respectively $\Lambda_1^k(\gamma)$- and $\Lambda_1^k(\gamma_1)$-concordant pairs with the operator $\mathfrak{A}(t_n)$. Moreover, recalling that the pair $\big(|\cdot|_{*\emptyset;t}; A(t)\big)$ is strictly $(*;\gamma_*|\varphi,\sigma)$-concordant, it is readily seen that the seminorm

$$\|\!|\cdot|\!\|_{*\bullet;n} = \sum_{j=1}^{\nu} |(w_j(-kA(t_n)))^* \cdot|_{*\emptyset;t_n+c_jk}$$

forms a $\Lambda_1^k(*;\gamma_*)$-concordant pair with $\mathfrak{A}(t_n)$. Apart from everything else, assuming that we have shown the estimate, with the operator $\mathfrak{B}_n$ given by (5.39), for all $t_n \in \breve{\Omega}_k$, $\chi \in \mathfrak{X}^*$, and $u \in \mathfrak{X}$,

$$|\langle \chi, \mathfrak{B}_n u\rangle| \leq C\|\!|\chi|\!\|_{*\bullet;n}\left(k^{\vartheta}\|\!|u|\!\|_{\bullet\emptyset;n} + k^{\vartheta_1}\|\!|u|\!\|_{\bullet I;n}\right), \tag{6.64}$$

it holds, with $\vartheta_0 = \min\{\vartheta, \vartheta_1\}$, $\gamma_0 = \vartheta_0 + \max\{\gamma - \vartheta, \gamma_1 - \vartheta_1\}$, and $\|\!|\cdot|\!\|_{\bullet;n} = k^{\vartheta-\vartheta_0}\|\!|\cdot|\!\|_{\bullet\emptyset;n} + k^{\vartheta_1-\vartheta_0}\|\!|\cdot|\!\|_{\bullet I;n}$,

$$|\langle \chi, \mathfrak{B}_n u\rangle| \leq Ck^{\vartheta_0}\|\!|\chi|\!\|_{*\bullet;n}\|\!|u|\!\|_{\bullet;n}.$$

Therefore, noting that by Lemma 1.2, the seminorm $\|\!|\cdot|\!\|_{\bullet;n}$ forms a $\Lambda_1^k(\gamma_0)$-concordant pair with $\mathfrak{A}(t_n)$ and taking the above reasonings into account, the claim follows by applying Theorems 2.12 and 2.17 with $\mathfrak{A}(t_n)$ substituted for $\mathfrak{A}_n$, with $\mathfrak{B}_n$ given by (5.39) and with $\gamma_{*0} = \gamma_*$.

It thus remains to show (6.64). To that end, note that as it follows from (5.39), the operator $\mathfrak{B}_n$ can be represented as $\mathfrak{B}_n = \mathfrak{B}_{1;n} + \mathfrak{B}_{2;n}$, where

$$\mathfrak{B}_{1;n} = \mathbf{w}^T(-kA(t_n))\,(\mathcal{D}_n - \mathcal{D}(t_n))\mathbf{v}(-kA(t_n))$$

and

$$\mathfrak{B}_{2;n} = -\mathbf{w}^T(-kA(t_n))\,k(\mathcal{D}_n - \mathcal{D}(t_n))\,(\mathfrak{i} + k\mathfrak{a}\mathcal{D}_n)^{-1}\mathfrak{a} \\ \times (\mathcal{D}_n - \mathcal{D}(t_n))\mathbf{v}(-kA(t_n)). \tag{6.65}$$

It is obvious, since $A(t)$ is under hypothesis $\mathbf{h\ddot{o}l}^{(\rightarrow)}(\vartheta, \vartheta_1; \mathcal{A}(t;\cdot,\cdot); |\cdot|_{\mathfrak{m};t}, \mathfrak{m} = \emptyset, 1, *\emptyset)$, that (6.64) holds with $\mathfrak{B}_{1;n}$ substituted for $\mathfrak{B}_n$. It therefore suffices to show (6.64) with $\mathfrak{B}_{2;n}$ in place of $\mathfrak{B}_n$, which is clearly achieved, with the aid of (6.65) and hypothesis $\mathbf{h\ddot{o}l}^{(\rightarrow)}(\vartheta, \vartheta_1; \mathcal{A}(t;\cdot,\cdot); |\cdot|_{\mathfrak{m};t}, \mathfrak{m} = \emptyset, 1, *\emptyset)$, if we prove that with $\mathbf{e}_j = \left(\delta_{j1}, \ldots, \delta_{j\nu}\right)^T$, $j = 1, \ldots, \nu$, for all $t_n \in \breve{\Omega}_k$ and $\mathbf{u} = \left(u_1, \ldots, u_\nu\right)^T \in \underbrace{\mathcal{D}om \times \cdots \times \mathcal{D}om}_{\nu}$,

$$\sum_{j=1}^{\nu} |\mathbf{e}_j^T (\mathfrak{i} + k\mathfrak{a}\mathcal{D}_n)^{-1} \mathfrak{a} k(\mathcal{D}_n - \mathcal{D}(t_n))\mathbf{u}|_{\bullet;t_n} \leq C \sum_{j=1}^{\nu} |u_j|_{\bullet;t_n}, \tag{6.66}$$

where it is denoted $|\cdot|_{\bullet;t} = k^{\vartheta}|\cdot|_{\emptyset;t} + k^{\vartheta_1}|\cdot|_{1;t}$. For showing (6.66), we use formula (5.49), with both sides multiplied from the right by $k(\mathcal{D}_n - \mathcal{D}(t_n))\mathbf{u}$, to obtain

$$\begin{aligned}(\mathfrak{i} + k\mathfrak{a}\mathcal{D}_n)^{-1} \mathfrak{a} k(\mathcal{D}_n - \mathcal{D}(t_n))\mathbf{u} &= (\mathfrak{i} + k\mathfrak{a}\mathcal{D}(t_n))^{-1} \mathfrak{a} k(\mathcal{D}_n - \mathcal{D}(t_n))\mathbf{u} \\ &- (\mathfrak{i} + k\mathfrak{a}\mathcal{D}(t_n))^{-1} \mathfrak{a} k(\mathcal{D}_n - \mathcal{D}(t_n)) (\mathfrak{i} + k\mathfrak{a}\mathcal{D}_n)^{-1} \mathfrak{a} k(\mathcal{D}_n - \mathcal{D}(t_n))\mathbf{u}.\end{aligned}$$

With this equality in mind, it is not hard to see, applying the kicking-back trick, that (6.66) will be verified if we show instead that for all $t_n \in \breve{\Omega}_k$ and $\mathbf{u} = \left(u_1, \ldots, u_\nu\right)^T \in \underbrace{\mathcal{D}om \times \cdots \times \mathcal{D}om}_{\nu}$,

$$\sum_{j=1}^{\nu} |\mathbf{e}_j^T (\mathfrak{i} + k\mathfrak{a}\mathcal{D}(t_n))^{-1} \mathfrak{a} k(\mathcal{D}_n - \mathcal{D}(t_n))\mathbf{u}|_{\bullet;t_n} \leq \frac{1}{2} \sum_{j=1}^{\nu} |u_j|_{\bullet;t_n}. \tag{6.67}$$

Since the entries of the matrix $(\mathfrak{i} - z\mathfrak{a})^{-1}\mathfrak{a}$ vanish as $|z| \to \infty$, (6.67) is, in fact, a consequence of the following fact. For any fixed rational function $\omega(z)$ which is (φ)-regular and satisfies the condition $\deg[\omega] \leq -1$, it holds, with some $\varepsilon > 0$, for all $k \in (0, K]$, $t \in [0, T]$, and $u \in \mathcal{D}om$,

$$|\omega(-kA(t))k(A(t + ck) - A(t))u|_{\bullet;t} \leq Ck^{\varepsilon}|u|_{\bullet;t}, \tag{6.68}$$

so that the factor Ck^{ε} on the right can be taken arbitrarily small at the expense of choosing suitably $K > 0$. Observe now that if $K > 0$ is sufficiently small, the negative fractional powers $(I + kA(t))^{-\tau}$, $\tau > 0$, of the operator $I + kA(t)$ are defined and bounded on $\mathfrak{X}$ for $k \in (0, K]$ (cf. Appendix A.3). Furthermore, since $\gamma_1 + \gamma_* < 1$, there exist $\gamma_1' > \gamma_1$ and $\gamma_*' > \gamma_*$ such that $\gamma_1' + \gamma_*' = 1$. Using the fact that the seminorms $|\cdot|_{1;t}$ and $|\cdot|_{*\emptyset;t}$ form respectively $(\gamma_1|\varphi, \sigma)$- and strictly $(*; \gamma_*|\varphi, \sigma)$-concordant pairs with $A(t)$, similarly to Lemma 1.9 it can be shown that for all $k \in (0, K]$, $t \in [0, T]$, $u \in \mathfrak{X}$, and $\chi \in \mathfrak{X}^*$,

$$k^{\gamma_1} \sup_{\|u\|=1} |(I + kA(t))^{-\gamma_1'} u|_{\emptyset;t} + k^{\gamma_*} \sup_{\|\chi\|_*=1} |((I + kA(t))^{-\gamma_*'})^* \chi|_{*\emptyset;t+ck} \leq C. \tag{6.69}$$

Also, it follows from Lemma 5.3 that

$$\|\omega_1(-kA(t))\| \leq C \quad \text{for all } k \in (0, K] \text{ and } t \in [0, T], \tag{6.70}$$

where $\omega_1(z) = (1 - z)\,\omega(z)$.

Now, with the aid of the identity[20]

$$\omega(-kA(t)) = (I + kA(t))^{-\gamma} \omega_1(-kA(t))(I + kA(t))^{-\gamma_*},$$

applying the estimates (6.62), (6.63), and (6.70), we therefore obtain, for all $k \in (0, K]$, $t \in [0, T]$, and $u \in \mathcal{Dom}$,

$$\begin{aligned} k^{\vartheta} &|\omega(-kA(t))k(A(t + ck) - A(t))u|_{\emptyset;t} \\ &\leq Ck^{1+\vartheta-\vartheta_\bullet-\gamma} \|\omega_1(-kA(t))\,(I + kA(t))^{-\gamma_*} k(A(t + ck) - A(t))u\| \\ &\quad \leq Ck^{1+\vartheta-\vartheta_\bullet-\gamma} \|(I + kA(t))^{-\gamma_*} k(A(t + ck) - A(t))u\| \\ &\qquad \leq Ck^{\vartheta-\vartheta_\bullet-\vartheta_{*\bullet}} \left(k^{\vartheta} |u|_{\emptyset;t} + k^{\vartheta_1} |u|_{1;t}\right). \end{aligned} \tag{6.71}$$

In a similar manner since $\gamma_1' + \gamma_*' = 1$, applying the identity

$$\omega(-kA(t)) = (I + kA(t))^{-\gamma_1'} \omega_1(-kA(t))(I + kA(t))^{-\gamma_*'},$$

and using (6.69) and (6.70) and the above argument, we get, for all $k \in (0, K]$, $t \in [0, T]$, and $u \in \mathcal{Dom}$,

$$\begin{aligned} k^{\vartheta_1} &|\omega(-kA(t))\,k(A(t + ck) - A(t))u|_{1;t} \\ &\leq Ck^{\vartheta_1+1-\gamma_1-\gamma_*} \left(k^{\vartheta} |u|_{\emptyset;t} + k^{\vartheta_1} |u|_{1;t}\right). \end{aligned} \tag{6.72}$$

Finally, (6.68) follows with $\varepsilon = \min\{\vartheta - \vartheta_\bullet - \vartheta_{*\bullet}, 1 - \gamma_1 - \gamma_*\}$ by combining (6.71) and (6.72), noting that $\varepsilon > 0$ since $\vartheta_\bullet + \vartheta_{*\bullet} < \vartheta$ and $\gamma_1 + \gamma_* < 1$. □

Theorem 6.14. *Suppose that $A(t) \in \mathcal{S}(\varphi, \sigma)$ and the RK method is both of type $V_I(\varphi)$ and of type $V_I^*(\varphi)$, for some $\varphi \in (0, \pi/2)$ and $\sigma \in \mathbb{R}$. Let further seminorms $|\cdot|_{\oplus;t}$ and $|\cdot|_{\mathfrak{m};t}$, $\mathfrak{m} = *\emptyset, *1, \emptyset$, form respectively $(\gamma_\oplus|\varphi, \sigma)$-,*

[20] We recall that $\gamma + \gamma_* = 1$.

$(*;\gamma_*|\varphi,\sigma)$-, $(*;\gamma_{*1}|\varphi,\sigma)$-, and strictly $(\gamma|\varphi,\sigma)$-concordant pairs with $A(t)$ for some $\gamma_\oplus,\gamma_*,\gamma_{*1},\gamma\in[0,1)$ such that $\gamma+\gamma_*=1$ and $\gamma+\gamma_{*1}<1$, and let the inequalities (6.62) and (6.63) hold for some $\mu_1\geq 0$ and $\vartheta_\bullet,\vartheta_{*\bullet}\in[0,1)$. Also, letting $\mathcal{A}(t;\cdot,\cdot)$ be a testing functional for $A(t)$, suppose that $A(t)$ satisfies the hypotheses $\mathbf{h\ddot{o}l}^{\oplus}_{(k)}[\vartheta_\oplus;|\cdot|_{\oplus;t}]$ and $\mathbf{h\ddot{o}l}^{(\leftarrow)}(\vartheta,\vartheta_1;\mathcal{A}(t;\cdot,\cdot);|\cdot|_{\mathfrak{m};t},\mathfrak{m}=\emptyset,*\emptyset,*1)$ for some $\vartheta_\oplus,\vartheta\in(0,1]$ and $\vartheta_1\in[0,1]$, where $\vartheta>\vartheta_\bullet+\vartheta_{*\bullet}$. Apart from everything else, suppose that (6.48) holds. Then the stability estimates (6.56) and (6.57) hold for any fixed $\mu_0>0$ sufficiently large and for any fixed $\xi\in[0,\xi_0)$, where $\xi_0=1+\min\{\vartheta-\gamma_*,\vartheta_1-\gamma_{*1}\}$, and they remain valid in the case $\xi=\xi_0$ if one adds $t_q^{-\gamma}\log(t_{q+1}/k)$ to the factor $t_q^{-\xi}$. Furthermore, if in addition $\vartheta_1>0$, the stability estimate (6.58) holds for any fixed $\mu_0>0$ sufficiently large and for any fixed $\xi_*\in[0,\xi_1)$, where $\xi_1=1-\gamma+\min\{\vartheta,\vartheta_1\}$. In the case $\xi_*=\xi_1$, this estimate retains its shape with $t_{n-l}^{-\xi_*}$ replaced by $t_{n-l}^{-\xi_*}+t_{n-l}^{-\gamma_{*2}}\log(t_{n-l+1}/k)$, where γ_{*2} is given by (2.74).

The proof is similar to that of Theorem 6.13.

In principle, in the statement of Theorem 6.13 we may withdraw the assumptions (6.62) and (6.63) since using the argument employed in the proof of Theorem 6.12 as well as hypothesis $\mathbf{h\ddot{o}l}^{\oplus}_{(k)}[\vartheta_\oplus;|\cdot|_t]$ leads to the fact that (2.54) holds with $\vartheta_0=\gamma_{*0}=0$ and $\gamma_0=\{1-\vartheta_\oplus,\gamma_\oplus\}$. Therefore, applying again Theorem 2.12, one can state an analogue of Theorem 6.13 in which the conditions (6.62) and (6.63) are removed but it is required instead that $\gamma_0<1-\gamma_*+\min\{\vartheta,\vartheta_1\}$, $\xi_*\leq 1-\gamma_0$, and $\xi\leq 1$ in addition to what is already accepted, where in the limiting cases $\xi_*=1-\gamma_0$ and $\xi=1$, the stability estimates are accordingly corrected. In a similar manner one can show a modification of Theorem 6.14 by applying Theorem 2.13.

We do not intend to examine methods of class S_{II} in the case with a variable operator for the reason that such a consideration would contain many additional technical difficulties while the results do not seem to be widely applicable. Now we turn to the study of methods of class S_{III}.

Theorem 6.15. *Suppose that $A(t)\in\mathcal{S}(\varphi,\sigma)$ and that the RK method is both of type $V_{III}(\varphi)$ and of type $V^*_{III}(\varphi)$ and satisfies the condition $\deg[\alpha(\cdot)-\alpha(\infty)]=-1$, for some $\varphi\in(0,\pi/2)$ and $\sigma\in\mathbb{R}$. Suppose further that a seminorm $|\cdot|_t$ forms a $(\gamma_1|\varphi,\sigma)$-concordant pair with the operator $A(t)$ and $A(t)$ itself satisfies hypothesis $\mathbf{h\ddot{o}l}^{\oplus}[\vartheta;|\cdot|_t]$, for some $\gamma_1\in[0,1)$ and $\vartheta\in(0,1]$. Finally, let $\omega(z)$ be a rational function which is (φ)-regular and subject to $\deg[\omega]\leq -1$. Then it holds, with $\mathfrak{V}_n=\omega(-kA(t_n))$ and $\mathfrak{V}_{-1}=\mathfrak{V}_0$, with any fixed $\mu_0>0$ sufficiently large, and with any fixed $\xi\in[0,1)$, for all*

$k \in (0, K]$ and $t_n, t_l \in \Omega_k$ such that $t_l \le t_n$,

$$\|(\mathfrak{A}(t_n) + \mu_0 I)^{\xi} \mathfrak{U}_{n-1,l} \mathfrak{V}_{l-1}\| \le C t_{n-l+1}^{-\xi}. \tag{6.73}$$

Also, if the operator $\mathfrak{R}$ in (5.19) is given by $\mathfrak{R} = \omega(-kA(0))$, then the solution Y_n of (5.19), (5.20) satisfies the stability estimate (6.57).

Proof. First of all note that by Lemma 5.3, it suffices to show the claim in the particular case $\omega(z) = (1 - z)^{-1}$ which is equivalent to the fact that $\mathfrak{V}_n = (I + kA(t_n))^{-1}$ for $t_n \in \breve{\Omega}_k$. From now on we therefore accept the last assumption.

Further, by Theorem 6.6 the operator $\mathfrak{A}(t_n)$ satisfies the Λ_2^k-condition, while using Theorem 6.9 yields that $\mathfrak{A}(t_n)$ substituted for $\mathfrak{A}_n$ satisfies hypothesis $\mathbf{H\ddot{O}L}^{(\to)}(\vartheta, 0; \|\!|\cdot\|\!|_{\mathfrak{m};n}, \mathfrak{m} = \emptyset, I, *\emptyset)$, where with some $\mu_0 > 0$,

$$\|\!|\cdot\|\!|_{\mathfrak{m};n} = \begin{cases} \|(\mathfrak{A}(t_n) + \mu_0 I)\cdot\| & \text{if} \quad \mathfrak{m} = \emptyset, \\ |(I + kA(t_n))^{-1}\cdot|_{t_n} & \text{if} \quad \mathfrak{m} = I, \\ \|((I + kA(t_n))^{-1})^{*}\cdot\|_{*} & \text{if} \quad \mathfrak{m} = *\emptyset, \end{cases}$$

since the seminorm $\|(\rho(-kA(t_n)))^{*}\cdot\|_{*}$ appearing in (6.43) is replaced by $\|((I + kA(t_n))^{-1})^{*}\cdot\|_{*}$, by Lemma 5.3 and duality. Observe that the triplet $(\|(\mathfrak{A}(t_n) + \mu_0 I)\cdot\|; \mathfrak{A}(t_n); (I + kA(t_n))^{-1})$ is $\Lambda_2^k(1)$-concordant, by Theorem 6.6, while the triplet $(|(I + kA(t_n))^{-1}\cdot|_{t_n}; \mathfrak{A}(t_n); (I + kA(t_n))^{-1})$ is $\Lambda_2^k(\gamma_1)$-concordant, by Theorem 6.5. Also, a slightly more delicate calculation than the one leading to (6.61) gives, again with the aid of (5.39) and (5.43), for all $t_n \in \breve{\Omega}_k$, $\chi \in \mathfrak{X}^*$, and $u \in \mathfrak{X}$,

$$|\langle \chi, \mathfrak{B}_n u\rangle| \le C \sum_{j=1}^{\nu} \|(w_j(-kA(t_n)))^{*}\chi\|_{*} \times \left(k^{\vartheta} \|(\mathfrak{A}(t_n) + \mu_0 I)\, u\| + \sum_{j=1}^{\nu} |v_j(-kA(t_n))\, u|_{t_n} \right),$$

where the seminorms $\sum_{j=1}^{\nu} \|(w_j(-kA(t_n)))^{*}\cdot\|_{*}$ and $\sum_{j=1}^{\nu} |v_j(-kA(t_n))\cdot|_{t_n}$ may be replaced, by Lemma 5.3, respectively by $\|((I + kA(t_n))^{-1})^{*}\cdot\|_{*}$ and by $|(I + kA(t_n))^{-1}\cdot|_{t_n}$, preserving the needed properties since $\deg[w_j] \le -1$ and $\deg[v_j] \le -1$ for $j = 1, \ldots, \nu$.

Furthermore, it is easily checked, using the assumed Hölder-continuity hypothesis for $A(t)$ and arguing as above, that for all $t_n, t_l \in \breve{\Omega}_k$,

$$\begin{aligned}
\|\mathfrak{V}_n^{-1}\mathfrak{V}_l\| &= \|(I+kA(t_n))\,(I+kA(t_l))^{-1}\| \\
&= \|I+k(A(t_l)-A(t_n))\,(I+kA(t_l))^{-1}\| \\
&\le 1+Ck\|(A(t_l)+\mu_1 I)\,(I+kA(t_l))^{-1}\| \\
&\quad +Ck\sup_{\|u\|=1}|(I+kA(t_l))^{-1}u|_{t_l} \\
&\le C. \qquad (6.74)
\end{aligned}$$

Therefore, using the above reasonings, the stability estimate (6.73) follows by Theorem 2.15 since the triplet $\big(\|(\mathfrak{A}(t_n)+\mu_0 I)^{\xi}\cdot\|;\mathfrak{A}(t_n);(I+kA(t_n))^{-1}\big)$ is $\Lambda_2^k(\xi|1)$-concordant, in view of Theorem 6.6.

To complete the proof, F_n and Y_0 are represented by

$$F_n = \mathfrak{V}_n\widetilde{F}_n \text{ and } Y_0 = \mathfrak{V}_{-1}\widetilde{y}^0,$$

with $\widetilde{F}_n = (I+kA(t_n))F_n$ and $\widetilde{y}^0 = (I+kA(0))\mathfrak{R}y^0$. Using now (5.41) and (5.43) gives

$$\|\widetilde{F}_n\| \le C\max_{j=1,\dots,\nu}\|f(t_{nj})\|, \qquad (6.75)$$

while, by (6.74), we have

$$\|\widetilde{y}^0\| \le C\|y^0\|. \qquad (6.76)$$

With the aid of (6.75) and (6.76), and the above observations, it is readily seen now that (6.57) follows by Theorem 2.19. □

Theorem 6.16. *Suppose that $A(t)\in\mathcal{S}(\varphi,\sigma)$ and that the RK method is both of type $V_{III}(\varphi)$ and of type $V_{III}^*(\varphi)$ and satisfies the condition $\deg[\alpha(\cdot)-\alpha(\infty)]=-1$, for some $\varphi\in(0,\pi/2)$ and $\sigma\in\mathbb{R}$. Let further $\mathcal{A}(t;\cdot,\cdot)$ be a testing functional for $A(t)$, let seminorms $|\cdot|_{\oplus;t}$ and $|\cdot|_{\mathfrak{m};t}$, $\mathfrak{m}=\emptyset,1,*\emptyset$, form respectively $(\gamma_\oplus|\varphi,\sigma)$-, $(1/2|\varphi,\sigma)$-, $(\gamma_1|\varphi,\sigma)$-, and strictly $(*;1/2|\varphi,\sigma)$-concordant pairs with $A(t)$, and let the operator $A(t)$ itself satisfy the hypotheses* $\mathbf{h\ddot{o}l}^{\oplus}_{(k)}[\vartheta_\oplus;|\cdot|_{\oplus;t}]$ *and* $\mathbf{h\ddot{o}l}^{(\rightarrow)}(\vartheta,\vartheta_1;\mathcal{A}(t;\cdot,\cdot);|\cdot|_{\mathfrak{m};t},\mathfrak{m}=\emptyset,1,*\emptyset)$, *for some $\vartheta,\vartheta_\oplus\in(0,1]$, $\vartheta_1\in[0,1]$, $\gamma_\oplus\in[0,1)$, and $\gamma_1\in[0,1/2)$. Also, suppose that the conditions (6.62) and (6.63) hold with $\gamma=\gamma_*=1/2$ and with some $\vartheta_\bullet,\vartheta_{*\bullet}\in[0,1)$ such that $\vartheta_\bullet+\vartheta_{*\bullet}<\vartheta$. Apart from everything else, let $\omega(z)$ be a (φ)-regular rational function. Then, with $\mathfrak{V}_n=\omega(-kA(t_n))$, for any fixed $\mu_0>0$ sufficiently large and for any fixed $\xi_*\in[0,\xi_0)$, where $\xi_0=1+\min\{\vartheta-1/2,\vartheta_1-\gamma_1\}$, if $\deg[\omega]\le-\xi_*$, we have, for all $k\in(0,K]$ and $t_n,t_l\in\Omega_k$ such that $t_l<t_n$,*

$$\|\mathfrak{V}_l^*(\mathfrak{A}^*(t_l)+\mu_0 I)^{\xi_*}\mathfrak{U}^*_{n-1,l+1}\|_* \le Ct_{n-l}^{-\xi_*}.$$

In the case $\xi_* = \xi_0$, *the last estimate is still in force with* $t_{n-l}^{-\xi_*}$ *replaced by* $t_{n-l}^{-\xi_*} + t_{n-l}^{-1/2}\log(t_{n-l+1}/k)$. *Furthermore, if in addition* $\vartheta_1 > 0$ *and the operator* $\mathfrak{R}$ *in (5.19) satisfies condition (6.48), then we have, for any fixed* $\mu_0 > 0$ *sufficiently large and for any fixed* $\xi \in [0, 1/2+\min\{\vartheta, \vartheta_1\})$ *such that* $\deg[\omega] \le -\xi$, *for all* $k \in (0, K]$ *and* $t_n, t_l \in \Omega_k$ *such that* $t_l \le t_n$, *that*

$$\|\mathfrak{V}_n(\mathfrak{A}(t_n) + \mu_0 I)^\xi \mathfrak{U}_{n-1,l}\| \le C t_{n-l+1}^{-\xi}$$

and

$$\|\mathfrak{V}_n(\mathfrak{A}(t_n) + \mu_0 I)^\xi Y_n\| \le C\left(t_{n+1}^{-\xi}\|y^0\| + k\sum_{l=0}^{n-1} t_{n-l}^{-\xi}\max_{j=1,\ldots,\nu}\|f(t_{lj})\|\right),$$

where Y_n *is the solution of (5.19), (5.20). In the case* $\xi = 1/2+\min\{\vartheta, \vartheta_1\}$, *both estimates remain valid if one adds* $t_q^{-\gamma_2}\log(t_{q+1}/k)$ *to* $t_q^{-\xi}$ *for* $q = n+1$ *and* $q = n-l$, *where*

$$\gamma_2 = \begin{cases} 1/2 & \text{if } \vartheta \le \vartheta_1, \\ \gamma_1 & \text{if } \vartheta > \vartheta_1. \end{cases}$$

Proof. In fact, the result can be shown with the aid of just the same reasonings as those used in the proofs of Theorems 6.12 and 6.13, applying as well Theorems 6.1, 6.1*, 6.6, 6.8, 2.12, and 2.17. Note that the fact that the pairs $(\|\mathfrak{V}_n(\mathfrak{A}(t_n)+\mu_0 I)^\xi\cdot\|; \mathfrak{A}(t_n))$ and $(\|\mathfrak{V}_n^*(\mathfrak{A}^*(t_n)+\mu_0 I)^{\xi_*}\cdot\|; \mathfrak{A}(t_n))$ are $\Lambda_2^k(\xi)$- and $\Lambda_2^k(*;\xi_*)$-concordant, respectively, can easily be verified by using Theorems 6.6 and 6.6* since the operator $\mathfrak{V}_n$ is permutable with $(\mathfrak{A}(t_n)+\mu_0 I)^\xi$ and $(\lambda I - k\mathfrak{A}(t_n))^{-1}$. □

The interested reader can state without difficulty an analogue of Theorem 6.16 in the situation when hypothesis $\mathbf{h\ddot{o}l}^{(\rightarrow)}(\ldots)$ is replaced by hypothesis $\mathbf{h\ddot{o}l}^{(\leftarrow)}(\ldots)$.

Remark 6.6. *In our considerations carried out within an abstract framework, we are forced to accept from time to time some differently stated hypotheses as independent ones, although they are not really such in actual practical applications. For example, if* $A(t)$ *is thought of as an approximation to a multidimensional elliptic partial differential operator, the above inequalities (6.62) and (6.63) are akin, in fact, to assuming hypothesis* $\mathbf{h\ddot{o}l}_{(k)}^{\oplus}(\ldots)$.

Remark 6.7. *Applying Lemma 1.11 shows that the assertions of Theorems 6.12 —–6.16 remain valid if instead of the seminorms* $\|(\mathfrak{A}(t_n)+\mu_0 I)^\xi\cdot\|$ *and* $\|(\mathfrak{A}^*(t_n)+\mu_0 I)^{\xi_*}\cdot\|_*$ *we use* $\|(\mathfrak{A}_n+\mu_0 I)^\xi\cdot\|$ *and* $\|(\mathfrak{A}_n^*+\mu_0 I)^{\xi_*}\cdot\|_*$, *respectively, under slightly stronger restrictions on* ξ *and* ξ_**, except Theorem 6.15 which needs no additional assumptions. More precisely, one then has to require in addition that* $\xi_* < \xi_0$, $\xi < \xi_1$ *in the statements of Theorems 6.13 and 6.16, that* $\xi < \xi_0$, $\xi_* < \xi_1$ *in the statement of Theorem 6.14, and that* $\xi_* < \min\{\vartheta, 1-\gamma_1\}$ *in the statement of Theorem 6.12. Also, as concerns the statement of Theorem 6.12 in the particular case where* $A(t)$ *satisfies hypothesis* **höl**$[\vartheta]$*, it should in addition be required that* $\xi < 1+\vartheta$.

Remark 6.8. *The results of Section 6.1 allow us to obtain stability estimates involving some different seminorms in place of* $\|(\mathfrak{A}(t_n)+\mu_0 I)^{\xi}\cdot\|$, $\|\mathfrak{B}_n(\mathfrak{A}(t_n)+\mu_0 I)^{\xi}\cdot\|$, $\|(\mathfrak{A}^*(t_l)+\mu_0 I)^{\xi_*}\cdot\|_*$*, and* $\|\mathfrak{B}_l^*(\mathfrak{A}^*(t_l)+\mu_0 I)^{\xi_*}\cdot\|_*$. *More precisely, as concerns, for instance, the first two mentioned seminorms,*[21] *let a seminorm* $]\![\cdot]\![_t$ *form a* $(\xi|\varphi,\sigma)$*-concordant pair with* $A(t)$*. Define the seminorm*

$$]\!]\!\cdot\!]\!]_n =]\![\psi(-kA(t_n))\cdot]\![_{t_n}, \tag{6.77}$$

where $\psi(z)$ *is a* (φ)*-regular rational function that is under the condition* $\deg[\psi] \le -\lfloor\xi\rfloor - 1$*, which is replaced by the condition* $\deg[\psi] \le -\xi$ *if* $\xi \in \mathbb{N}$ *and if the seminorm* $]\![\cdot]\![_t$ *forms a* $(\xi|\xi|\varphi,\sigma)$*-concordant pair with* $A(t)$*. Then, with* $]\!]\!\cdot\!]\!]_n$ *substituted for* $\|(\mathfrak{A}(t_n)+\mu_0 I)^{\xi}\cdot\|$ *or for* $\|\mathfrak{B}_n(\mathfrak{A}(t_n)+\mu_0 I)^{\xi}\cdot\|$*, the assertions of Theorems 6.10, 6.11, 6.12, 6.13, 6.14, and 6.16 remain valid, assuming in addition, if the RK method is of class* S_{II} *or of class* S_{III}*, that* $\deg[\psi] \le -\xi(\varpi+1)$*, where* $\varpi = -\deg[\alpha(\cdot)-\alpha(\infty)]$ *and the last nonstrict inequality is replaced by the corresponding strict one if* ξ *is an integer. Moreover, if (6.21) holds, then for* $\xi = 0$ *it is enough to accept that* $\deg[\psi] \le 0$ *instead of all the above restrictions on* $\deg[\psi]$*. With the same substitution, the assertion of Theorem 6.15 remains valid provided that* $\deg[\psi] \le 0$ *if* $\xi = 0$ *and* $\deg[\psi] \le 1 - 2\xi$ *otherwise.*

The above assertion is, in fact, a consequence of Theorems 6.1, 6.3, and 6.5.

Note that in virtue of the above analysis of concrete families of RK methods (see Section 5.2), Theorems 6.10, 6.12, 6.13, and 6.14 are applicable for all Radau IA, IIA and Lobatto IIIC methods as well as for all RK methods whose Butcher array is specified by (5.26), provided that condition (5.27) holds. At the same time all Gauss-Legendre methods of odd order can serve as examples of application of Theorems 6.11, 6.15, and 6.16. Also, Gauss-Legendre methods of even order and Lobatto IIIA, IIIB methods of odd order lead to rational approximations which are appropriate for dealing with the assertion of Theorem 6.10 and rational schemes on the basis of

[21] The other two seminorms can be replaced in a similar manner.

Lobatto IIIA, IIIB methods of even order can be used in the format of Theorem 6.11.

6.4 Comments and bibliographical remarks

Section 6.1: As concerns Theorems 6.1, 6.3, and 6.5, what the reader meets here is their maiden presentation, while their particular versions, that is, Theorems 6.2, 6.4, and 6.6 are actually present in our old work [16, 18, 22]. The proof of Theorem 6.1 combines some ideas from [37] with those employed in [18] (see also [35]).
Section 6.2: Apparently, Theorems 6.7 and 6.8 are new. At the same time our earlier work [16, 19, 25, 26] contains assertions that are quite close to Theorem 6.9. This group of results is, however, essentially based on applying the hypotheses like $\mathbf{h\ddot{o}l}^{\oplus}[\vartheta; |\cdot|_t]$, which is of limited use in practice. (See also the comment to Section 6.3.)
Section 6.3: In the situation when $A(t) \equiv A$ is independent of t,[22] under the assumption of $A(\varphi)$-stability and under the condition $|r(\infty)| < 1$, a stability estimate which is equivalent to (6.49) with $\xi = 0$, is first stated, apparently, in Brenner & Thomée [52].[23] (For later references, we also mention Larsson, Thomée, and Wahlbin [100] and Lubich & Nevanlinna [104].) It is worth noting, for fairness' sake, that the same achievement (with $\xi = 0$) is presented in a Hilbert space framework by Le Roux [102] whose techniques are, nevertheless, extended to the Banach case. It turns out that the restriction $|r(\infty)| < 1$ is not crucial, and it is avoided in our work [10, 11, 12, 16, 18, 22], where the corresponding non-strict inequality is allowed as well. Later, a similar result, for $\xi = 0$ only, is shown in Crouzeix *et al.* [61] and Palencia [127], by using more straightforward techniques. The general case $\xi \geq 0$ has been examined by the author in [10, 11, 12, 16, 18, 22], and Theorems 6.10 and 6.11 arise from there. Also, we mention two recent papers of Hansbo [82, 83] who in particular presents certain strong stability type estimates. As concerns her results for single step methods, it is easy, however, to deduce them directly from [16, 18, 22]. In the matter of the assertions collected in the present chapter, it is seen that if $\xi > 0$, we have to distinguish between the classes S_I, S_{II} on the one hand and S_{III} on the other hand since for the last class of methods additional damping is required. At the same time, when using general seminorms of the form

[22] One actually handles then the rational method (5.30).

[23] In principle, the paper [52] deals with rational approximations of general (not necessarily holomorphic) C_0 semigroups.

(6.77), the role of a damping factor played by the operator $\psi(-kA(t_n))$ is distinguishable in a different manner for these classes (see Remark 6.8).

Turning now to the history of the question in the situation with a variable operator, we mention, first of all, certain early results in Gudovich & Terteryan [78], Sobolevskii [149, 152], and Thomée & Wahlbin [163], which deal with concrete problems or concrete narrow families of discretization methods. For results of general character, we refer to our work [16, 19, 25, 26] and to González & Palencia [73, 74]. We remark that the papers [73, 74] which have appeared later contain no improvements in the properly Hölder case in comparison with our results.[24] The assertions of the present Theorems 6.12 and 6.15 follow in essence our investigation in [16, 19, 25, 26] while Theorems 6.13, 6.14, and 6.16 are presented for the first time here. We remark that the assertions of Theorems 6.12 and 6.15 are of limited use in practice and the same can be said regarding all results of González & Palencia [73, 74]. The point is that those results are based on putting to use Hölder-continuity type hypotheses which cannot be verified for partial differential operators and their discrete approximations in some useful function spaces, for instance, in those with maximum norm, unless we are dealing with the one-dimensional case. At the same time the assertions of Theorems 6.13, 6.14, and 6.16 are free of that disadvantage.

[24]In fact, the situation considered in [73, 74] is even restricted to the case where $\xi = 0$ and the condition $|r(\infty)| < 1$ is assumed.

Chapter 7

Convergence Estimates

Since stability estimates have been shown, the question of convergence is easily settled if the input data of the starting differential problem (5.1) are sufficiently smooth. In applications connected with PDE problems, we meet, however, the phenomenon of order reduction when the order of convergence observed is smaller than the classical one, unless certain compatibility conditions are imposed on the input data. Apparently, Crouzeix [59] is the first to examine this phenomenon in detail in the context of Hilbert spaces. Below we study different possibilities, depending on the smoothness of the input data, for showing error estimates by using Banach space techniques. In what follows we are particularly interested in dealing with input data not being smooth enough, that is, when the classical order of consistancy is not achieved.

In this chapter, given numbers $0 \leq \tau_1 \leq \tau_2$ and $\kappa \in [1, \infty]$, we denote by $L_\kappa(\tau_1, \tau_2; \mathfrak{X})$ the Banach space of strongly measurable functions $\psi(s) : [\tau_1, \tau_2] \to \mathfrak{X}$ with the norm

$$\|\psi(.)\|_{L_\kappa(\tau_1,\tau_2;\mathfrak{X})} = \begin{cases} \left(\int_{\tau_1}^{\tau_2} \|\psi(s)\|^\kappa ds\right)^{1/\kappa} & \text{if } 1 \leq \kappa < \infty, \\ \operatorname{ess\,sup}_{s\in[\tau_1,\tau_2]} \|\psi(s)\| & \text{if } \quad \kappa = \infty. \end{cases}$$

7.1 Preliminaries

It is known (see Appendix B) that under certain conditions there exists a unique solution of problem (5.1),[1] where the operator $A(t)$ depends on t. Sometimes it is also possible to point out conditions under which the solution

[1]By solution we always mean classical solution (cf. Appendix B).

has the desired order of regularity (see, e.g., Sobolevskii [148]). However for the reasons explained in Appendix B, we never use such conditions below. Instead we posit the unique solvability of the starting problem (5.1) as well as the fact that the solution is as smooth as needed in order to achieve the desired order of convergence. At the same time we keep the conditions that ensure the unique solvability and stability of the discrete problem (5.19), (5.20) so that our main assumptions will be that $A(t)$ is (φ, σ)-sectorial[2] for some $\varphi \in (0, \pi/2)$, $\sigma \in \mathbb{R}$ and satisfies certain Hölder-continuity type conditions, among those stated in Chapter 5. Acting along these lines, we show error estimates in terms of the solution. On the other hand, in the autonomous case $A(t) \equiv A$, we are able, by using the concept of extended solution of problem (5.1) (see Appendix B), to obtain error estimates in terms of the data.

Note further that the coefficient matrix $\mathfrak{a} = (a_{jl})_{j,l=1}^{\nu}$ and the vector $\mathbf{b}^T = (b_1, \ldots, b_\nu)^T$ are associated with the quadrature formulas

$$\int_0^k \psi(s)\, ds \approx k \sum_{j=1}^{\nu} b_j \psi(c_j k) \tag{7.1}$$

and

$$\int_0^{c_j k} \psi(s)\, ds \approx k \sum_{l=1}^{\nu} a_{jl} \psi(c_l k), \quad j = 1, \ldots, \nu, \tag{7.2}$$

where $\psi(s)$ may be thought of as an $(\mathfrak{X})$-valued function.[3] It is well known (see, e.g., [53, 65]) that stating the simplifying assumptions $\mathcal{B}(p)$ and $\mathcal{C}(q)$ leads to the fact that the quadrature formulas (7.1) and (7.2) are precise for all polynomials of degree $p-1$ and $q-1$, respectively. For subsequent use we denote

$$Q_n(\psi) = -\left(\int_{t_n}^{t_{n+1}} \psi(s)\, ds - k \sum_{j=1}^{\nu} b_j \psi(t_n + c_j k) \right)$$

and

$$Q_{nj}(\psi) = -\left(\int_{t_n}^{t_n + c_j k} \psi(s)\, ds - k \sum_{l=1}^{\nu} a_{jl} \psi(t_n + c_l k) \right), \quad j = 1, \ldots, \nu.$$

---

[2]We recall that Dom $A(t) = \mathcal{D}om$ is independent of t, which is accepted throughout Part II.

[3]It is just the case in what follows.

Lemma 7.1. *Given $p, q \geq 1$, for any $\psi(t) : [0,T] \to \mathfrak{X}$ sufficiently smooth, the simplifying assumption $\mathcal{B}(p)$ implies*

$$\|Q_n(\psi)\| \leq Ck^p \int_{t_n}^{t_{n+1}} \|\psi^{(p)}(s)\|\, ds, \tag{7.3}$$

while $\mathcal{C}(q)$ implies

$$\|Q_{nj}(\psi)\| \leq Ck^q \int_{t_n}^{t_{nj}} \|\psi^{(q)}(s)\|\, ds, \quad j = 1, \ldots, \nu. \tag{7.4}$$

Proof. Clearly, it suffices to show (7.3) only. By Taylor expansion we have

$$\psi(t) = \sum_{l=0}^{p-1} \frac{(t-t_n)^l}{l!} \psi^{(l)}(t_n) + \int_{t_n}^{t} \frac{(t-s)^{p-1}}{(p-1)!} \psi^{(p)}(s)\, ds. \tag{7.5}$$

With this formula in mind, (7.3) easily follows noting that the value of $Q_n(\psi)$ remains unchanged when substituting $\psi(t) - \sum_{l=0}^{p-1} \frac{(t-t_n)^l}{l!} \psi^{(l)}(t_n)$ for $\psi(t)$, since the quadrature formula (7.1) is precise for all polynomials of degree $p-1$. □

Further, we denote

$$H_{qj} = \frac{1}{q!} \left(\sum_{l=1}^{\nu} a_{jl} c_l^q - \frac{c_j^{q+1}}{q+1} \right).$$

Lemma 7.2. *Let, given $q \geq 1$, the simplifying assumption $\mathcal{C}(q)$ hold. Then, for any $\psi(t) : [0,T] \to \mathfrak{X}$ sufficiently smooth,*[4]

$$Q_{nj}(\psi) = k^{q+1} H_{qj} \psi^{(q)}(t_n) + k^{q+1} \int_{t_n}^{t_{n+1}} \mathcal{P}_j\left(\frac{s-t_n}{k}\right) \psi^{(q+1)}(s)\, ds,$$
$$j = 1, \ldots, \nu,$$

where $\mathcal{P}_j(s)$ is a bounded Peano kernel.

Proof. In fact, the claim follows in the same manner as in the proof of the preceding lemma, using (7.5) with $q+1$ in place of p, noting that the quadrature formula (7.2) is precise for all polynomials of degree $q-1$, and observing that $Q_{nj}(\psi_q) = k^{q+1} H_{qj} \psi^{(q)}(t_n)$, where $\psi_q(t) = \frac{(t-t_n)^q}{q!} \psi^{(q)}(t_n)$. □

[4] Cf., e.g., Lubich & Ostermann [105]

Lemma 7.3. *Let the RK method in question be of order p and of stage order $q \le p-2$. Then the order condition holds*

$$\sum_{j=1}^{\nu} b_j H_{qj} = 0.$$

This result can be established in the standard manner by using Taylor expansion (cf. Ostermann & Roche [123]).

Now we derive suitable representations of the error, assuming that $\mathfrak{R} = I$.[5] When making this assumption, we are distinguishing between the autonomous and non-autonomous cases, and by doing so, we show that this distinction is caused by the accepted compromise between our intentions and real possibilities. In the first (autonomous) case we show error estimates in terms of the data, which are realistic; however similar estimates in terms of the solution are also possible. In the general (non-autonomous) case we estimate the error only in terms of the solution, and we cannot satisfactorily presently demonstrate anything different. In both cases, autonomous and non-autonomous, we now assume that problem (5.1) possesses a unique solution $y(t)$.

Let $A(t) \equiv A$ be independent of t.[6] Suppose that $A \in \mathcal{S}(\varphi, \sigma)$ and the RK method is of type $C(\varphi)$ for some $\varphi \in (0, \pi/2)$ and $\sigma \in \mathbb{R}$. In this case we have the extended solution of (5.19), (5.20) given by

$$Y_{n+1} = r(-kA)Y_n + k\sum_{j=1}^{\nu} w_j(-kA)\, f(t_{nj}) \text{ for } t_n \in \breve{\Omega}_k, \quad Y_0 = y^0.$$

Also, for the solution $y(t)$ of (5.1) at nodal points we use the representation

$$y(t_{n+1}) = r(-kA)y(t_n) + k\sum_{j=1}^{\nu} w_j(-kA)\, f(t_{nj}) - \phi_n \quad \text{for } t_n \in \breve{\Omega}_k,$$

where ϕ_n is the truncation error. Therefore, for the global error $E_n = Y_n - y(t_n)$, we have

$$E_{n+1} = r(-kA)E_n + \phi_n,$$

which further gives, since $E_0 = 0$,

$$E_n = \sum_{j=0}^{n-1} r(-kA)^{n-j-1}\phi_j \quad \text{for } t_n \in \Omega_k. \tag{7.6}$$

[5]In this chapter more general expressions for $\mathfrak{R}$ will be used only in the very particular case where $A(t) \equiv A$ and $f(t) \equiv 0$.

[6]In this case the unique solvability of (5.1) will, in fact, follow from our subsequent suppositions.

Turning next to the general case where $A(t)$ depends on t, we assume that the conditions of Theorem 5.1 are satisfied so that the discrete problem (5.19), (5.20) is uniquely solvable. For $y(t)$ the solution of (5.1), we have, using the above functionals $Q_n(\cdot)$ and $Q_{nj}(\cdot)$, that

$$y(t_{n+1}) = y(t_n) - k\sum_{j=1}^{\nu} b_j(A(t_{nj})y(t_{nj}) - f(t_{nj})) - Q_n(y_t)$$

and

$$y(t_{nj}) = y(t_n) - k\sum_{l=1}^{\nu} a_{jl}(A(t_{nl})y(t_{nl}) - f(t_{nl})) - Q_{nj}(y_t), \quad j = 1, \dots, \nu.$$

Subtracting these equalities from the respective Runge-Kutta equations in (5.19) and (5.20), we get, with E_n just the same as above and with $E_{nj} = Y_{nj} - y(t_{nj})$,

$$E_{n+1} = E_n - k\sum_{j=1}^{\nu} b_j A(t_{nj})E_{nj} + Q_n(y_t) \tag{7.7}$$

and

$$E_{nj} = E_n - k\sum_{l=1}^{\nu} a_{jl}A(t_{nl})E_{nl} + Q_{nj}(y_t), \quad j = 1, \dots, \nu. \tag{7.8}$$

Therefore, solving (7.8) for $\mathbf{E}_n = \big(E_{n1}, \dots, E_{n\nu}\big)^T$ and assuming that the expressions $(\mathfrak{i} + k\mathfrak{a}\mathcal{D}_n)^{-1}\mathbf{Q}_n(y_t)$ and $\mathcal{D}_n\,(\mathfrak{i} + k\mathfrak{a}\mathcal{D}_n)^{-1}\mathbf{Q}_n(y_t)$ are defined, where $\mathbf{Q}_n(\cdot) = \big(Q_{n1}(\cdot), \dots, Q_{n\nu}(\cdot)\big)^T$, similarly to (5.54) we find

$$\mathbf{E}_n = (\mathfrak{i} + k\mathfrak{a}\mathcal{D}_n)^{-1}\mathbf{e} \otimes E_n + (\mathfrak{i} + k\mathfrak{a}\mathcal{D}_n)^{-1}\mathbf{Q}_n(y_t),$$

which further gives, inserted into (7.7),

$$\begin{aligned} E_{n+1} &= (I - k\mathbf{b}^T\mathcal{D}_n(\mathfrak{i}+k\mathfrak{a}\mathcal{D}_n)^{-1}\mathbf{e})E_n - k\mathbf{b}^T\mathcal{D}_n(\mathfrak{i}+k\mathfrak{a}\mathcal{D}_n)^{-1}\mathbf{Q}_n(y_t) + Q_n(y_t) \\ &= \mathfrak{U}_n E_n - k\mathbf{b}^T\mathcal{D}_n(\mathfrak{i} + k\mathfrak{a}\mathcal{D}_n)^{-1}\mathbf{Q}_n(y_t) + Q_n(y_t). \end{aligned}$$

Using this recurrence formula, since $E_0 = 0$, it follows that for $t_n \in \Omega_k$,

$$E_n = \sum_{j=0}^{n-1} \mathfrak{U}_{n-1,j+1}\big(- k\mathbf{b}^T\mathcal{D}_j(\mathfrak{i} + k\mathfrak{a}\mathcal{D}_j)^{-1}\mathbf{Q}_j(y_t) + Q_j(y_t)\big). \tag{7.9}$$

In conclusion, we recall the well-known fact (cf., e.g., Sinestrari [146]) that if $A(t) \in \mathcal{S}(\varphi, \sigma)$ for some $\varphi \in (0, \pi/2)$ and $\sigma \in \mathbb{R}$, we then define for any fixed $t \in [0,T]$, by using formula (B.2), a family of operators $e^{-sA(t)}$, $s > 0$, which possesses all the properties of holomorphic semigroups.[7] For further reference we now show a result of a preparatory character which is involved with the operator $e^{-sA(t)}$.

Lemma 7.4. *Let $A(t) \in \mathcal{S}(\varphi,\sigma)$ and let a seminorm $|\cdot|_t$ form a $(\xi|\varphi,\sigma)$-concordant pair with $A(t)$, for some $\varphi \in (0,\pi/2)$, $\sigma \in \mathbb{R}$, and $\xi \geq 0$. Then, if $T < \infty$, we have, with $K > 0$ sufficiently small, for all $k \in (0,K]$, $t,s \in [0,T]$, and $u \in \mathfrak{X}$,*

$$|e^{-sA(t)}u|_t \leq Cs^{-\xi}\|u\|, \tag{7.10}$$

and, in particular, for any fixed $\eta \in \mathbb{N} \cup \{0\}$,

$$\|A(t)^\eta e^{-sA(t)}\| \leq Cs^{-\eta}. \tag{7.11}$$

If $\sigma \geq 0$ in addition, then the above holds in the case $T = K = \infty$ as well.

Proof. Let Γ_s be the contour coinciding with $\partial\big(\Sigma_\varphi \cup \mathcal{D}(1/s)\big)$. Furthermore, going over to the shifted operator $A(t) - \sigma I$ if $\sigma < 0$, it is sufficient to consider the case $\sigma \geq 0$ only. Assuming therefore the last restriction and using next formula (B.2) with Γ_s in place of Ξ, we have, for all $s > 0$,

$$e^{-sA(t)} = \frac{1}{2\pi i}\int_{\Gamma_s} e^{-s\lambda}(\lambda I - A(t))^{-1}d\lambda.$$

Now (7.10) follows when integrating by parts and acting further as in the proof of Theorem 2.1.[8] With regard to the second stated estimate (7.11), it is the particular case of (7.10) where $\xi = \eta$ and $|\cdot|_t = \|A(t)^\xi \cdot\|$. □

7.2 Autonomous equations

In this section we examine convergence in the case with a constant operator. (Therefore we write A instead of $A(t)$.) As above, we then work with the extended solution of (5.19), (5.20) specified by (5.30).

We begin with the homogeneous case $f(t) \equiv 0$, for which the extended discrete solution is specified by

$$Y_{n+1} = r(-kA)Y_n \quad \text{for } t_n \in \breve{\Omega}_k, \qquad Y_0 = \mathfrak{R}y^0, \tag{7.12}$$

[7] Except strong continuity at 0 if Dom $A(t)$ is not dense in $\mathfrak{X}$

[8] In fact, this technique can be thought of as a slight modification of a well-known argument in the theory of holomorphic semigroups (cf., e.g., Reed & Simon [133, the proof of Theorem X.52]).

and it is compared to the extended solution of the original Cauchy problem (5.1) given by $y(t) = e^{-tA}y^0$ for $t > 0$, where the operator e^{-tA}, $t > 0$, is defined by formula (B.2).

First we show an auxiliary assertion.

Lemma 7.5. *Let $A \in \mathcal{S}(\varphi, \sigma)$ and let the RK method be of order p and $A(\varphi)$-stable, for some $\varphi \in (0, \pi/2)$, $\sigma \in \mathbb{R}$, and $p \in \mathbb{N}$. Then, if $T < \infty$, with*

$$\widetilde{r}_n(z) = (r(z)^n - e^{nz}) \left(\frac{1-z}{z}\right)^p, \tag{7.13}$$

it holds, with $K > 0$ sufficiently small, for all $k \in (0, K]$ and $t_n \in \Omega_k$,

$$\|\widetilde{r}_n(-kA)\| \leq C.$$

In the case $\sigma \geq 0$ one can take $T = K = \infty$.

Proof. Let $r_1(z)$ be the stability function of an A-stable method of order $\geq p$, with the property $r_1(\infty) = 0$. (For example, one can take a ν-stage Radau IA method with $\nu \geq (p+1)/2$.) Then, clearly, it suffices to show the desired estimate with $\widetilde{r}_n^{(j)}(-kA)$, $j = 1, 2$, in place of $\widetilde{r}_n(-kA)$, where

$$\widetilde{r}_n^{(1)}(z) = (r(z)^n - r_1(z)^n) \left(\frac{1-z}{z}\right)^p \text{ and } \widetilde{r}_n^{(2)}(z) = (r_1(z)^n - e^{nz}) \left(\frac{1-z}{z}\right)^p.$$

Note that there exists a fixed $d_1 > 1$ such that with some $q \geq 1$,

$$|r(z) - r(\infty)| \leq C|z|^{-q} \text{ and } |r_1(z)| \leq 1 \quad \text{for all } z \notin \operatorname{Int} \mathcal{D}(d_1).$$

Since $|r(\infty)| \leq 1$, this estimate further implies that $|r(z)^n| \leq e^{cn|z|^{-q}}$ and, thus, $|\widetilde{r}_n^{(1)}(z)| \leq Ce^{cn|z|^{-q}}$, for all $n \geq 0$ and $z \notin \operatorname{Int} \mathcal{D}(d_1)$. Also, in view of the fact that $|r(-z)| \leq 1$ and $|r_1(-z)| \leq 1$ for $z \in \Sigma_\varphi$, it is not hard to see that $|\widetilde{r}_n^{(1)}(-z)| \leq Ce^{-cn|z|^{-q}}$ for all $n \geq 0$ and $z \in \Sigma_{\varphi_1}$ such that $|z| \geq 1$, for any fixed $\varphi_1 \in (0, \varphi)$. Next, fixing arbitrarily $\varphi_1 \in (0, \varphi)$, using the expansion

$$r(z)^n - r_1^n(z) = (r(z) - r_1(z)) \sum_{l=0}^{n-1} r(z)^{n-1-l} r_1(z)^l$$

as well as the estimates, for some $d > 0$,

$$|r(z) - r_1(z)| \leq C|z|^{p+1} \quad \text{for all } z \in \mathcal{D}(d),$$

$$|r(-z)| \le Ce^{-c|z|} \text{ and } |r_1(-z)| \le Ce^{-c|z|} \quad \text{for all } z \in \Sigma_{\varphi_1} \cap \mathcal{D}(1),$$

and taking the condition $r(0) = r'(0) = r_1(0) = r_1'(0) = 1$ into account, it follows that

$$|\widetilde{r}_n^{(1)}(z)| \le Cn\,|z|e^{cn|z|} \quad \text{for all } z \in \mathcal{D}(d)$$

and

$$|\widetilde{r}_n^{(1)}(-z)| \le Cn\,|z|e^{-cn|z|} \quad \text{for all } z \in \Sigma_{\varphi_1} \cap \mathcal{D}(1).$$

It is therefore seen that the function $\widetilde{r}_n^{(1)}(z)$ satisfies the conditions of Lemma 5.6, which yields that $\|\widetilde{r}_n^{(1)}(-kA)\| \le C$.

Similarly, it can be shown, applying Lemma 5.5, that $\|\widetilde{r}_n^{(2)}(-kA)\| \le C$, and the proof is now complete. □

The first assertion on convergence is as follows.

Theorem 7.1. *Let $A \in \mathcal{S}(\varphi, \sigma)$ and let the RK method be of order p and $A(\varphi)$-stable, for some $\varphi \in (0, \pi/2)$, $\sigma \in \mathbb{R}$, and $p \in \mathbb{N}$. Let also $\mathfrak{R} = I$. Then, if $T < \infty$ and $y^0 \in \operatorname{Dom} A^{j+1}$ with some integer $1 \le j \le p$, for Y_n given by (7.12) and for $y(t)$ the extended solution of (5.1) with $f(t) \equiv 0$, it holds, with $K > 0$ sufficiently small, for all $k \in (0, K]$ and $t_n \in \Omega_k$,*

$$\|Y_n - y(t_n)\| \le Ck^j t_n \|A^{j+1}y^0\|. \tag{7.14}$$

In the case $\sigma \ge 0$ one can take $T = K = \infty$.

Proof. It follows from Theorems 6.10 and 6.11 and Lemma 7.4 that with $K > 0$ sufficiently small, for all $k \in (0, K]$ and $t_n \in \Omega_k$,

$$\|e^{-t_n A}\| + \|r(-kA)^n\| \le C. \tag{7.15}$$

Furthermore, since the method is of order p, by Taylor expansion, for any fixed $\varphi_1 \in (0, \varphi)$ the function $\widetilde{r}(z) = z^{-j-1}(r(z) - e^z)$, where j comes from the statement of the theorem, satisfies, when substituted for $\omega(z)$, the conditions of Lemma 5.4 and the condition $\widetilde{r}(\infty) = 0$. It therefore follows that for all $u \in \operatorname{Dom} A^{j+1}$,

$$\|\left(r(-kA) - e^{-kA}\right)u\| \le k^{j+1}\|\widetilde{r}(-kA)\|\,\|A^{j+1}u\| \le Ck^{j+1}\|A^{j+1}u\|.$$

With this inequality in mind, since $Y_n = r(-kA)^n y^0$, using (7.15) and the identity

$$\begin{aligned} Y_n - y(t_n) &= \left(r(-kA)^n - e^{-t_n A}\right) y^0 \\ &= \sum_{l=1}^{n} \left(r(-kA) - e^{-kA}\right) e^{-t_{n-l}A} r(-kA)^{l-1} y^0, \end{aligned} \tag{7.16}$$

a simple estimation yields, for all $t_n \in \Omega_k$,

$$\|Y_n - y(t_n)\| \leq Ck^{j+1}n\|A^{j+1}y^0\| = Ck^j t_n\|A^{j+1}y^0\|,$$

which shows (7.14).

The more precise wording for $\sigma \geq 0$ is obvious. □

Theorem 7.1 can be widely used since the property of $A(\varphi)$-stability is established for very many discretization methods. For further progress we have to distinguish between the above introduced classes of methods.

First we introduce one more helpful notion.

Definition 7.1. *We say that a rational function $\omega(z)$* approaches *a function $\eta(z)$ with* order *m if*

$$\omega(z) = \eta(z) + \mathcal{O}(|z|^m) \quad as\ |z| \to 0.$$

Our next result concerns methods of class S_I.

Theorem 7.2. *Suppose that $A \in \mathcal{S}(\varphi, \sigma)$, the RK method is of order p and $A_I(\varphi)$-stable, and a seminorm $]\![\cdot]\!]_t$ forms a $(\xi|\varphi, \sigma)$-concordant pair with A, for some $p \in \mathbb{N}$, $\varphi \in (0, \pi/2)$, $\sigma \in \mathbb{R}$, and $\xi \geq 0$. Suppose further, for some non-negative integers j, ς, and m subject to $j \leq p$ and $\xi < \varsigma + j$, that $y^0 \in \operatorname{Dom} A^j$ and $\mathfrak{R} = \omega(-kA)$, where $\omega(z)$ is a rational function which is (φ)-regular, approaches e^{mz} with order p, and satisfies the condition $\deg[\omega] \leq -\varsigma$. Then, if $T < \infty$, for Y_n specified by (7.12) and for $y(t)$ the extended solution of (5.1) with $f(t) \equiv 0$, we have, with $K > 0$ sufficiently small, for all $k \in (0, K]$ and $t_n \in \Omega_k$ with $n \geq 1$, that*

$$]\![Y_n - y(t_{n+m})]\!]_{t_n} \leq Ck^p t_{n+m}^{-(\xi+p-j)}\|A^j y^0\|. \tag{7.17}$$

The result remains valid even if $\xi = m + j$ provided that, in addition to above, the seminorm $]\![\cdot]\!]_t$ forms a $(\xi|\xi|\varphi, \sigma)$-concordant pair with A if $\xi \geq 1$ and (6.21) is satisfied with $]\![\cdot]\!]_t$ substituted for $|\cdot|_t$ if $\xi = 0$. In the case $\sigma \geq 0$, the above statement holds for $T = \infty$ and one can then take $K = \infty$.

Proof. Noting that all the involved operators are pairwise permutable, we have, with $n_1 = \lfloor n/2 \rfloor$ and $n_2 = n - n_1$,

$$\begin{aligned}
Y_n - y(t_{n+m}) &= r(-kA)^n \omega(-kA) y^0 - e^{-t_{n+m}A} y^0 \\
&= \left(r(-kA)^{n_1} - e^{-t_{n_1}A}\right) r(-kA)^{n_2} \omega(-kA) y^0 \\
&\quad + \left(r(-kA)^{n_2} - e^{-t_{n_2}A}\right) e^{-t_{n_1}A} \omega(-kA) y^0 \\
&\quad + \left(\omega(-kA) - e^{-t_m A}\right) e^{-t_n A} y^0 \\
&=: G_1 + G_2 + G_3.
\end{aligned}$$

With $\widetilde{r}_n(z)$ given by (7.13), G_1 and G_2 can also be represented by

$$G_1 = (-k)^p A^{p-j}(I + kA)^{-p} r(-kA)^{n_2} \omega(-kA)\, \widetilde{r}_{n_1}(-kA) A^j y^0$$

and

$$G_2 = (-k)^p A^{p-j}(I + kA)^{-p} e^{-t_{n_1} A} \omega(-kA)\, \widetilde{r}_{n_2}(-kA) A^j y^0.$$

It is obvious that the claim follows if we show that for all $k \in (0, K]$, $t_n \in \Omega_k$ with $n \geq 1$, and $t \in [0, T]$,

$$]\![G_1]\!\![_t +]\![G_2]\!\![_t +]\![G_3]\!\![_t \leq C k^p t_n^{-(\xi+p-j)} \|A^j y^0\|. \tag{7.18}$$

Observe that instead of estimating $]\![G_1]\!\![_t$ as stated in (7.18) it suffices to show that with $G = (-k)^p A^{p-j}(I + kA)^{-p} r(-kA)^n \omega(-kA)\, \widetilde{r}_n(-kA) A^j y^0$, for all $k \in (0, K]$, $t_n \in \Omega_k$, and $t \in [0, T]$,

$$]\![G]\!\![_t \leq C k^p t_{n+1}^{-(\xi+p-j)} \|A^j y^0\|. \tag{7.19}$$

To that end, we note that since the method is $A_I(\varphi)$-stable, the function $\rho(-z)$ has no zeros in Σ_{φ_1} for any fixed $\varphi_1 \in (0, \varphi)$ and it holds $\deg[\rho^{-1}] = 1$. Putting now $\omega_0(z) = (1 - z)^{\varsigma+j-p} \rho(z)^{j-p} \omega(z)$, it follows from the above observations that the function $\omega_0(z)$ is (φ)-regular and for any fixed $\varphi_1 \in (0, \varphi)$ this function is bounded in Σ_{φ_1} with $\omega_0(z) = \omega_0(\infty) + \mathcal{O}(|z|^{-1})$ as $|z| \to \infty$, for $z \in \Sigma_{\varphi_1}$. In virtue of Lemma 5.4, the operator $\omega_0(-kA)$ is defined and uniformly bounded on $\mathfrak{X}$, that is,

$$\|\omega_1(-kA)\| \leq C \quad \text{for all } k \in (0, K]. \tag{7.20}$$

Now, recalling that $\mathfrak{A} = A\rho(-kA)$, applying Theorem 6.10 in the case $f(t) \equiv 0$ and Remark 6.8, and using the fact that all the involved operators are pairwise permutable, we get, since $\xi < m + j$,[9] with n_1 and n_2 defined as above, for all $t \in [0, T]$ and $t_n \in \Omega_k$,

$$\begin{aligned}]\![G]\!\![_t &= k^p]\![(I + kA)^{-\varsigma-j} r(-kA)^{n_1} \mathfrak{A}^{p-j} r(-kA)^{n_2} \omega_0(-kA) \widetilde{r}_n(-kA) A^j y^0]\!\![_t \\ &\leq C k^p t_{n+1}^{-\xi} \|\mathfrak{A}^{p-j} r(-kA)^{n_2} \omega_0(-kA)\, \widetilde{r}_n(-kA)\, A^j y^0\| \\ &\leq C k^p t_{n+1}^{-(\xi+p-j)} \|\omega_0(-kA)\, \widetilde{r}_n(-kA)\, A^j y^0\|, \end{aligned}$$

which further leads to (7.19) in view of (7.20) and Lemma 7.5.

Finally, the needed estimate (7.18) for $]\![G_2]\!\![_t$ and $]\![G_3]\!\![_t$ follows by actually using the same argument, with the aid of Lemma 7.4 and of the estimate

[9] Or $\xi = m + j$ and the extra conditions on $]\![\cdot]\!\![_t$ are assumed as pointed out in the statement of the theorem.

$\|\rho(-kA)^{p-j}\| + \|(kA)^{p-j}\rho(-kA)^{p-j}\| + \|\omega_1(-kA)\| \leq C$,[10] where $\omega_1(z) = z^{-p}(\omega(z) - e^{mz})$.

It is also seen from the above reasonings that in the case $\sigma \geq 0$ the result is still valid for $T = \infty$, with no restriction on k from above. □

For methods of class S_{II} or of class S_{III} we have the following convergence result.

Theorem 7.3. *Let the conditions of Theorem 7.2 be satisfied with the difference that the RK method is now $A_{II}(\varphi)$- or $A_{III}(\varphi)$-stable and the condition $\xi < \varsigma + j$ (or the condition $\xi \leq m + j$ with the corresponding extra requirements) is replaced by the condition $\xi \leq (\varsigma + p)/(\varpi + 1) - p + j$, with $\varpi = -\deg[\alpha(\cdot) - \alpha(\infty)]$, where the nonstrict inequality is replaced by the corresponding strict one if ξ is an integer, unless $\xi = 0$ and (6.21) holds. Then, if $T < \infty$, for Y_n and $y(t)$ specified the same as in the conditions of Theorem 7.2, the error estimate (7.17) holds in the considered case as well, and the result is still valid for $T = K = \infty$ if $\sigma \geq 0$.*

The proof is similar to that of Theorem 7.2, applying Theorems 6.10 and 6.11 and Remark 6.8.

Theorem 7.3 can be applied for all Gauss-Legendre and Lobatto IIIA, IIIB methods while Theorem 7.2 is applicable for all other concrete families of methods discussed in Section 5.2.

Further, we examine inhomogeneous problems of the form (5.1) but continuing in the case of a constant operator. We therefore handle the rational method (5.30) now. In the inhomogeneous case we meet a phenomenon called order reduction. This phenomenon shows itself by the fact that if the discretization method is of high order, we have to impose certain compatibility requirements on the solution or on the data[11] to achieve the optimal order of accuracy otherwise a lower order only is observed. Below we consider different situations, including those for which the order reduction phenomenon occurs.

Theorem 7.4. *Suppose that $A \in S(\varphi, \sigma)$ and the RK method is of type $C(\varphi)$, of order p, and of strict order q, for some $\varphi \in (0, \pi/2)$, $\sigma \in \mathbb{R}$, and $p, q \in \mathbb{N}$ subject to $p \geq q$. Suppose further that $\mathfrak{R} = I$, $T < \infty$, problem (5.1) possesses an extended solution $y(t)$ of order $p + 1$,*[12] *and*

---

[10]Which follows by Lemma 5.4, similarly to (7.20)

[11]These restrictions are often thought of as artificial. In fact, there is no common opinion of what are artificial conditions. Many authors try to avoid conditions of the form $y^{(l)}(t) \in \operatorname{Dom} A^j$ with $j \geq 2$ while the situation with $j = 1$ is thought of as quite acceptable.

[12]Which is simultaneously a solution of (5.1), see Appendix B.

$A^{\max\{0,p-m-l\}}f^{(l)}(\cdot) \in L_\infty(0,t;\mathfrak{X})$ for $l = q,\dots,p$ and $t \in [0,T]$, with some $m = 0,\dots,p-q$. Then, if Y_n is the extended solution of (5.19), (5.20), we have, with $y^0_{(p+1)}$ coming from formula (B.5) stated for $p+1$ in place of p and with $K > 0$ sufficiently small, for all $k \in (0,K]$ and $t_n \in \Omega_k$, that[13]

$$\|Y_n - y(t_n)\| \le Ck^p t_n \|y^0_{(p+1)}\| + Ck^p \|(t_n - \cdot) f^{(p+1)}(\cdot)\|_{L_1(0,t_n;\mathfrak{X})} + Ck^{p-m} t_n \sum_{l=q}^{p} \|A^{\max\{0,p-m-l\}} f^{(l)}(\cdot)\|_{L_\infty(0,t_n;\mathfrak{X})}. \tag{7.21}$$

In the case $\sigma \ge 0$, the above holds for $T = \infty$ as well and one can then take $K > 0$ arbitrarily large but finite if $m \ge 1$ and $K = \infty$ if $m = 0$.

Proof. Our argument will be based on using formula (7.6). An essential ingredient of this effort is finding a suitable estimate of the quantity $\|\phi_n\|$, where ϕ_n is the truncation error appearing in (7.6).

Since $\phi_n = r(-kA)y(t_n) - y(t_{n+1}) + k\mathbf{w}^T(-kA)\mathbf{\Phi}_n$, where $\mathbf{\Phi}_n$ is given by (5.29), we have, by Taylor expansions for $y(t)$ and $f(t)$,

$$\phi_n = r(-kA)\, y(t_n) - \sum_{l=0}^{p} \frac{k^l}{l!} y^{(l)}(t_n) + kW_{0;n} + R_n, \tag{7.22}$$

where

$$W_{J;n} = \sum_{j=1}^{\nu} w_j(-kA) \sum_{l=J}^{p-1} \frac{(c_j k)^l}{l!} f^{(l)}(t_n) \quad \text{for } J = 0,\dots,p-1 \tag{7.23}$$

and

$$\begin{aligned} R_n &= k\sum_{j=1}^{\nu} w_j(-kA) \int_{t_n}^{t_{nj}} \frac{(t_{nj}-s)^{p-1}}{(p-1)!} f^{(p)}(s)\, ds \\ &\quad - \int_{t_n}^{t_{n+1}} \frac{(t_{n+1}-s)^p}{p!} y^{(p+1)}(s)\, ds. \end{aligned} \tag{7.24}$$

Now we introduce the functions

$$h_p(z) = \frac{p!}{z^{p+1}} \left(r(z) - \sum_{j=0}^{p} \frac{z^j}{j!} \right)$$

[13]Therefore, the classical order is observed in the case $m = 0$ only.

and

$$h_l(z) = \frac{l!}{z^{l+1}} \left(r(z) - \sum_{j=0}^{l} \frac{z^j}{j!} \right) - \sum_{j=1}^{\nu} c_j^l w_j(z), \quad l = 0, \ldots, p-1. \tag{7.25}$$

It can easily be checked (cf., e.g., Brenner, Crouzeix, and Thomée [50, Lemma 2]) that

$$h_l(z) = 0 \quad \text{for } l = 0, \ldots, q-1 \tag{7.26}$$

and

$$h_l(z) = \mathcal{O}(|z|^{p-l}) \text{ as } z \to 0 \quad \text{for } l = q, \ldots, p, \tag{7.27}$$

since the method is of order p and of strict order q. With the aid of the equality $-Ay = y' + f$, we derive from (7.22), with $W_{J;n}$ defined by (7.23),

$$\begin{aligned} \phi_n - R_n &= kQ_0(-kA)\,(-A)y(t_n) - \sum_{l=1}^{p} \frac{k^l}{l!} y^{(l)}(t_n) + kW_{0;n} \\ &= kQ_0(-kA)y'(t_n) - \sum_{l=1}^{p} \frac{k^l}{l!} y^{(l)}(t_n) + kW_{1;n} - kh_0(-kA)f(t_n), \end{aligned}$$

where we denote

$$Q_J(z) = h_J(z) + \sum_{j=1}^{\nu} c_j^J w_j(z), \quad J = 0, \ldots, p-1.$$

Applying repeatedly the trick based on using the relationship $-Ay^{(l)} = y^{(l+1)} - f^{(l)}$ for $l = 1, \ldots, p$, we finally obtain a new representation for $\phi_n - R_n$ as follows

$$\begin{aligned} \phi_n - R_n &= k^2 Q_1(-kA)\,(-A)y'(t_n) - \sum_{l=2}^{p} \frac{k^l}{l!} y^{(l)}(t_n) \\ &\quad + kW_{1;n} - kh_0(-kA)f(t_n) = \cdots \\ &= -\sum_{l=0}^{p} \frac{k^{l+1}}{l!} h_l(-kA) f^{(l)}(t_n) + \frac{k^{p+1}}{p!} h_p(-kA) y^{(p+1)}(t_n). \end{aligned} \tag{7.28}$$

Putting further for short $l' = \max\{0, p-m-l\}$, $l = q, \ldots, p$ and denoting $\widetilde{h}_l(z) = z^{-l'} h_l(z)$ for $l = q, \ldots, p$, we note that the functions $\widetilde{h}_l(z)$ are (φ)-regular[14] and subject to $\deg[\widetilde{h}_l] \le 0$, in view of (7.27). Applying therefore Lemma 5.3, we find, for all $k \in (0, K]$, that

$$\|\widetilde{h}_l(-kA)\| \le C, \quad l = q, \ldots, p. \tag{7.29}$$

[14] We recall that $w_j(z)$, $j = 1, \ldots, \nu$, possess this property.

Using now (7.28) instead of (7.22) yields, with the aid of (7.26),

$$\begin{aligned}\phi_n &= -k^{p+1-m}\sum_{l=q}^{p}\frac{(-1)^{l'}}{l!}k^{\max\{l+m-p,0\}}\widetilde{h}_l(-kA)A^{l'}f^{(l)}(t_n)\\ &\quad+\frac{k^{p+1}}{p!}\widetilde{h}_p(-kA)y^{(p+1)}(t_n)+R_n, \end{aligned}\tag{7.30}$$

which further implies, by (7.29),

$$\|\phi_n\|\le Ck^{p+1-m}\sum_{l=q}^{p}\|A^{l'}f^{(l)}(t_n)\|+Ck^{p+1}\|y^{(p+1)}(t_n)\|+\|R_n\|. \tag{7.31}$$

At the same time since $\|w_j(-kA)\|\le C$, a trivial estimation based on using (7.24) gives

$$\|R_n\|\le Ck^{p+1}\max_{0\le s\le t_{n+1}}\|f^{(p)}(s)\|+Ck^p\int_{t_n}^{t_{n+1}}\|y^{(p+1)}(s)\|\,ds. \tag{7.32}$$

Next, since problem (5.1) possesses an extended solution of order $p+1$, applying formula (B.5) with $p+1$ in place of p and noting that $\|e^{-tA}\|\le C$ for $0<t\le T$, it follows that

$$\|y^{(p+1)}(t_n)\|\le C\|y^0_{(p+1)}\|+C\int_0^{t_n}\|f^{(p+1)}(s)\|\,ds \tag{7.33}$$

and

$$\begin{aligned}\int_{t_n}^{t_{n+1}}\|y^{(p+1)}(s)\|\,ds &\le Ck\|y^0_{(p+1)}\|+C\int_{t_n}^{t_{n+1}}\left(\int_0^t\|f^{(p+1)}(s)\|\,ds\right)dt\\ &\le Ck\|y^0_{(p+1)}\|+C\int_{t_n}^{t_{n+1}}(t_{n+1}-s)\,\|f^{(p+1)}(s)\|\,ds\\ &\quad+Ck\int_0^{t_n}\|f^{(p+1)}(s)\|\,ds. \end{aligned}\tag{7.34}$$

Combining (7.31), (7.32), (7.33), and (7.34) thus yields

$$\begin{aligned}\|\phi_n\| &\le Ck^{p+1}\|y^0_{(p+1)}\|+Ck^{p+1-m}\sum_{l=q}^{p}\|A^{l'}f^{(l)}(.)\|_{L_\infty(0,t_{n+1};\mathfrak{X})}\\ &\quad+Ck^p\int_{t_n}^{t_{n+1}}(t_{n+1}-s)\,\|f^{(p+1)}(s)\|\,ds+Ck^{p+1}\int_0^{t_n}\|f^{(p+1)}(s)\|\,ds. \end{aligned}\tag{7.35}$$

Finally, by a trivial calculation we have

$$k\sum_{j=0}^{n-1}\int_0^{t_j}\|f^{(p+1)}(s)\|\,ds=\sum_{j=1}^{n-1}t_{n-j}\int_{t_{j-1}}^{t_j}\|f^{(p+1)}(s)\|\,ds$$

$$\leq\sum_{j=1}^{n-1}\int_{t_{j-1}}^{t_j}(t_n-s)\,\|f^{(p+1)}(s)\|\,ds\leq\int_0^{t_n}(t_n-s)\,\|f^{(p+1)}(s)\|\,ds. \quad (7.36)$$

With the aid of this inequality and by inserting (7.35) into (7.6) the claim is shown because $\|r(-kA)^j\|\leq C$ for $j=0,\ldots,T/k$, by Theorems 6.10 and 6.11. □

Clearly, the assertion of Theorem 7.4 remains valid if, instead of the norm $\|\cdot\|$, the error $Y_n-y(t_n)$ is measured with some seminorm which forms a $(0|\varphi,\sigma)$-concordant pair with A. The next result shows that the classical order may still be observed when using stronger seminorms for an estimation of the error.

Theorem 7.5. *Let the conditions of Theorem 7.4 be satisfied for $T<\infty$,*[15] *$\varphi\in(0,\pi/2)$, $\sigma\in\mathbb{R}$, and non-negative integers p,q,m such that $p\geq q\geq 1$ and $0\leq m\leq p-q$, and let Y_n and $y(t)$ be defined as in that theorem. Also, suppose that a seminorm $]\![\cdot]\!\![_t$ forms a $(\xi|\varphi,\sigma)$-concordant pair with the operator A for some $\xi\in(0,1]$ and that a rational function $\omega(z)$ is (φ)-regular and satisfies the condition $\deg[\omega]\leq-\xi(\varpi+1)$ in which $\leq$ should be replaced by $<$ if $\xi=1$, unless $|r(\infty)|<1$ and the pair $(]\![\cdot]\!\![_t;A)$ is $(1|1|\varphi,\sigma)$-concordant, where*

$$\varpi=\begin{cases}0 & \textit{if } |r(\infty)|<1,\\ -\deg[\alpha(\cdot)-\alpha(\infty)] & \textit{otherwise.}\end{cases}$$

Then, with $K>0$ sufficiently small (with $K=\infty$ if $\sigma\geq 0$) and with any fixed $\kappa\in[1/(1-\xi),\infty]$, we have, for all $k\in(0,K]$ and $t_n\in\Omega_k$, $n\geq 1$,

$$]\![\omega(-kA)\,(Y_n-y(t_n))]\!\![_{t_n}\leq Ck^pt_n^{1-\xi}\|y^0_{(p+1)}\|$$
$$+\,Ck^{p-m}t_n^{1-\xi}\sum_{l=q+1}^{p}\|A^{\max\{0,p-m-l\}}f^{(l)}(\cdot)\|_{L_\infty(0,t_n;\mathfrak{X})}$$
$$+\,Ck^pt_n^{1-\xi-1/\kappa}\|(t_n-\cdot)f^{(p+1)}(\cdot)\|_{L_\kappa(0,t_n;\mathfrak{X})},$$

[15] It is admissible that $T=\infty$ if $\sigma\geq 0$.

where, unless $\xi = 1$, the factor $t_n^{1-\xi-1/\kappa}$ is replaced by $\ell_n := \log(t_{n+1}/k)$ if $\kappa = 1/(1-\xi)$, while, in the extreme case $\kappa = \infty$ and $\xi = 1$, the factors $t_n^{1-\xi-1/\kappa}$ and $t_n^{1-\xi}$ are replaced by ℓ_n^2 and ℓ_n, respectively.

Proof. We demonstrate our argument only for $\xi < 1$ since the below proof barely changes in the case $\xi = 1$.

Having therefore assumed that $\xi < 1$, for any scalar positive-valued function $\psi(s)$, $s \geq 0$, with the aid of a simple calculation it follows that with $\widetilde{\psi}_n(s) = (t_n - s)\psi(s)$, for all $t_n \in \Omega_k$,

$$
\begin{aligned}
k\sum_{j=0}^{n-1} t_{n-j}^{-\xi} \int_0^{t_j} \psi(s)\,ds &= \sum_{j=1}^{n-1} k \sum_{l=1}^{n-j} t_l^{-\xi} \int_{t_{j-1}}^{t_j} \psi(s)\,ds \\
&\leq C\sum_{j=1}^{n-1} t_{n-j}^{1-\xi} \int_{t_{j-1}}^{t_j} \psi(s)\,ds \leq C\sum_{j=0}^{n-1} t_{n-j}^{-\xi} \int_{t_j}^{t_{j+1}} \widetilde{\psi}_n(s)\,ds, \qquad (7.37)
\end{aligned}
$$

and further, applying Hölder's inequality for integrals and sums, consecutively,

$$
\begin{aligned}
\sum_{j=0}^{n-1} t_{n-j}^{-\xi} \int_{t_j}^{t_{j+1}} \widetilde{\psi}_n(s)\,ds &\leq k\sum_{j=0}^{n-1} t_{n-j}^{-\xi} k^{-1/\kappa} \left(\int_{t_j}^{t_{j+1}} \widetilde{\psi}_n(s)^\kappa ds\right)^{1/\kappa} \\
&\leq Ct_n^{1-\xi-1/\kappa} \left(\sum_{j=0}^{n-1} \int_{t_j}^{t_{j+1}} \widetilde{\psi}_n(s)^\kappa ds\right)^{1/\kappa} \\
&= Ct_n^{1-\xi-1/\kappa} \left(\int_0^{t_n} \widetilde{\psi}_n(s)^\kappa ds\right)^{1/\kappa}. \qquad (7.38)
\end{aligned}
$$

Also, under the accepted conditions, by Theorems 6.10 and 6.11 we have, for all $t \in [0,T]$, $u \in \mathfrak{X}$, and $0 \leq j \leq n-1$, that

$$
]\![\omega(-kA)r(-kA)^{n-j-1}u]\![_t \leq Ct_{n-j}^{-\xi}\|u\|. \qquad (7.39)
$$

Finally, the claim follows by using formula (7.6) in combination with (7.35), (7.37), (7.38), and (7.39). □

It appears that sometimes the same order of convergence is achieved under reduced regularity requirements in comparison with what is stated above. We will give only one particular assertion which, however, enables us to show the main idea.

Theorem 7.6. *Suppose that the conditions of Theorem 7.4 are satisfied for* $m = 0$ *and for some* $\varphi \in (0, \pi/2)$, $\sigma \in \mathbb{R}$, $p, q \in \mathbb{N}$ *subject to* $q \leq p-1$, *and* $T < \infty$ ($T \leq \infty$ *if* $\sigma \geq 0$), *with the difference that it is now assumed that* $f^{(q)}(0) \in \operatorname{Dom} A^{p-q-1}$ *instead of the condition* $A^{p-q} f^{(q)}(\cdot) \in L_\infty(0, t; \mathfrak{X})$ *for* $t \in [0, T]$.[16] *Let furthermore*

$$\deg[\widetilde{h}_q] \leq \deg[\rho], \tag{7.40}$$

where $\widetilde{h}_q(z)$ *is given by (7.25) for* $l = q$. *Then the error estimate (7.21) holds in which the summand* $Ck^p t_n \|A^{p-q} f^{(q)}(\cdot)\|_{L_\infty(0,t_n;\mathfrak{X})}$ *is replaced by* $Ck^p \|A^{p-q-1} f^{(q)}(0)\|$. *Also, if* $q = p-1$, *one can substitute* $Ck^p t_n \|f^{(p)}(0)\|$ *for the summand* $Ck^p t_n \|f^{(p)}(\cdot)\|_{L_\infty(0,t_n;\mathfrak{X})}$.

Proof. Clearly, by (7.6) and (7.30) it suffices to find a different estimate for the expression $\widehat{E}_n = \sum_{j=0}^{n-1} r(-kA)^{n-j-1} \widehat{\phi}_j$, where

$$\widehat{\phi}_j = k^{p+1} \frac{(-1)^{p-q+1}}{q!} A \widetilde{h}_q(-kA)\, A^{p-q-1} f^{(q)}(t_j).$$

Using now (7.40) and the fact that the function $\rho(z)^{-1}$ is (φ)-regular, by Lemma 5.3 we have, with $\widehat{h}_q(z) = \frac{(-1)^{p-q+1}}{q!} \rho(z)^{-1} \widetilde{h}_q(z)$,

$$\|\widehat{h}_q(-kA)\| \leq C. \tag{7.41}$$

Denoting next $Z_j = k^p \widehat{h}_q(-kA) A^{p-q-1} f^{(q)}(t_j)$, observing that $kA\rho(-kA) = \alpha(-kA) = I - r(-kA)$, and using the identity

$$\begin{aligned} \widehat{E}_n &= \sum_{j=0}^{n-1} r(-kA)^{n-j-1} kA\, \rho(-kA)\, Z_j \\ &= Z_n - r(-kA)^n Z_0 + \sum_{j=0}^{n-1} r(-kA)^{n-j-1} (Z_j - Z_{j+1}), \end{aligned}$$

we find, with the aid of (7.15) and (7.41),

$$\begin{aligned} \|\widehat{E}_n\| &\leq \|Z_n\| + C\|Z_0\| + \sum_{j=0}^{n-1} \|Z_j - Z_{j+1}\| \\ &\leq Ck^p \left(\|A^{p-q-1} f^{(q)}(0)\| + \|A^{p-q-1} f^{(q)}(t_n)\| \right) \\ &\quad + Ck^p \sum_{j=0}^{n-1} \int_{t_j}^{t_{j+1}} \|A^{p-q-1} f^{(q+1)}(s)\|\, ds. \end{aligned}$$

[16] We therefore leave the restrictions on $f^{(l)}(t)$, $l = q+1, \ldots, p-1$, as stated before.

Therefore the claim follows by combining this inequality with the estimate

$$\|A^{p-q-1} f^{(q)}(t_n)\| \leq \|A^{p-q-1} f^{(q)}(0)\| + \int_0^{t_n} \|A^{p-q-1} f^{(q+1)}(s)\| \, ds.$$

Furthermore, note that

$$\|f^{(p)}(\cdot)\|_{L_\infty(0,t_n;\mathfrak{X})} \leq \|f^{(p)}(0)\| + \int_0^{t_n} \|f^{(p+1)}(s)\| \, ds.$$

Using this result for an estimation in (7.31) and (7.32) and taking (7.36) into account, it is seen, with the aid of the argument used in the proof of Theorem 7.4, that the more precise wording for $q = p - 1$ is true as well. □

It is not hard to see that for methods of type $C_I(\varphi)$ or of type $C_{III}(\varphi)$ it always holds

$$\deg[\widetilde{h}_l] \leq -1 = \deg[\rho], \quad l = q, \ldots, p.$$

In conclusion, we remark that all RK methods belonging to the concrete families discussed in Section 5.2 are of type $C(\varphi)$ for any $\varphi \in (0, \pi/2)$. All of these methods are therefore used in the format of Theorems 7.4, 7.5, and 7.6. The corresponding magnitudes of their orders and stage orders can be found in Section 5.2.

7.3 Non-autonomous equations

In the present section we show error estimates in terms of the solution for the case with a variable operator. We assume throughout that $T < \infty$, problem (5.1) possesses a unique solution $y(t)$[17] and that $\mathfrak{R} = I$ in (5.19). In our consideration we concentrate on examining the situation where the solution of (5.1) is of low space regularity, and more precisely, it is assumed that at most $y^{(l)}(t) \in \mathcal{D}om$ for $t \in [0, T]$ and for some $l \geq 1$.

We start by showing the following result.

Theorem 7.7. *Suppose that $A(t) \in \mathcal{S}(\varphi, \sigma)$ and $A(t)$ satisfies hypothesis* **lip**, *while the RK method is of type $V_I(\varphi)$, of order p, and of stage order q, for some $\varphi \in (0, \pi/2)$, $\sigma \in \mathbb{R}$, and $p, q \in \mathbb{N}$ such that $q \leq p - 2$. Let also $A(\cdot)\, y^{(q+2)}(\cdot), y^{(p+1)}(\cdot) \in L_1(0, T; \mathfrak{X})$. Then, for the error $E_n = Y_n - y(t_n)$*

[17] Hereinafter $y(t)$ has the same meaning.

of the numerical solution by (5.19), (5.20), it holds, with $K > 0$ sufficiently small and $\mu_1 \geq 0$ sufficiently large, for all $k \in (0, K]$ and $t_n \in \Omega_k$,

$$\|E_n\| \leq Ck^{q+2}\|(A(0) + \mu_1 I)\, y^{(q+1)}(0)\| \\ + Ck^{q+2}\|(A(\cdot) + \mu_1 I)\, y^{(q+2)}(\cdot)\|_{L_1(0,t_n;\mathfrak{X})} + Ck^p\|y^{(p+1)}(\cdot)\|_{L_1(0,t_n;\mathfrak{X})}. \quad (7.42)$$

Proof. Similarly to (5.43) we conclude that with $\mu_1 \geq 0$ sufficiently large, the operator $(\mathcal{D}_n - \mathcal{D}(t_n))\,(\mathfrak{i} + k\mathfrak{a}\mathcal{D}_n)^{-1} \otimes (A(t_n) + \mu_1)^{-1}$ is defined on $\mathfrak{X}$ and[18]

$$\|(\mathcal{D}_n - \mathcal{D}(t_n))\,(\mathfrak{i} + k\mathfrak{a}\mathcal{D}_n)^{-1} \otimes (A(t_n) + \mu_1)^{-1}\| \leq Ck. \quad (7.43)$$

Using therefore the representation

$$\mathcal{D}_n(\mathfrak{i}+k\mathfrak{a}\mathcal{D}_n)^{-1}\otimes(A(t_n)+\mu_1)^{-1} = \mathcal{D}(t_n)\,(\mathfrak{i}+k\mathfrak{a}\mathcal{D}(t_n))^{-1}\otimes(A(t_n)+\mu_1)^{-1} \\ + (\mathfrak{i} + k\mathfrak{a}\mathcal{D}(t_n))^{-1}(\mathcal{D}_n - \mathcal{D}(t_n))(\mathfrak{i} + k\mathfrak{a}\mathcal{D}_n)^{-1} \otimes (A(t_n) + \mu_1)^{-1}, \quad (7.44)$$

which is, in fact, an analogue of (5.44), it follows, with the aid of (5.35) and (7.43), that the operator $\mathcal{D}_n(\mathfrak{i} + k\mathfrak{a}\mathcal{D}_n)^{-1} \otimes (A(t_n) + \mu_1)^{-1}$ is defined on $\mathfrak{X}$ and

$$\|\mathcal{D}_n(\mathfrak{i} + k\mathfrak{a}\mathcal{D}_n)^{-1} \otimes (A(t_n) + \mu_1)^{-1}\| \leq C. \quad (7.45)$$

Also, it follows from (6.56) with $\xi = 0$ that for $t_n, t_j \in \Omega_k$ such that $t_j \leq t_n$,

$$\|\mathfrak{U}_{n-1,j+1}\| \leq C. \quad (7.46)$$

Now, by (7.9) and Lemma 7.2, we have $E_n = E_n^{(1)} + E_n^{(2)} + E_n^{(3)}$, where, with $\mathbf{H} = \left(H_{q1}, \ldots, H_{q\nu}\right)^T$,

$$E_n^{(1)} = -k^{q+2}\sum_{j=0}^{n-1} \mathfrak{U}_{n-1,j+1}\mathbf{b}^T\mathcal{D}_j(\mathfrak{i} + k\mathfrak{a}\mathcal{D}_j)^{-1}\mathbf{H}y^{(q+1)}(t_j),$$

$$E_n^{(2)} = -k^{q+2}\sum_{j=0}^{n-1} \mathfrak{U}_{n-1,j+1}\mathbf{b}^T\mathcal{D}_j(\mathfrak{i}+k\mathfrak{a}\mathcal{D}_j)^{-1}\int_{t_j}^{t_{j+1}} \mathcal{P}_j\left(\frac{s - t_j}{k}\right) y^{(q+2)}(s)\, ds,$$

and

$$E_n^{(3)} = \sum_{j=0}^{n-1} \mathfrak{U}_{n-1,j+1}Q_j(y_t).$$

[18] We recall that the operator $A(t)$ is now Lipschitz-continuous.

It will be seen from below that these expressions are well defined. By the triangle inequality it thus suffices to show that (7.42) holds with $\|E_n^{(j)}\|$, $j = 1, 2, 3$, in place of $\|E_n\|$.

First of all we observe that this fact easily follows in respect of $\|E_n^{(3)}\|$ since we have by (7.3) and (7.46),

$$\|E_n^{(3)}\| \leq Ck^p \int_0^{t_n} \|y^{(p+1)}(s)\| \, ds.$$

Further, using (7.46) and (7.45) and noting that by the assumed hypothesis of Lipschitz-continuity, similarly to (6.47),

$$\|(A(t) + \mu_1 I)\,(A(s) + \mu_1 I)^{-1}\| \leq C \quad \text{for all } t, s \in [0, T], \tag{7.47}$$

we find, since $A(t)$ is closed, that

$$\begin{aligned} \|E_n^{(2)}\| &\leq Ck^{q+2} \sum_{j=0}^{n-1} \|(A(t_n) + \mu_1 I) \int_{t_j}^{t_{j+1}} \mathcal{P}_j\left(\frac{s - t_j}{k}\right) y^{(q+2)}(s) \, ds\| \\ &\leq Ck^{q+2} \int_0^{t_n} \|(A(s) + \mu_1 I)\, y^{(q+2)}(s)\| \, ds. \end{aligned}$$

This result shows (7.42) with $\|E_n^{(2)}\|$ substituted for $\|E_n\|$.

It thus remains to estimate suitably $\|E_n^{(1)}\|$. To that end, we find from (7.44) that $E_n^{(1)} = E_n^{(4)} + E_n^{(5)}$, where

$$E_n^{(4)} = -k^{q+2} \sum_{j=0}^{n-1} \mathfrak{U}_{n-1,j+1} A(t_j)\, \mathbf{w}^T(-kA(t_j))\, \mathbf{H} y^{(q+1)}(t_j)$$

and

$$\begin{aligned} E_n^{(5)} &= -k^{q+1} \sum_{j=0}^{n-1} \mathfrak{U}_{n-1,j+1} \mathbf{b}^T (\mathfrak{i} + k\mathfrak{a}\mathcal{D}(t_j))^{-1} \\ &\quad \times (\mathcal{D}_j - \mathcal{D}(t_j))\, (\mathfrak{i} + k\mathfrak{a}\mathcal{D}_j)^{-1} \mathbf{H} y^{(q+1)}(t_j). \end{aligned}$$

It follows from (7.46), (5.35), and (7.43) that

$$\|E_n^{(5)}\| \leq Ck^{q+3} \sum_{j=0}^{n-1} \|(A(t_j) + \mu_1 I)\, y^{(q+1)}(t_j)\|.$$

However, by the following identity stated for $(\mathfrak{X})$-valued functions,

$$k\psi(t_j) = \int_{t_j}^{t_{j+1}} \big(\psi(s) - (t_{j+1} - s)\,\psi'(s)\big)\,ds,$$

we have instead, using (7.47) and the fact that $A(t)$ is closed,

$$\begin{aligned}\|E_n^{(5)}\| &\leq C\sum_{l=q+1}^{q+2} k^{l+1}\sum_{j=0}^{n-1}\int_{t_j}^{t_{j+1}} \|(A(s)+\mu_1 I)\,y^{(l)}(s)\|\,ds\\ &= C\sum_{l=q+1}^{q+2} k^{l+1}\int_0^{t_n} \|(A(s)+\mu_1 I)\,y^{(l)}(s)\|\,ds.\end{aligned}$$

This fact further gives that

$$\begin{aligned}\|E_n^{(5)}\| &\leq Ck^{q+2}\|(A(0)+\mu_1 I)\,y^{(q+1)}(0)\|\\ &\quad +Ck^{q+2}\int_0^{t_n} \|(A(s)+\mu_1 I)\,y^{(q+2)}(s)\|\,ds,\end{aligned}$$

and (7.42) holds with $\|E_n^{(5)}\|$ substituted for $\|E_n\|$. It thus remains to show (7.42) when substituting $\|E_n^{(4)}\|$ for $\|E_n\|$.

Using now the identity

$$\mathbf{w}^T(-kA(t_j)) = \mathbf{b}^T - kA(t_j)\mathbf{b}^T\mathfrak{a}(\mathfrak{i}+k\mathfrak{a}\mathfrak{D}(t_j))^{-1},$$

and taking Lemma 7.3 into account, we find

$$E_n^{(4)} = k^{q+2}\sum_{j=0}^{n-1}\mathfrak{U}_{n-1,j+1}kA(t_j)\mathbf{b}^T\mathfrak{a}(\mathfrak{i}+k\mathfrak{a}\mathfrak{D}(t_j))^{-1}\mathbf{H}A(t_j)y^{(q+1)}(t_j). \quad (7.48)$$

Since $\deg[\rho] = -1$, $\rho(z)^{-1}$ is (φ)-regular, and the entries of the matrix $\mathfrak{a}(\mathfrak{i}-z\mathfrak{a})^{-1}$ vanish as $|z|\to\infty$, by Lemma 5.3 it follows that with $\eta(z) = \rho(z)^{-1}\mathbf{b}^T\mathfrak{a}(\mathfrak{i}-z\mathfrak{a})^{-1}\mathbf{H}$,

$$\|\eta(-kA(t))\| \leq C \quad \text{for } k\in(0,K] \text{ and } t\in[0,T]. \quad (7.49)$$

Next, using the equalities

$$\mathbf{w}^T(-kA(t_j))\mathbf{H} = \rho(-kA(t_j))\,\eta(-kA(t_j))$$

and

$$kA(t_j)\,\rho(-kA(t_j)) = k\mathfrak{A}(t_j) = I - \mathfrak{U}_j - k\mathfrak{B}_j$$

instead of (7.48) we have, with $Z_j = k^{q+2}\eta(-kA(t_j))\, A(t_j)\, y^{(q+1)}(t_j)$, that

$$E_n^{(4)} = \sum_{j=0}^{n-1} \mathfrak{U}_{n-1,j+1} k\mathfrak{A}(t_j) Z_j = E_n^{(6)} + E_n^{(7)},$$

where

$$\begin{aligned} E_n^{(6)} &= \sum_{j=0}^{n-1} (\mathfrak{U}_{n-1,j+1} - \mathfrak{U}_{n-1,j})\, Z_j \\ &= Z_n - \mathfrak{U}_{n-1,0} Z_0 + \sum_{j=0}^{n-1} \mathfrak{U}_{n-1,j+1}(Z_j - Z_{j+1}) \end{aligned}$$

and

$$E_n^{(7)} = -k \sum_{j=0}^{n-1} \mathfrak{U}_{n-1,j+1} \mathfrak{B}_j Z_j.$$

Since $A(t)$ is under hypothesis **lip**, it follows from (6.61) that $\|\mathfrak{B}_j\| \le C$, which further implies

$$\|E_n^{(7)}\| \le C k^{q+3} \sum_{j=0}^{n-1} \|A(t_j)\, y^{(q+1)}(t_j)\|.$$

Therefore $\|E_n^{(7)}\|$ is finally estimated just the same as $\|E_n^{(5)}\|$ and it remains to prove that (7.42) is fulfilled with $\|E_n^{(6)}\|$ in place of $\|E_n\|$.

Applying again hypothesis **lip** for $A(t)$, it holds

$$\|(A(t_j) - A(t_{j+1}))\, y^{(q+1)}(t_j)\| \le Ck \|(A(t_j) + \mu_1 I)\, y^{(q+1)}(t_j)\|$$

and, by Lemma 6.1,

$$\|\eta(-kA(t_j)) - \eta(-kA(t_{j+1}))\| \le Ck.$$

Using now the last two estimates together with (7.47), (7.49), and the above argument and applying the identity

$$\begin{aligned} Z_j - Z_{j+1} &= k^{q+2} \big(\eta(-kA(t_j)) - \eta(-kA(t_{j+1}))\big)\, A(t_j)\, y^{(q+1)}(t_j) \\ &\quad + k^{q+2} \eta(-kA(t_{j+1}))\, \big(A(t_j) - A(t_{j+1})\big)\, y^{(q+1)}(t_j) \\ &\quad + k^{q+2} \eta(-kA(t_{j+1}))\, A(t_{j+1})\, \big(y^{(q+1)}(t_j) - y^{(q+1)}(t_{j+1})\big), \end{aligned}$$

a simple estimation gives

$$\begin{aligned}\|Z_j - Z_{j+1}\| &\le Ck^{q+3}\|(A(t_j)+\mu_1 I)\, y^{(q+1)}(t_j)\| \\ &+Ck^{q+2}\int_{t_j}^{t_{j+1}} \|(A(s)+\mu_1 I)\, y^{(q+2)}(s)\|\, ds. \quad (7.50)\end{aligned}$$

Therefore, noting that by (7.47),

$$\begin{aligned}\|A(t_n)\, y^{(q+1)}(t_n)\| &\le \|A(0)\, y^{(q+1)}(0)\| + \int_0^{t_n} \|A(t_n)\, y^{(q+2)}(s)\|\, ds \\ &\le C\|(A(0)+\mu_1 I)\, y^{(q+1)}(0)\| + C\int_0^{t_n} \|(A(s)+\mu_1 I)\, y^{(q+2)}(s)\|\, ds,\end{aligned}$$

with the aid of (7.46), (7.49), and (7.50) it follows from the above representation for $E_n^{(6)}$ that

$$\begin{aligned}\|E_n^{(6)}\| &\le Ck^{q+2}\|(A(0)+\mu_1 I)\, y^{(q+1)}(0)\| \\ &+Ck^{q+2}\int_0^{t_n} \|(A(s)+\mu_1 I)\, y^{(q+2)}(s)\|\, ds \\ &+Ck^{q+3}\sum_{j=0}^{n-1} \|(A(t_j)+\mu_1 I)\, y^{(q+1)}(t_j)\|.\end{aligned}$$

This completes the proof since the last term on the right is estimated just the same as $\|E_n^{(5)}\|$. □

It is natural to expect that the order of convergence reduces under weaker requirements of space regularity. In the following statement no space regularity restrictions are imposed on higher derivatives of the exact solution.

Theorem 7.8. *Let the operator $A(t)$ and the RK method satisfy the conditions of Theorem 7.7 for some $\varphi \in (0, \pi/2)$, $\sigma \in \mathbb{N}$, and $p, q \in \mathbb{N}$ such that $q \le p-1$. Let also the coefficient matrix $\mathfrak{a}$ be non-singular. Then, for the error E_n of the numerical solution of (5.1) by (5.19), (5.20), it holds, with $K > 0$ sufficiently small, for all $k \in (0, K]$ and $t_n \in \Omega_k$,*

$$\begin{aligned}\|E_n\| &\le Ck^{q+1}\|y^{(q+1)}(0)\| + Ck^{q+1}\|y^{(q+2)}(\cdot)\|_{L_1(0,t_n;\mathfrak{X})} \\ &+Ck^p\|y^{(p+1)}(\cdot)\|_{L_1(0,t_n;\mathfrak{X})}. \quad (7.51)\end{aligned}$$

Proof. Since $\mathfrak{a}$ is non-singular, we have

$$\mathfrak{D}_j(\mathfrak{i}+k\mathfrak{a}\mathfrak{D}_j)^{-1}\mathbf{Q}_j(y_t) = \mathfrak{D}_j(\mathfrak{i}+k\mathfrak{a}\mathfrak{D}_j)^{-1}\mathfrak{a}\widetilde{\mathbf{Q}}_j(y_t),$$

where $\widetilde{\mathbf{Q}}_j(\cdot) = \mathfrak{a}^{-1}\mathbf{Q}_j(\cdot)$. Also, by (5.43) it follows that

$$\|(\mathfrak{D}_j - \mathfrak{D}(t_j))\,(\mathfrak{i} + k\mathfrak{a}\mathfrak{D}_j)^{-1}\mathfrak{a}\| \leq C.$$

With the aid of these observations as well as of (5.44) and of Lemma 7.2, taking into account the estimates (5.35), (7.3), and (7.46) and using formula (7.9) it is not hard to see that the claim will follow if we show (7.51) with

$$\|\widetilde{E}_n\| = \|k\sum_{j=0}^{n-1} \mathfrak{U}_{n-1,j+1}\mathbf{b}^T\mathfrak{D}(t_j))\,(\mathfrak{i} + k\mathfrak{a}\mathfrak{D}(t_j))^{-1}\mathfrak{a}\widetilde{\mathbf{Q}}_j(y_t)\| \tag{7.52}$$

in place of $\|E_n\|$. This aim can however be achieved by using the argument employed in the proof of Theorem 7.7 and by applying the equality

$$\mathbf{b}^T\mathfrak{D}(t_j))\,(\mathfrak{i} + k\mathfrak{a}\mathfrak{D}(t_j))^{-1}\mathfrak{a}\widetilde{\mathbf{Q}}_j(y_t) = A(t_j))\,\rho(-kA(t_j))\,\mathbf{G}^T(-kA(t_j))\,\widetilde{\mathbf{Q}}_j(y_t),$$

where $\mathbf{G}^T(z) = \rho(z)^{-1}\mathbf{b}^T(\mathfrak{i} - z\mathfrak{a})^{-1}\mathfrak{a}$, since

$$\|\mathbf{G}(-kA(t_j))\| \leq C,$$

by Lemma 5.3. □

Next we show an analogue of both Theorem 7.7 and Theorem 7.8 when the operator $A(t)$ is under a different Hölder-continuity type hypothesis.

Theorem 7.9. *Let $A(t) \in \mathcal{S}(\varphi,\sigma)$ and let the RK method be both of type $V_I(\varphi)$ and of type $V_I^*(\varphi)$ as well as of order p and of stage order q, for some $\varphi \in (0,\pi/2)$, $\sigma \in \mathbb{R}$, and $p, q \in \mathbb{N}$ such that $q \leq p-2$. Let also $\mathcal{A}(t;\cdot,\cdot)$ be a testing functional for $A(t)$. Suppose further that seminorms $|\cdot|_t$ and $|\cdot|_{\mathfrak{m};t}$, $\mathfrak{m} = \emptyset, *\emptyset$, form respectively $(\gamma_0|\varphi,\sigma)$-, $(\gamma|\varphi,\sigma)$-, and $(*;\gamma_*|\varphi,\sigma)$-concordant pairs with $A(t)$ and that $A(t)$ itself satisfies the hypotheses* $\mathbf{h\ddot{o}l}^{\oplus}_{(k)}[\vartheta_0; |\cdot|_t]$ *and* $\mathbf{lip}(\mathcal{A}(t;\cdot,\cdot); |\cdot|_{\mathfrak{m};t}, \mathfrak{m} = \emptyset, *\emptyset)$, *with some $\gamma_0, \gamma, \gamma_* \in [0,1)$ and $\vartheta_0 \in (0,1]$. Then, if $A(\cdot)y^{(q+2)}(\cdot), y^{(p+1)}(\cdot) \in L_1(0,T;\mathfrak{X})$, the error of the numerical solution of (5.1) by (5.19), (5.20) satisfies the estimate (7.42). Furthermore, if the coefficient matrix $\mathfrak{a}$ is non-singular, then the estimate (7.51) holds as well, with the same regularity requirements on $y(t)$ as in the conditions of Theorem 7.8.*[19]

The proof is similar to those of Theorems 7.7 and 7.8, with the aid of Theorem 6.13.

Also, we give some related convergence results. We first show how the error can be estimated in stronger norms.

[19] One can then assume that $q \leq p-1$ instead of the above restriction $q \leq p-2$.

Theorem 7.10. *Let the operator $A(t)$ and the RK method satisfy the conditions of Theorem 7.7 for some $\varphi \in (0, \pi/2)$, $\sigma \in \mathbb{N}$, and $p, q \in \mathbb{N}$ such that $q \leq p-1$. Also, suppose that a seminorm $]\!\![\cdot]\!\![_t$ forms a $(\xi|1|\varphi, \sigma)$-concordant pair with the operator $A(t)$ for some $\xi \in (0, 1]$ and that a rational function $\omega(z)$ is (φ)-regular and satisfies the condition* $\deg[\omega] \leq -1$. *Then, if $A(\cdot)y^{(q+1)}(\cdot), y^{(p+1)}(\cdot) \in L_\kappa(0, T; \mathfrak{X})$ for some $\kappa \in [1/(1-\xi), \infty]$, the error E_n of the numerical solution of (5.1) by (5.19), (5.20) satisfies the estimate, with some $\mu_1 \geq 0$ and with $K > 0$ sufficiently small, for all $k \in (0, K]$ and $t_n \in \Omega_k$,*

$$\begin{aligned}]\!\![\omega(-kA(t_n))E_n]\!\![_{t_n} &\leq Ck^{q+1}\|(A(.) + \mu_1 I)y^{(q+1)}(.)\|_{L_\kappa(0,t_n;\mathfrak{X})} \\ &\quad + Ck^p\|y^{(p+1)}(.)\|_{L_\kappa(0,t_n;\mathfrak{X})}, \end{aligned}$$

where the right-hand side of this inequality is taken $\log(2 + 1/k)$ *times in the case $\kappa = 1/(1-\xi)$.*

Proof. First of all it follows from Theorem 6.12 and Remark 6.8 that for all $t_n, t_j \in \Omega_k$ such that $t_j \leq t_n$ and $u \in \mathfrak{X}$,

$$]\!\![\omega(-kA(t_n))\mathfrak{U}_{n-1,j+1}u]\!\![_{t_n} \leq Ct_{n-j}^{-\xi}\|u\|.$$

Using therefore this estimate together with (7.45), (7.3), and (7.4), a direct estimation on the basis of (7.9) shows the claim, with the aid of the argument leading to (7.38). □

One more direction for possible extension is performed in the next assertion.

Theorem 7.11. *Let the operator $A(t)$ and the RK method be under the conditions of Theorem 7.7 for some $\varphi \in (0, \pi/2)$, $\sigma \in \mathbb{N}$, and $p, q \in \mathbb{N}$ such that $q \leq p-1$, and let the coefficient matrix $\mathfrak{a}$ be non-singular. Let there further be given seminorms $|\cdot|_\bullet$ and $|\cdot|_{*\bullet;t}$ of which the former one is independent of t and has domain $\subseteq \mathfrak{X}$ while the latter one forms a $(*;\gamma_*|\varphi, \sigma)$-concordant pair with $A(t)$ for some $\gamma_* \in [0, 1)$ so that we have, with some $\mu_1 \geq 0$ sufficiently large, for all $t \in [0, T]$, $\chi \in \mathfrak{X}^*$, and $u \in$* Dom $|\cdot|_\bullet$, *that*

$$|\langle \chi, A(t)\,(A(t) + \mu_1 I)^{-1}u\rangle| \leq C|((A(t) + \mu_1 I)^{-1})^*\chi|_{*\bullet;t}|u|_\bullet.$$

Then if $y^{(p+1)}(\cdot) \in L_1(0, T; \mathfrak{X})$ and $|y^{(q+1)}(\cdot)|_\bullet \in L_\kappa(0, T; \mathbb{R})$ for some $\kappa \in [1/(1-\gamma_*), \infty]$, the error E_n of the numerical solution satisfies the estimate, with $K > 0$ sufficiently small, for all $k \in (0, K]$ and $t_n \in \Omega_k$,

$$\begin{aligned} \|E_n\| &\leq Ck^{q+1}\left(\|\,|y^{(q+1)}(\cdot)|_\bullet\|_{L_\kappa(0,t_n;\mathbb{R})} + \|y^{(q+1)}(\cdot)\|_{L_\kappa(0,t_n;\mathfrak{X})}\right) \\ &\quad + Ck^p\|y^{(p+1)}(.)\|_{L_1(0,t_n;\mathfrak{X})}, \end{aligned} \tag{7.53}$$

where the first term on the right-hand side of this inequality is taken $\log(2 + 1/k)$ *times in the case* $\kappa = 1/(1-\gamma_*)$.

Proof. As in the proof of Theorem 7.8, using (5.44), (7.3), and (7.4), it suffices to show the claim with $\|\widetilde{E}_n\|$ defined by (7.52) in place of $\|E_n\|$. By (7.53) we have, with the aid of the above argument,

$$\|\widetilde{E}_n\| \le Ck^{q+1} \sum_{j=0}^{n-1} \sup_{\|\chi\|_*=1} \left|((A(t_j) + \mu_1 I)^{-1})^* \mathfrak{U}^*_{n-1,j+1}\chi\right|_{*\bullet;t_j} \times \int_{t_j}^{t_{j+1}} |y^{(q+1)}(s)|_\bullet \, ds.$$

It thus remains to apply Theorem 6.12 (more precisely, the stability estimate (6.58)), Remark 6.8, and the reasoning leading to (7.38). □

As concerns the last two theorems, note that similar assertions are established under the Hölder-continuity type hypotheses that are assumed in the statement of Theorem 7.9.

Further we present some error estimates that are intended for use when the RK method is of class S_{III}. We do no more than to show two separate results.

Theorem 7.12. *Suppose that the operator* $A(t)$ *is under hypothesis* **lip** *and* $A(t) \in \mathcal{S}(\varphi, \sigma)$ *while the RK method satisfies the condition* $\deg[\alpha(\cdot) - \alpha(\infty)] = -1$ *and is both of type* $V_{III}(\varphi)$ *and of type* $V^*_{III}(\varphi)$ *as well as of order* p *and of stage order* q, *for some* $\varphi \in (0, \pi/2)$, $\sigma \in \mathbb{N}$, *and* $p, q \in \mathbb{N}$ *such that* $q \le p - 1$. *Then, if* $y^{(p+1)}(\cdot), A(\cdot)y^{(p+1)}(\cdot) \in L_1(0,T;\mathfrak{X})$, *the error* E_n *of the numerical solution of problem (5.1) by (5.19), (5.20) satisfies the estimate, with* $K > 0$ *sufficiently small, for all* $k \in (0, K]$ *and* $t_n \in \Omega_k$,

$$\|E_n\| \le Ck^{q+1} \left(\|(I + kA(0))y^{(q+1)}(0)\| + \|(I + kA(\cdot))y^{(q+2)}(\cdot)\|_{L_1(0,t_n;\mathfrak{X})} \right) + Ck^p \|(I + kA(\cdot))y^{(p+1)}(\cdot)\|_{L_1(0,t_n;\mathfrak{X})}.$$

Proof. In fact, the result can be shown in a manner similar to that used in the proof of Theorem 7.8. To achieve success we also use the estimates, for $0 \le j \le n-1$,

$$\|\mathfrak{U}_{n-1,j+1}(I + kA(t_j))^{-1}\| \le C$$

and

$$\|(\mathfrak{D}_j - \mathfrak{D}(t_j))\,(\mathfrak{i} + k\mathfrak{a}\mathfrak{D}_j)^{-1}(I + kA(t_j)^{-1}\| \le C$$

of which the former one follows by Theorem 6.15 while the latter one can be proved in the same way as (5.43). □

Theorem 7.13. *Suppose that* $A(t) \in \mathcal{S}(\varphi, \sigma)$ *and the RK method is both of type* $V_{III}(\varphi)$ *and of type* $V_{III}^*(\varphi)$ *as well as of order* p *and of stage order* q *and satisfies the condition* $\deg[\alpha(\cdot) - \alpha(\infty)] = -1$, *for some* $\varphi \in (0, \pi/2)$, $\sigma \in \mathbb{R}$, *and* $p, q \in \mathbb{N}$ *such that* $q \leq p - 2$. *Suppose further that seminorms* $|\cdot|_{\mathfrak{m};t}$, $\mathfrak{m} = \emptyset, *\emptyset$, *and* $|\cdot|_t$ *form respectively* $(1/2|\varphi, \sigma)$-, $(*; 1/2|\varphi, \sigma)$-, *and* $(\gamma_0|\varphi, \sigma)$*-concordant pairs with* $A(t)$ *and that* $A(t)$ *satisfies the hypotheses* $\mathbf{h\ddot{o}l}^{\oplus}_{(k)}[\vartheta_1; |\cdot|_t]$ *and* $\mathbf{lip}(\mathcal{A}(t;\cdot,\cdot); |\cdot|_{\mathfrak{m};t}, \mathfrak{m} = \emptyset, *\emptyset)$, *with* $\mathcal{A}(t;\cdot,\cdot)$ *a testing functional for* $A(t)$ *and with some* $\gamma_0 \in [0, 1)$ *and* $\vartheta_1 \in (0, 1]$. *Let also the conditions (6.62) and (6.63) hold with* $\gamma = \gamma_* = 1/2$ *and with some* $\vartheta_\bullet, \vartheta_{*\bullet} \in [0, 1)$ *such that* $\vartheta_\bullet + \vartheta_{*\bullet} < 1$. *Then the error estimate (7.42) holds under the same regularity requirements on* $y(t)$ *as in the conditions of Theorem 7.7. If in addition the coefficient matrix* $\mathfrak{a}$ *is non-singular, then the estimate (7.51) holds as well*[20] *under the regularity requirements on* $y(t)$ *as in the conditions of Theorem 7.8.*

The proof is similar to those of Theorems 7.7 and 7.8, noting that the stability estimate (7.46) is in force under the accepted conditions, as it follows from Theorem 6.16.

It is worth noting that the conditions of Theorems 7.7, 7.8, 7.9, 7.10, and 7.11 are fulfilled for all Radau IA, IIA and Lobatto IIIC methods as well as for all RK methods which satisfy (5.26), (5.27). Besides, Theorems 7.12 and 7.13 are applicable, for instance, for all Gauss-Legendre of odd order. The magnitudes of orders and stage orders for these methods are pointed out in Section 5.2.

7.4 Working against order reduction

Now the reader sees that the classical order of Runge-Kutta methods is not in general achieved if the operator in question is unbounded and the data are not sufficiently smooth. Therefore the question arises if there really are any means of avoiding the phenomenon of order reduction or at least of raising the order of convergence. In the present section we discuss some possibilities along those lines. For simplicity we restrict ourselves to the case where $A(t) \in \mathcal{S}(\varphi, \sigma)$ with $\sigma > 0$, which yields that the operator $A(t)$ has a bounded inverse with a uniform estimate $\|A(t)^{-1}\| \leq C$ for $t \in [0, T]$.

We begin with a discussion in the case where $A(t) \equiv A$ is independent of t. Let the RK method be of order p and of strict order $q \leq p - 1$. We associate to problem (5.1) the following auxiliary problem, with $j = p - q - 1$

[20] We then assume that $q \leq p - 1$.

or with $j = p - q$,

$$\begin{aligned} z' + Az &= (-1)^j A^{-j} f^{(j)}, \quad 0 < t \leq T, \\ z(0) &= y^0 + \sum_{l=1}^{j} (-1)^l A^{-l} f^{(l-1)}(0). \end{aligned} \tag{7.54}$$

The point is that the solutions of (5.1) and (7.54) are connected by

$$y(t) = z(t) - \sum_{l=1}^{j} (-1)^l A^{-l} f^{(l-1)}(t). \tag{7.55}$$

Therefore, having obtained an approximation for $z(t)$ and using next the last relation, it is easy to get an approximation to $y(t)$. At the same time it is essential that one can approximately calculate $z(t)$ by a suitable RK method, achieving the optimal order of accuracy ($= p$) under reduced space regularity requirements on the time derivatives of $f(t)$. Acting so, we have the following.

Theorem 7.14. *Let $T = \infty$, $j = p - q$ or $j = p - q - 1$, $A \in \mathcal{S}(\varphi, \sigma)$ and let the RK method in question be of type $C(\varphi)$, of order p, and of strict order q, for some $\varphi \in (0, \pi/2)$, $\sigma > 0$, and $p, q \in \mathbb{N}$ such that $q \leq p - 1$. Suppose further that the function $f(t)$ is $p+j+1$ times strongly continuously differentiable and problem (5.1) possesses an extended solution of order $p+1$. Finally, let Z_n be the solution by the corresponding rational method (5.30) applied to (7.54) with $\mathfrak{R} = I$, and let Y_n be defined by Z_n through*

$$Y_n = Z_n - \sum_{l=1}^{j} (-1)^l A^{-l} f^{(l-1)}(t_n).$$

Then, assuming in addition, if $j = p - q - 1$, that $f^{(q)}(t) \in \operatorname{Dom} A$ for $t \in [0, T]$, the error $Y_n - y(t_n)$ of the numerical solution of the original Cauchy problem (5.1) satisfies the estimate, with $K > 0$ sufficiently small, for all $k \in (0, K]$ and $t_n \in \Omega_k$,

$$\begin{aligned} \|Y_n - y(t_n)\| \leq{} & Ck^p t_n \|y^0_{(p+1)}\| + Ck^p t_n \sum_{l=1}^{j} \|A^{-l} f^{(p+l)}(0)\| \\ & + Ck^p t_n \sum_{l=q}^{p} \|A^{p-l-j} f^{(j+l)}(\cdot)\|_{L_\infty(0,t_n;\mathfrak{X})} \\ & + Ck^p \|(t_n - \cdot) A^{-j} f^{(p+j+1)}(\cdot)\|_{L_1(0,t_n;\mathfrak{X})}, \end{aligned}$$

where $y^0_{(p+1)}$ comes from formula (B.5) stated for $p+1$ in place of p.

Proof. Since problem (5.1) possesses an extended solution of order $p+1$, it is not hard to see that the same is true for problem (7.54) and

$$z^0_{(p+1)} = y^0_{(p+1)} + \sum_{l=1}^{j} (-1)^l A^{-l} f^{(p+l)}(0),$$

which results from (7.55) by using (B.4). The claim thus follows by direct application of Theorem 7.4 to problem (7.54). □

Note that the expression on the right-hand side of the last error estimate contains a term of the form $Ck^p \| Ay^{(p-1)}(.)\|_{L_\infty(0,t_n;\mathfrak{X})}$ if $j = p-q-1$ while no positive powers of A appear in the case $j = p-q$.

Further we turn to the case with a variable operator. We will show only that, modifying the RK method in a suitable way, it is possible to raise the order of converegence at least by 1, without imposing stronger restrictions of space regularity on the exact solution but requiring instead more regularity in time. Note that by further extensions of these techniques in order to get higher orders of convergence under the original requirements of space regularity we are led to the necessity of imposing certain (difficult for verification) extra conditions on the operator $A(t)$. Anyway we leave such extensions beyond our present consideration.

Let the RK method in question be of order p and of stage order $q \le p-2$. This method is applied to the auxiliary problem[21]

$$z' + A(t)z = -\frac{d}{dt}\left(A(t)^{-1} f(t)\right), \quad 0 < t \le T, \tag{7.56}$$

$$z(0) = y^0 - A(0)^{-1} f(0). \tag{7.57}$$

For our aims it is important that $y(t)$ can be expressed by $y(t) = z(t) + A(t)^{-1} f(t)$. We state here only one result illustrating the main idea. Further possibilities can be studied on the basis of the assertions presented in Section 7.3.

Theorem 7.15. *Let $T < \infty$, $\mathfrak{R} = I$ and let the operator $A(t)$ and the RK method satisfy the conditions of Theorem 7.7*[22] *for some $\varphi \in (0, \pi/2)$, $\sigma > 0$, and $p, q \in \mathbb{N}$ such that $q \le p-2$. Suppose further that $A(t)$ is $p+1$ times strongly continuously differentiable on $\mathcal{D}om = \mathrm{Dom}\, A(t)$ and that the functions $y(t)$ and $f(t)$ are sufficiently smooth with respect to t. Apart from everything else, let Z_n be the discrete solution by the RK method applied to*

[21] We recall that $A(t)^{-1}$ is uniformly bounded, by our assumptions.

[22] In particular these assumptions mean that $\mathrm{Dom}\, A(t)$ is independent of t.

the auxiliary problem (7.56) and let $Y_n = Z_n + A^{-1}(t_n)f(t_n)$. *Then it holds, with* $K > 0$ *sufficiently small, for all* $k \in (0, K]$ *and* $t_n \in \Omega_k$,

$$\begin{aligned}\|Y_n - y(t_n)\| &\le Ck^{q+2}\sum_{l=1}^{q+2}\|y^{(l)}(0)\| + Ck^{q+2}\|y^{(q+3)}(\cdot)\|_{L_1(0,t_n;\mathfrak{X})}\\ &\quad + Ck^p\sum_{l=q+3}^{p+1}\|y^{(l)}(0)\| + Ck^p\|y^{(p+2)}(\cdot)\|_{L_1(0,t_n;\mathfrak{X})}.\end{aligned}$$

Proof. By Theorem 7.7, we have[23]

$$\begin{aligned}\|Y_n - y(t_n)\| = \|Z_n - z(t_n)\| \le Ck^{q+2}\|A(0)\, z^{(q+1)}(0)\|\\ + Ck^{q+2}\|A(\cdot)\, z^{(q+2)}(\cdot)\|_{L_1(0,t_n;\mathfrak{X})} + Ck^p\|z^{(p+1)}(\cdot)\|_{L_1(0,t_n;\mathfrak{X})}.\end{aligned} \tag{7.58}$$

Now, with the aid of the equalities

$$z(t) = y(t) - A(t)^{-1}f(t) = -A(t)^{-1}y'(t)$$

and

$$(A(t)^{-1})' = -A(t)^{-1}A'(t)A(t)^{-1}$$

as well as of the estimate[24]

$$\|A^{(l)}(t)A(t)^{-1}\| \le C \quad \text{for } l = 1, \dots, p+1,$$

by Leibniz' rule it follows that for $m = q+1, q+2$,

$$\|A(t)\, z^{(m)}(t)\| = \|A(t)\, \left(A(t)^{-1}y'(t)\right)^{(m)}\| \le C\sum_{l=1}^{m+1}\|y^{(l)}(t)\|$$

and

$$\|z^{(p+1)}(t)\| \le C\sum_{l=1}^{p+2}\|y^{(l)}(t)\|.$$

Therefore combining the last two estimates with (7.58) and observing that for $t \in [0, T]$,

$$\sum_{l=1}^{m}\|y^{(l)}(t)\| \le C\sum_{l=1}^{m}\|y^{(l)}(0)\| + C\|y^{(m+1)}(\cdot)\|_{L_1(0,t;\mathfrak{X})}, \quad m = q+2, p+1,$$

the desired result follows. □

[23] One can write $A(s)$ instead of $A(s) + \mu_1 I$ since $A(s)^{-1}$ is uniformly bounded.

[24] This estimate obtains by the $(p+1)$-tuple strongly continuous differentiability of $A(t)$ on $\mathcal{D}om$.

Note that applying the RK method directly to problem (5.1) and using Theorem 7.8 does not allow us to be confident that $\|Y_n - y(t_n)\| \sim \mathcal{O}(k^{q+2})$ unless we make certain assumptions on the behaviour of $A(t)y^{(q+2)}(t)$ (cf. Theorem 7.7).

7.5 Comments and bibliographical remarks

Section 7.2: The homogeneous case seems to have been thoroughly investigated by now even in the framework of Banach spaces. Apparently, Le Roux [102] is the first who derived error estimates in the case with sectorial operator. It is remarkable that her result, although presented for Hilbert spaces, admits an extension to a Banach space setting. Such an extension is stated in Brenner & Thomée [52], where an analogue of the present Theorem 7.1 for general C_0 semigroups is also contained. Analogues of Theorem 7.2 in the case $\xi = 0$, stated for methods of class S_I as well, are found in [102] and [52]. Regarding the general case $\xi \geq 0$, the results related to Theorem 7.2 are presented in our work [37]. Although the paper [37] deals with the case $m = 0$ only, its techniques are trivially extended in order to be applicable for $m \geq 1$ too. Apart from everything else, we give here a similar result for methods of the classes S_{II} and S_{III} (see Theorem 7.3). Note that for both classes of methods the presence of a damping factor, whose role is played by the operator $\omega(-kA)$, is essential even for $\xi = 0$ if the initial condition y^0 is not smooth enough. The idea of damping, when deriving error estimates for non-smooth data, goes back to the work of Luskin & Rannacher [109] and Rannacher [132] carried out for Hilbert spaces and some special methods. Also, some nonsmooth data error estimates in a Banach space setting are contained in Hansbo [82, 83], for A-stable methods only when dealing with the classes S_{II} and S_{III}.[25] It is easily seen from the above considerations that showing nonsmooth data error estimates reduces, through applying Lemma 7.5, to making use of certain strong stability type estimates (cf., e.g., the proof of Theorem 7.2) while the role of damping in stability analysis has already been demonstrated in the preceding chapter.

In the inhomogeneous case one should expect extra difficulties which result from the fact that the solution of the original Cauchy problem (5.1) has to satisfy certain compatibility conditions in order to obtain the optimal order of converegence otherwise one meets an order reduction. It turns out that a suitable choice of method allows one to reduce the requirements of

[25] For methods of these classes, the requirement of A-stability implies the condition $\deg[\alpha(\cdot) - \alpha(\infty)] = -1$.

space regularity, which are very often thought of as artificial. This possibility is studied in detail for the first time by Crouzeix [59] in the context of Hilbert spaces; unfortunately, his techniques cannot be extended to the Banach case. For equations in a Banach space, a related result is first given in Brenner, Crouzeix, and Thomée [50]. Their achievement has been further extended in Ostermann & Roche [123, 124] and in Lubich & Ostermann [105]. It follows from their work that sometimes it is natural to expect even fractional orders of convergence. The proofs of Theorems 7.4 and 7.6 follow in essence the paper [50].

Section 7.3: The question of convergence of Runge-Kutta methods, applied to linear non-autonomous differential equations in a Banach space, does not seem to be studied thoroughly up to now. We mention however a paper of González & Ostermann [71] which contains results that are close to the assertion of the above Theorem 7.7. At the same time Theorem 7.7 is stated under somewhat less restrictive requirements than in [71]. Some further developments are given in the statements of Theorems 7.8, 7.10, and 7.11. Theorem 7.12 presents a convergence result for methods of class S_{III} as well. The mentioned results, including that of the paper [71], are not helpful for applications connected with multidimensional problems in spaces with maximum norm while the assertions of Theorems 7.9 and 7.13 are intended for use in such situations.

Section 7.4: The question of working against order reduction has been extensively discussed in the literature (see, e.g., Abarbanel, Gottlieb, and Carpenter [1], Calvo & Palencia [55], Carpenter et al. [57], Keeling [92], Pathria [130], and Sanz-Serna, Verwer, and Hundsdorfer [140]). Our consideration in the present section has no pretensions to being a complete exposition of the problem.

Chapter 8

Variable Stepsize Approximations

Our aim here is to extend some of the previous results to the case of non-uniform time partitions. In this chapter we never speak about discrete semigroups since the transition operator depends essentially on the corresponding time moment. In our analysis below we nevertheless distinguish between two cases: commutative and non-commutative. In the commutative case, although the transition operator varies with time, it is always a (varying) function of one fixed operator. This circumstance allows us to use the operator calculus methods. In this sense the situation is similar to that covered by autonomous equations and uniform grids. In the non-commutative case we make use of certain perturbation techniques.

Throughout this chapter any set of nodal points $\{t_0, t_1, \ldots, t_N\}$, $N \geq 1$, such that $t_0 = 0 < t_1 < \ldots < t_N = T$ is called a non-uniform grid (non-uniform partition) in the segment $[0, T]$ and is denoted by $\mho_k$. The situation when $T = \infty$ is also covered by our consideration; but in this case $\mho_k$ consists of infinitely many nodal points. The parameter k is now not a number but a sequence (finite or infinite), and more precisely, k is defined as $k = \{k_0, \ldots, k_{N-1}\}$, where $k_n = t_{n+1} - t_n$ is the variable stepsize. Hereinafter, given a grid $\mho_k$ with $N+1$ nodal points, it is always accepted that $k_N = k_{N-1}$ and $t_{N+1} = t_N$. As above, we distinguish any objects that may depend on discrete time t_n by n as a subscript. Also, we denote $\breve{\mho}_k = \mho_k \setminus \{t_N\}$ and

$$\partial Z_n = k_n^{-1}(Z_{n+1} - Z_n) \quad \text{for } t_n \in \breve{\mho}_k, \tag{8.1}$$

for any discrete function $Z_n : \breve{\mho}_k \to \mathfrak{X}$.

Further, let there be a non-uniform grid $\mho_k$ and let the above RK method (5.19), (5.20) be used for discretization of problem (5.1) on $\mho_k$, with $\mathfrak{R} = I$.[1] Since Runge-Kutta discretization applies locally, that is, separately at each time moment, we then handle a discrete problem which takes, in fact, the same form as the above discrete problem (5.19), (5.20), with the minor difference that k is replaced by k_n and, in particular, t_{nj} are now given by $t_{nj} = t_n + c_j k_n$. It is not hard to see that the statement of the main theorem on representability of the starting problem (5.19), (5.20) in the form (1.1)[2] is in force with k_n in place of k, provided that ∂Y_n in (1.1) is now understood in the new sense, as specified by (8.1). At the same time, if $A(t) \equiv A$ is independent of t, similarly to the case with a constant stepsize, we then consider problem (1.1) with

$$\mathfrak{A}_n = k_n^{-1}\alpha(-k_n A), \quad \mathfrak{Q}_n = I, \tag{8.2}$$

and

$$F_n = \mathbf{w}^T(-k_n A)\mathbf{\Phi}_n = \sum_{j=1}^{\nu} w_j(-k_n A) f(t_{nj}), \tag{8.3}$$

as a generalized form of (5.19), (5.20). As in Chapter 5, we then deal with the rational method (5.30), with k_n substituted for k, which defines uniquely a discrete function $Y_n : \mho_k \to \mathfrak{X}$. This function is called the **extended solution** by the RK method on $\mho_k$ in the autonomous case. Regardless of the way in which $\mathfrak{A}_n$ appears, we always assume that $\mathfrak{A}_N = \mathfrak{A}_{N-1}$, where $N+1$ is the number of nodes of $\mho_k$.

Below we derive stability and error estimates for the solution or if $A(t) \equiv A$ then for the extended solution of the Runge-Kutta procedure (5.19), (5.20) in the situation with a variable stepsize. In fact, we take the same conditions on the operator $A(t)$ as in the case with a constant stepsize. At the same time, using the assumption that $A(t) \in \mathcal{S}(\varphi, \sigma)$, we assume in addition that $\sigma \geq 0$, for simplicity. Also, we are not concerned with all of the above introduced classes of methods S_I, S_{II}, S_{III}, dealing with class S_I only. In the case with a time-variable, Hölder-continuous operator, we take some extra restrictions on the used families of non-uniform grids.

---

[1] For the sake of simplicity, no correction of the initial condition y^0 is introduced in this chapter.

[2] That is, of Theorem 5.1

8.1 The commutative case

Here we consider the case where the operator $A(t) \equiv A$ does not depend on t. In this situation, for any $t_n, t_l \in \mho_k$ the corresponding transition operators $\mathfrak{U}_n$ and $\mathfrak{U}_l$ commute, and one can therefore use an operator calculus formula as the main tool of analysis.

First of all we give some useful comments. Let the RK method be $A_I(\varphi)$-stable with some $\varphi \in (0, \pi/2)$ and let φ_1 be any fixed number such that $\varphi_1 \in (0, \varphi)$. For subsequent reference we note that there exist constants $d > 0$ sufficiently small and $d_1 > 0$ sufficiently large such that the function $r(z)$ has no poles in $\mathcal{D}(d)$ and outside of $\operatorname{Int}\mathcal{D}(d_1)$,[3] and the following inequalities hold

$$|r(z)| \leq e^{c|z|} \quad \text{for all } z \in \mathcal{D}(d), \tag{8.4}$$

$$\max\left\{(1+|z|)^{-1}, |r(-z)|\right\} \leq e^{-c|z|} \quad \text{for all } z \in \Sigma_{\varphi_1} \cap \mathcal{D}(d_1), \tag{8.5}$$

and

$$\max\left\{(1+|z|)^{-1}, |r(z)|\right\} \leq \varepsilon \quad \text{for all } z \notin \operatorname{Int}\mathcal{D}(d_1), \tag{8.6}$$

with some $\varepsilon \in (|r(\infty)|, 1)$. Note that at the expense of choosing d_1 sufficiently large, ε can be taken as close to $|r(\infty)|$ as desired.

Next we show certain auxiliary results. In the corresponding statements we let the operator $A(t)$ depend upon t for the reason that this assumption will be needed for our subsequent purposes.

Lemma 8.1. *Let $A(t) \in \mathcal{S}(\varphi, \sigma)$ and let the RK method be $A_I(\varphi)$-stable, with some $\varphi \in (0, \pi/2)$ and $\sigma \geq 0$. Then, for any finite sequence of positive numbers $h_1, \ldots, h_n$, it holds, for all $t \in [0, T]$,*

$$\left\| \prod_{j=1}^{n} r(-h_j A(t)) \right\| \leq C.$$

Proof. Without loss of generality it can be assumed that $h_n \geq \cdots \geq h_1$. Accepting therefore the last condition, we further use the following identity

$$\prod_{j=1}^{n} r(-h_j A(t)) = r(\infty)^n I + \sum_{j=1}^{n} r(\infty)^{n-j} G_j(t), \tag{8.7}$$

where

$$G_j(t)) = (r(-h_j A(t)) - r(\infty) I) \prod_{l=1}^{j-1} r(-k_l A(t)). \tag{8.8}$$

[3]Clearly, the same is true for the function $\alpha(z)$.

Denoting next

$$\tau_j = \sum_{l=1}^{j} h_l, \tag{8.9}$$

we let $\Gamma^{(j)}$ be the contour coinciding with the boundary of the set $\mathcal{D}(\tau_j^{-1}d) \cup \Sigma_{\varphi_1}$, where d is the same as in (8.4) and φ_1 comes from the statement of Lemma A.1. Note that this lemma can be applied now with $R = 0$ since $\sigma \geq 0$. Clearly, $\Gamma^{(j)}$ can be decomposed into two parts as follows $\Gamma^{(j)} = \Gamma_1^{(j)} \cup \Gamma_2^{(j)}$, where $\Gamma_1^{(j)} = \partial\mathcal{D}(d\tau_j^{-1}) \cap \big(- \Sigma_{\pi-\varphi_1}\big)$ and $\Gamma_2^{(j)} = \{z \in \mathbb{C} : \arg z = \pm\varphi_1,\ \tau_j^{-1}d \leq |z| \leq \infty\}$. Using then the operator calculus formula (A.10) with $A(t)$ in place of $\mathfrak{L}$ and with $\Gamma^{(j)}$ in place of $\Xi_{R+\varepsilon}$, we get, for $1 \leq j \leq n$,

$$\begin{aligned} G_j(t) &= \frac{1}{2\pi i}\int_{\Gamma^{(j)}} (r(-h_j z) - r(\infty)) \prod_{l=1}^{j-1} r(-h_l z)\,(zI - A(t))^{-1}\,dz \\ &= G_1^{(j)} + G_2^{(j)}, \end{aligned} \tag{8.10}$$

where $G_l^{(j)} = (2\pi i)^{-1}\int_{\Gamma_l^{(j)}} \cdots$, $l = 1, 2$. Since $A(t) \in \mathcal{S}(\varphi, \sigma)$, using Lemma A.1 yields, with the aid of (8.4), that

$$\|G_1^{(j)}\| \leq C \int_{\Gamma_1^{(j)}} e^{c\tau_j|z|}|z|^{-1}|dz| \leq C. \tag{8.11}$$

Further, applying again Lemma A.1 and taking into account (8.5), (8.6), and the obvious inequality

$$|r(-z) - r(\infty)| \leq C(1 + |z|)^{-1} \quad \text{for all } z \in \Sigma_{\varphi_1},$$

we obtain

$$\begin{aligned} \|G_2^{(j)}\| &\leq C\int_{\tau_j^{-1}d}^{d_1 h_j^{-1}} e^{-c\tau_{j-1}x}x^{-1}\,dx + C\int_{d_1 h_j^{-1}}^{\infty} (1 + h_j x)^{-1}x^{-1}\,dx \\ &=: I_1 + I_2. \end{aligned} \tag{8.12}$$

It is not hard to see that for $h_j \leq \tau_{j-1}$,

$$I_1 \leq C\int_{\tau_j^{-1}d}^{h_j^{-1}d_1} e^{-c\tau_j x}x^{-1}\,dx \leq C\int_d^{\infty} e^{-\frac{b}{2}x}x^{-1}\,dx = C, \tag{8.13}$$

while for $h_j \geq \tau_{j-1}$,

$$I_1 \leq C \int_{\tau_j^{-1} d}^{h_j^{-1} d_1} x^{-1}\, dx \leq C \int_{d/2}^{d_1} x^{-1}\, dx = C. \tag{8.14}$$

Now, combining (8.10)-(8.14) and using the obvious fact that $I_2 \leq C$, with the aid of the triangle inequality it follows that $\|G_j(t)\| \leq C$ for $1 \leq j \leq n$, which together with (8.7) gives, after a trivial estimation,

$$\left\| \prod_{j=1}^{n} r(-h_j A(t)) \right\| \leq C(1 + |r(\infty)| + \cdots + |r(\infty)|^n).$$

Clearly, this inequality shows the claim since $|r(\infty)| < 1$. □

Lemma 8.2. *Let the conditions of Lemma 8.1 be satisfied for some $\varphi \in (0, \pi/2)$ and $\sigma \geq 0$. Then for any non-decreasing finite sequence of positive numbers $0 < h_1 \leq \cdots \leq h_n$, it holds, for all $t \in [0,T]$ and $l = 1, \ldots, n$,*

$$\left\| \alpha(-h_l A(t)) \prod_{\substack{j \neq l, \\ 1 \leq j \leq n}} r(-h_j A(t)) \right\| \leq C \left(h_l \sum_{j=0}^{n-1} \varepsilon^j \tau_{n-j}^{-1} + |r(\infty)|^n \right),$$

where ε comes from (8.6) and τ_j is given by (8.9).

Proof. Since $\deg[\rho] = -1$, it follows from Lemma 5.3 that

$$\|(I + hA(t))\rho(-hA(t))\| \leq C \quad \text{for all } h > 0,$$

which further yields, for all $h > 0$ and $u \in \mathfrak{X}$,

$$\begin{aligned} \|\alpha(-hA(t))u\| &= h\|(I + hA(t))\rho(-hA(t))A(t)\,(I + hA(t))^{-1}u\| \\ &\leq Ch\|A(t)\,(I + hA(t))^{-1}u\|. \end{aligned} \tag{8.15}$$

Let now Γ be the contour coinciding with $\partial\Sigma_{\varphi_1}$.[4] Using therefore the operator calculus formula (A.10) in which the path of integration is reduced to Γ since the integrand is bounded at 0, we have

$$h_l A(t)\,(I + h_l A(t))^{-1} \prod_{\substack{j \neq l, \\ 1 \leq j \leq n}} r(-h_j A(t)) = r(\infty)^{n-1} I + \overline{G}^{(l)}, \tag{8.16}$$

[4] As above, φ_1 comes from the statement of Lemma A.1. Also, recall that the estimate (A.1) in this statement holds with $R = 0$ since $\sigma \geq 0$.

where

$$\overline{G}^{(l)} = \frac{1}{2\pi i}\int_{\Gamma}\left(\frac{h_l z}{1+h_l z}\prod_{\substack{j\neq l,\\ 1\le j\le n}} r(-h_j z) - r(\infty)^{n-1}\right)(zI - A(t))^{-1}\,dz. \tag{8.17}$$

Let next $\overline{\Gamma}_1 = \{z\in\mathbb{C} : \arg z = \pm\varphi_1,\ 0\le |z|\le d_1 h_1^{-1}\}$, $\overline{\Gamma}_2 = \partial\mathcal{D}(d_1 h_1^{-1})\cap\left(-\Sigma_{\pi-\varphi_1}\right)$ and let $\overline{\Gamma} = \overline{\Gamma}_1\cup\overline{\Gamma}_2$ be oriented clockwise. Since the norm of the integrand on the right-hand side of (8.17) is $O(|z|^{-2})$ as $|z|\to\infty$, after a tolerable deformation of the path of integration we use formula (8.17) with $\overline{\Gamma}$ in place of Γ. Therefore, taking into account that $\int_{\overline{\Gamma}}(zI-A(t))^{-1}\,dz = 0$ and applying Cauchy's Theorem we obtain, instead of (8.17),

$$\overline{G}^{(l)} = \frac{h_l}{2\pi i}\int_{\overline{\Gamma}} z(1+h_l z)^{-1}\prod_{\substack{j\neq l,\\ 1\le j\le n}} r(-h_j z)(zI-A(t))^{-1}\,dz. \tag{8.18}$$

Denote next by $\overline{G}_1^{(l)}$ and $\overline{G}_2^{(l)}$ the right-hand side of this equality with $\overline{\Gamma}_1$ and $\overline{\Gamma}_2$, respectively, in place of $\overline{\Gamma}$. With the aid of Lemma A.1, (8.5), and (8.6) we get

$$\begin{aligned}
\|\overline{G}_1^{(l)}\| &\le Ch_l\int_{\overline{\Gamma}_1} |(1+h_l|z|)^{-1}\prod_{\substack{j\neq l,\\ 1\le j\le n}} |r(-h_j z)|\,|dz| \\
&\le Ch_l\left(\int_0^{d_1 h_n^{-1}} e^{-c\tau_n x}\,dx + \sum_{j=1}^{n-1}\varepsilon^j\int_{d_1 h_{n+1-j}^{-1}}^{d_1 h_{n-j}^{-1}} e^{-c\tau_{n-j}x}\,dx\right) \\
&\le Ch_l\int_0^{\infty}\sum_{j=0}^{n-1}\varepsilon^{j-1}e^{-c\tau_{n-j}x}\,dx \le Ch_l\sum_{j=0}^{n-1}\varepsilon^j\tau_{n-j}^{-1}.
\end{aligned} \tag{8.19}$$

At the same time by Lemma A.1 and (8.6) it follows that

$$\begin{aligned}
\|\overline{G}_2^{(l)}\| &\le Ch_l\int_{\overline{\Gamma}_2}(1+h_l|z|)^{-1}\prod_{\substack{j\neq l,\\ 1\le j\le n}} |r(-h_j z)|\,|dz| \\
&\le Ch_l\varepsilon^n\int_{\overline{\Gamma}_2}|dz| \le Ch_l\varepsilon^{n-1}\tau_1^{-1}.
\end{aligned} \tag{8.20}$$

It thus remains to combine (8.15), (8.16), (8.18), (8.19), and (8.20). □

We further extend both Lemma 8.1 and Lemma 8.2 in the following way.

Lemma 8.3. *Let the conditions of Lemma 8.1 be satisfied for some $\varphi \in (0, \pi/2)$ and $\sigma \geq 0$. Let also seminorms $|\cdot|_t$ and $|\cdot|_{*;t}$ form respectively $(\xi|1|\varphi, \sigma)$- and $(*; \xi_*|1|\varphi, \sigma)$-concordant pairs with $A(t)$, for some $\xi, \xi_* \in [0, 1]$, and let $\omega(z)$ be a rational function which is (φ)-regular and satisfies the condition $\deg[\omega] \leq -\max\{\xi, \xi_*\}$. Then for any non-decreasing finite sequence of positive numbers $0 < h_1 \leq \cdots \leq h_n$, it holds, for all $t \in [0, T]$ and $l = 1, \ldots, n$,*

$$\sup_{\|u\|=1} \Big|\omega(-h_l A(t)) \prod_{\substack{j \neq l, \\ 1 \leq j \leq n}} r(-h_j A(t))\, u\Big|_t \leq C \sum_{j=0}^{n-1} \varepsilon^j \tau_{n-j}^{-\xi} + C h_l^{-\xi} |r(\infty)|^n$$

and

$$\sup_{\|\chi\|_*=1} \Big|\omega(-h_l A(t))^* \prod_{\substack{j \neq l, \\ 1 \leq j \leq n}} r(-h_j A(t))^* \chi\Big|_{*;t} \leq C \sum_{j=0}^{n-1} \varepsilon^j \tau_{n-j}^{-\xi_*} + C h_l^{-\xi_*} |r(\infty)|^n,$$

where ε comes from (8.6) and τ_j is given by (8.9).

The proof is similar to that of Lemma 8.1 and of Lemma 8.2 in the cases $\xi = 0$ and $\xi \in (0, 1]$, respectively.

Next we introduce some notions which are needed to state the below restrictions on the used families of grids.

Definition 8.1. *With any finite sequence $\mathcal{K}_n = (k_j)_{j=1}^n$, we associate the reordered sequence $\mathcal{H}_n = (h_j)_{j=1}^n$ given by $h_j = k_{\iota(j)}$, where the bijection $\iota(\cdot)$ is chosen so that*

$$h_1 \leq \cdots \leq h_n$$

and

$$\iota(j) < \iota(j+1) \text{ whenever } h_j = h_{j+1}.$$

The finite sequence $\mathcal{H}_n$ will then be called the ordered sequence *for $\mathcal{K}_n$.*

Definition 8.2. *Let $\mathfrak{T}$ be a family of grids in $[0, T]$. The family $\mathfrak{T}$ is called* exponentially balanced with basis $Q \geq 1$ *if there exists a constant $C > 0$ such that for any grid $\mho_k \in \mathfrak{T}$ with $N + 1$ nodal points*[5] *and for any indices l, n subject to $0 \leq l \leq n < N$,*

$$\sum_{j=l}^{n} k_j \leq C Q^{n-l+1-m} \sum_{j=1}^{m} h_j \quad \textit{for } m = 1, \ldots, n - l + 1,$$

[5] We recall that $k = \{k_0, \ldots, k_{N-1}\}$.

where the finite sequence $\mathcal{H}_{n-l+1} = (h_j)_{j=1}^{n-l+1}$ *is the ordered sequence for* $\widetilde{\mathcal{K}}_{n-l+1} = (k_{j+l-1})_{j=1}^{n-l+1}$.

Definition 8.3. *A family of grids* $\mathcal{T}$ *is called* locally balanced with range $Q \geq 1$ *if* $Q^{-1}k_{n+1} \leq k_n \leq Qk_{n+1}$ *whenever* $t_n, t_{n+1} \in \breve{\mho}_k$*, for all* $\mho_k \in \mathcal{T}$.

Let $\mathcal{T}$ be a locally balanced family of grids with range $Q \geq 1$. Given a grid $\mho_k \in \mathcal{T}$ with $N+1$ nodal points and given indices l, n such that $0 \leq l \leq n < N$, let $\widetilde{\mathcal{K}}_{n-l+1} = (k_{j+l-1})_{j=1}^{n-l+1}$ and let $\mathcal{H}_{n-l+1} = (h_j)_{j=1}^{n-l+1}$ be the ordered sequence for $\widetilde{\mathcal{K}}_{n-l+1}$. Then, for any $m = 1, \ldots, n-l+1$,

$$\sum_{j=l}^{n} k_j - \sum_{j=1}^{m} h_j \leq (n-l+1-m)h_{n-l+1}, \tag{8.21}$$

while since $\mathcal{T}$ is locally balanced with range Q,

$$\sum_{j=1}^{m} h_j \geq h_{n-l+1} \sum_{j=1}^{m} Q^{-(n-l+1-j)} \geq Q^{-(n-l+1-m)} h_{n-l+1}. \tag{8.22}$$

Combining (8.21) and (8.22) therefore yields, with any fixed $Q_1 > Q$, for $m = 1, \ldots, n-l+1$,

$$\sum_{j=l}^{n} k_j \leq (n-l+2-m)Q^{n-l+1-m} \sum_{j=1}^{m} h_j \leq CQ_1^{n-l+1-m} \sum_{j=1}^{m} h_j,$$

which means that $\mathcal{T}$ is, in fact, exponentially balanced with basis $Q_1 > Q$, where Q_1 can be chosen as close to Q as desired.

As a simple example, we note that the family of grids defined by $k_n = Q^n k_0$, $n = 0, 1, \ldots$, $Q = \text{const} > 1$, where $k_0 > 0$ is a free parameter, is obviously locally balanced with range Q.

Yan [166] defines the so-called quasi-quasiuniform assumption on the variable stepsize which is as follows. $\mathcal{T}$ is called a family of quasi-quasiuniform grids if, whenever $t_n, t_{n+1} \in \breve{\mho}_k$, for all $\mho_k \in \mathcal{T}$,

$$k_n \leq k_{n+1} \tag{8.23}$$

and

$$ck_{n+1} \leq k_n \leq Ct_{n+1}/(n+1). \tag{8.24}$$

However it turns out that any family of increasing quasi-quasiuniform grids is, in fact, exponentially balanced with basis $Q > 1$, where Q can be chosen as close to 1 as desired.

To see this fact, we note first that, as it follows from (8.23), for all $\mho_k \in \mathfrak{T}$ and $t_l, t_n \in \breve{\mho}_k$ with $1 \le l \le n$,

$$\frac{1}{l}\sum_{j=0}^{l-1} k_j \le \frac{1}{n-l+1}\sum_{j=l}^{n} k_j,$$

which further implies

$$t_{n+1}/(n+1) \le (t_{n+1} - t_l)/(n-l+1).$$

Obviously, the last inequality holds for $l = 0$ as well. Therefore, instead of (8.24) we actually have, whenever $t_l, t_{n+1} \in \breve{\mho}_k$ and $t_l \le t_n$, for all $\mho_k \in \mathfrak{T}$,

$$ck_{n+1} \le k_n \le (t_{n+1} - t_l)/(n-l+1).$$

With this inequality in mind, we obtain, whenever $t_l < t_n$ and $1 \le m \le n-l+1$,

$$\begin{aligned}\sum_{j=l}^{n} k_j &\le \sum_{j=l}^{n-1} k_j + Ck_{n-1} \le \bigl(1 + C/(n-l)\bigr)\sum_{j=l}^{n-1} k_j \\ &\le \cdots \le \prod_{q=m}^{n-l}(1 + C/q)\sum_{j=l}^{l+m-1} k_j \le \left(\frac{n-l+1}{m}e\right)^{C}\sum_{j=l}^{l+m-1} k_j, \quad (8.25)\end{aligned}$$

and, clearly, $\sum_{j=l}^{n} k_j$ admits the same bound for $l = n$ as well. Also, letting Q be any fixed number > 1, it holds, with any fixed $d \ge 0$,

$$\left(\frac{n-l+1}{m}\right)^{d} \le CQ^{n-l+1-m}. \quad (8.26)$$

Finally, combining (8.25), (8.26), and the fact that any $\mho_k \in \mathfrak{T}$ is an increasing grid shows the claim.

Definition 8.4. *$\mathfrak{T}$ is called a* family of quasi-uniform grids *if there exists a constant C such that*

$$\max_{t_n \in \breve{\mho}_k} k_n \le C \min_{t_n \in \breve{\mho}_k} k_n \quad \textit{for all } \mho_k \in \mathfrak{T}.$$

If $\mathfrak{T}$ is a family of quasi-uniform grids, then, for any $\mho_k \in \mathfrak{T}$, $t_l, t_n \in \breve{\mho}_k$ with $l \le n$, and $1 \le m \le n-l+1$, letting, as above, $\mathcal{H}_{n-l+1} = (h_j)_{j=1}^{n-l+1}$

be the ordered sequence for $\widetilde{\mathcal{K}}_{n-l+1} = (k_{j+l-1})_{j=1}^{n-l+1}$,

$$\begin{aligned}\sum_{j=l}^{n} k_j &\leq C(n-l+1)h_{n-l+1} \leq C\frac{n-l+1}{m}\frac{h_{n-l+1}}{h_1}\sum_{j=1}^{m} h_j \\ &\leq C\frac{n-l+1}{m}\sum_{j=1}^{m} h_j.\end{aligned}$$

Using this estimate and the particular case of (8.26), where the exponent d on the left equals 1, we obtain that any family of quasi-uniform grids is exponentially balanced with basis $Q > 1$, where Q can be taken as close to 1 as required.

The above discussion and our subsequent consideration in this chapter shows that it suffices to deal only with exponentially balanced families of grids.

Now we state a stability result for problems with a constant operator.

Theorem 8.1. *Let the operator $A(t) \equiv A$ be independent of t, let $A \in \mathcal{S}(\varphi, \sigma)$, and let the RK method be of type $C_I(\varphi)$, for some $\varphi \in (0, \pi/2)$ and $\sigma \geq 0$. Let also $\mathfrak{T}$ be an arbitrary family of grids and let $T = \infty$. Then, for all grids $\mho_k \in \mathfrak{T}$, for Y_n the extended solution by the RK method on $\mho_k$, it holds*

$$\|Y_n\| \leq C\left(\|y^0\| + \sum_{l=0}^{n-1} k_l \max_{1\leq j\leq \nu} \|f(t_{lj})\|\right) \quad \text{for } t_n \in \mho_k. \tag{8.27}$$

Furthermore, if in addition $\mathfrak{T}$ is an exponentially balanced family of grids with basis Q such that

$$Q < |r(\infty)|^{-1}, \tag{8.28}$$

it then holds, for any fixed $\xi \in [0,1]$, for all $\mho_k \in \mathfrak{T}$ and for all $t_n \in \mho_k$,

$$\|\mathfrak{A}_n^{\xi} Y_n\| \leq C\left(t_{n+1}^{-\xi}\|y^0\| + \sum_{l=0}^{n-1} k_l (t_{n+1} - t_{l+1})^{-\xi} \max_{1\leq j\leq \nu} \|f(t_{lj})\|\right). \tag{8.29}$$

Proof. First of all by Lemma 5.3 we obtain since $\deg[w_j] \leq 0$,

$$\|\omega_j(-kA)\| \leq C, \quad j = 1, \ldots, \nu, \tag{8.30}$$

uniformly with respect to $k > 0$. Therefore, if $\mathfrak{T}$ is an arbitrary family of grids, (8.27) follows by applying Lemma 8.1 and (8.30) for an estimation on the basis of the representation, for $t_n \in \breve{\mho}_k$,

$$Y_n = r(-k_{n-1}A)\cdots r(-k_0A)y^0 + \sum_{l=0}^{n-1} k_l r(-k_{n-1}A)\cdots r(-k_{l+1}A)F_l. \quad (8.31)$$

Let now $\mathfrak{T}$ be an exponentially balanced family of grids with basis Q satisfying (8.28). We then have, for all $\mho_k \in \mathfrak{T}$ and $t_n, t_l \in \breve{\mho}_k$ such that $t_l \le t_n$,

$$|r(\infty)|^{n-l+1} < Q^{-(n-l+1)} \le Ck_n(t_{n+1}-t_l)^{-1} \quad (8.32)$$

and

$$\left(\sum_{m=1}^{n-l+1-j} h_m\right)^{-1} \le CQ^j(t_{n+1}-t_l)^{-1}, \quad j = 0,\dots,n-l, \quad (8.33)$$

where $(h_m)_{m=1}^{n-l+1}$ is the ordered sequence for $(k_{m+l-1})_{m=1}^{n-l+1}$. Select further some $\varepsilon \in (|r(\infty)|, Q^{-1})$. Applying then Lemma 8.2[6] and taking (8.32) and (8.33) into account yields that for $t_l \le t_n$,

$$\begin{aligned}\|\mathfrak{A}_n \prod_{j=l}^{n-1} r(-k_jA)\| &\le C(t_{n+1}-t_l)^{-1}\left(1+\varepsilon Q+\cdots+(\varepsilon Q)^{n-l}\right)\\ &\le C(t_{n+1}-t_l)^{-1}. \quad (8.34)\end{aligned}$$

Therefore, by Lemma 8.1, (8.34), and the convexity inequality (A.18), we get, for any fixed $\xi \in [0,1]$, for $t_l \le t_n$,

$$\|\mathfrak{A}_n^\xi \prod_{j=l}^{n-1} r(-k_jA)\| \le C(t_{n+1}-t_l)^{-\xi}.$$

Using now this inequality for an estimation on the basis of (8.31), with both sides multiplied from the left by $\mathfrak{A}_n^\xi$, leads to (8.29). □

Remark 8.1. *Let $T = \infty$ and let the operator $A(t) \equiv A$ and the RK method satisfy the conditions of Lemma 8.1 for some $\varphi \in (0,\pi/2)$ and $\sigma \ge 0$. Let also a seminorm $[\![\cdot]\!]_t$ form a $(\xi|1|\varphi,\sigma)$-concordant pair with A for some $\xi \in [0,1]$ and let $\omega(z)$ be a rational function which is (φ)-regular and satisfies the condition $\deg[\omega] \le -\xi$. Then the estimate (8.27) holds with $[\![Y_n]\!]_{t_n}$ substituted for $\|Y_n\|$ if $\xi = 0$ and if $\mathfrak{T}$ is an arbitrary family of grids,*

[6] We recall that ε in the statement of Lemma 8.2 can be chosen as close to $|r(\infty)|$ as desired.

and the estimate (8.29) holds with $]\!\![\omega(-k_nA)Y_n]\!\![_{t_n}$ *substituted for* $\|\mathfrak{A}_n^\xi Y_n\|$ *if* $\xi \in (0,1]$ *and if* $\mathfrak{T}$ *is an exponentially balanced family of grids with basis* Q *satisfying (8.28).*

This result follows from Lemma 8.3.

8.2 The non-commutative case

In this section we proceed from the supposition that the operator $A(t)$ may depend on t. Throughout the section it is assumed that $T < \infty$.

First of all we state the following Gronwall's lemma type result.

Lemma 8.4. *Let* $\mathfrak{T}$ *be a family of non-uniform grids in the segment* $[0,T]$ *satisfying the conditions, with some* $K > 0$*, for all* $\mho_k \in \mathfrak{T}$ *and* $t_n \in \breve{\mho}_k$*,*

$$k_n \le K \tag{8.35}$$

and

$$k_n \le Ck_{n+1}. \tag{8.36}$$

Let further $z_{n,l}$ *and* $w_{n,l}$ *be nonnegative solutions of the respective inequalities,*[7] *for all* $\mho_k \in \mathfrak{T}$ *and* $t_l, t_n \in \mho_k$ *with* $t_l \le t_n$*,*

$$z_{n,l} \le C(t_{n+1} - t_l)^{-\zeta_1} + C\sum_{q=l}^{n-1} k_q z_{n,q+1}(t_{q+1} - t_l)^{-\zeta_2} \tag{8.37}$$

and

$$w_{n,l} \le C(t_{n+1} - t_l)^{-\zeta_1} + C\sum_{q=l}^{n-1} k_q (t_{n+1} - t_{q+1})^{-\zeta_2} w_{q,l},$$

with some $\zeta_1, \zeta_2 \in [0,1)$*. Then, if* K *is sufficiently small, we have, for all* $\mho_k \in \mathfrak{T}$ *and* $t_l, t_n \in \mho_k$ *with* $t_l \le t_n$*, that*

$$z_{n,l} \le C(t_{n+1} - t_l)^{-\zeta_1} \tag{8.38}$$

and

$$w_{n,l} \le C(t_{n+1} - t_l)^{-\zeta_1}. \tag{8.39}$$

In the particular case $\zeta_1 = 0$*, condition (8.36) is not needed.*

[7] Actually, both $z_{n,l}$ and $w_{n,l}$ depend on $\mho_k$.

Proof. We show only the former stated estimate (8.38), noting that (8.39) is proved by using the same reasoning.

Denoting $\tilde{z}_{n,l} = (t_{n+1} - t_l)^{\zeta_1} z_{n,l}$, it follows from (8.37) that for all $\mho_k \in \mathfrak{T}$ and $t_l, t_n \in \mho_k$ with $t_l \leq t_n$,

$$\tilde{z}_{n,l} \leq C + C(t_{n+1} - t_l)^{\zeta_1} \sum_{q=l}^{n-1} k_q (t_{n+1} - t_q)^{-\zeta_1} \tilde{z}_{n,q+1} (t_{q+1} - t_l)^{-\zeta_2} \tag{8.40}$$

since

$$(t_{n+1} - t_{q+1})^{-\zeta_1} \leq C(t_{n+1} - t_q)^{-\zeta_1} \quad \text{for } t_{q+1} \leq t_n,$$

by (8.36). Let further $T_1 \in (0, T/2]$ be a fixed number which will be specified later and let $K > 0$ be subject to $K \leq T_1$. Given $t_n \in \mho_k$, we put, for $j \geq 1$,

$$\tau_j = \big[\max\{0, t_{n+1} - jT_1\}, t_{n+1} - (j-1)T_1\big) \text{ and } Z_j = \max_{t_l \in \tau_j} \tilde{z}_{n,l},$$

and denote by $t'_{(j)}$ and $t''_{(j)}$ the largest and smallest nodes, respectively, of $\mho_k \cup \{t_{n+1}\}$ such that $\tau_j \subseteq [t'_{(j)}, t''_{(j)}]$, unless τ_j is void. By (8.40) we therefore have, for $m \geq 1$,

$$Z_m \leq C + C \sum_{j=1}^{m} Z_j \times \max_{t_l \in \tau_m} \left((t_{n+1} - t_l)^{\zeta_1} \sum_{\substack{l \leq q \leq n-1, \\ t_{q+1} \in \tau_j}} k_q (t_{n+1} - t_q)^{-\zeta_1} (t_{q+1} - t_l)^{-\zeta_2} \right). \tag{8.41}$$

Taking (8.35) into account, recalling that $K \leq T_1$, and noting that $t_l \leq t'_{(j)} \leq t''_{(j)} \leq t_{n+1}$ if $t_l \notin \tau_j$, a simple calculation now gives, with $\Delta_{x_0} = [x_0, x_0 + 3T_1]$, for $j \geq 1$ and $t_l \leq t_n$,

$$\begin{aligned} \sum_{\substack{l \leq q \leq n-1, \\ t_{q+1} \in \tau_j}} k_q (t_{n+1} - t_q)^{-\zeta_1} (t_{q+1} - t_l)^{-\zeta_2} &\leq \int_{t'_{(j)}}^{t''_{(j)}} (t_{n+1} - x)^{-\zeta_1} (x - t_l)^{-\zeta_2} dx \\ &\leq \max_{x_0 \in [0, T - 2T_1]} \int_{[t_l, t_{n+1}] \cap \Delta_{x_0}} (t_{n+1} - x)^{-\zeta_1} (x - t_l)^{-\zeta_2} dx \\ &\leq C(t_{n+1} - t_l)^{-\zeta_1} \big(T_1^{1-\zeta_1} + T_1^{1-\zeta_2}\big). \end{aligned}$$

Combining this estimate with (8.41), we obtain, for $m \geq 1$,

$$Z_m \leq C + C\big(T_1^{1-\zeta_1} + T_1^{1-\zeta_2}\big) \sum_{j=1}^{m} Z_j,$$

where no constant C depends on T_1. Taking now T_1 sufficiently small, it follows that

$$Z_m \leq C + \frac{1}{2}\sum_{j=1}^{m} Z_j,$$

which further implies that $Z_m \leq C$ since $m \leq \lfloor T/T_1 \rfloor + 1$. Clearly, this inequality shows (8.38), by the definition of Z_j and $\widetilde{z}_{n,l}$.

In conclusion, note that, applying the same argument, it is not necessary to use (8.36) if $\zeta_1 = 0$. □

Now we show stability in the case with a variable operator. It is convenient to denote, given a grid $\mho_k$, for any $t \in [0,T]$ and $t_l, t_n \in \mho_k$,

$$\mathfrak{A}_l(t) = k_l^{-1}\alpha(-k_l A(t)), \quad \mathfrak{U}_l(t) = I - k_l\mathfrak{A}_l(t), \quad \mathfrak{U}_{n,l}(t) = \mathfrak{U}_n(t)\cdots\mathfrak{U}_l(t),$$

dealing with the RK method (5.19), (5.20) applied to (5.1) on $\mho_k$, and

$$\mathfrak{U}_l = I - k_l\mathfrak{A}_l, \quad \mathfrak{U}_{n,l} = \mathfrak{U}_n\cdots\mathfrak{U}_l,$$

dealing with problem (1.1) considered on $\mho_k$.[8] Using the above notation, similarly to (1.4) the solution Y_n of (1.1) in the non-uniform case is represented as

$$Y_n = \mathfrak{U}_{n-1,0}\, y^0 + \sum_{l=0}^{n-1} k_l\mathfrak{U}_{n-1,l+1}F_l. \tag{8.42}$$

Also, for subsequent use we point out the identity, for $t_n, t_l \in \mho_k$ such that $t_l \leq t_n$,

$$\mathfrak{U}_{n-1,l} = \mathfrak{U}_{n-1,l}(t_l) + \sum_{q=l}^{n-1} k_q\mathfrak{U}_{n-1,q+1}(\mathfrak{A}_q(t_l) - \mathfrak{A}_q)\,\mathfrak{U}_{q-1,l}(t_l). \tag{8.43}$$

Henceforth, given a non-uniform grid $\mho_k$, by the **discrete solution**, for brevity, we mean the solution by the RK method (5.19), (5.20) applied to problem (5.1) on $\mho_k$. Note that all the below statements contain assumptions that ensure the representability of the starting RK problem in the form (1.1) in which $\mathfrak{A}_n$, $\mathfrak{B}_n$, and F_n are specified by (5.38), (5.39), and (5.41), with k_n substituted for k. In what follows we always assume that $\mathfrak{A}_n$, $\mathfrak{B}_n$, and F_n are defined as specified and we use the representation (8.42) for the discrete solution Y_n without explicit mention.

Our first result on stability is as follows.

[8] It is assumed that $\mathfrak{U}_{n,l}(t) = \mathfrak{U}_{n,l} = I$ if $n < l$.

Theorem 8.2. *Suppose $A(t) \in S(\varphi,\sigma)$, a seminorm $|\cdot|_t$ forms a $(\gamma_1|\varphi,\sigma)$-concordant pair with $A(t)$, and $A(t)$ saisfies hypothesis* $\mathbf{h\ddot{o}l}^{\oplus}[\vartheta;|\cdot|_t]$, *for some $\varphi \in (0,\pi/2)$, $\sigma \geq 0$, $\gamma_1 \in [0,1)$, and $\vartheta \in (0,1]$. Let further the RK method be of type $V_I(\varphi)$ and let $\mathcal{T}$ be an exponentially balanced family of grids with basis $Q > 1$ satisfying (8.28). Then, with $K > 0$ sufficiently small and with any fixed $\xi \in [0,1)$, for all grids $\mho_k \in \mathcal{T}$ such that $\max_{t_n\in\mho_k} k_n \leq K$ and for all $t_n \in \mho_k$, the discrete solution Y_n satisfies the estimate*[9]

$$\|\mathfrak{A}_n(t_n)^{\xi} Y_n\| \leq C t_{n+1}^{-\xi}\|y^0\| + C\sum_{l=0}^{n-1} k_l (t_{n+1}-t_{l+1})^{-\xi} \max_{1\leq j\leq \nu} \|f(t_{lj})\|. \quad (8.44)$$

Proof. Similarly to (6.61) and (6.59) we have, with some $\mu_0 \geq 0$, for all $\mho_k \in \mathcal{T}$, $t_n \in \mho_k$, and $u \in \mathfrak{X}$,

$$\|\mathfrak{B}_n u\| \leq C k_n^{\vartheta}\|(\mathfrak{A}_n(t_n)+\mu_0 I)\, u\| + C\sum_{j=1}^{\nu} |v_j(-k_n A(t_n))\, u|_{t_n} \quad (8.45)$$

and

$$\|F_n\| \leq C \max_{1\leq j\leq \nu} \|f(t_{nj})\|. \quad (8.46)$$

Next, by the convexity inequality (A.18) and Lemma 8.1 it holds, for $t_l \leq t_n$,

$$\begin{aligned}\|\mathfrak{A}_n(t_n)^{\xi}\mathfrak{U}_{n-1,l}(t_l)\| &\leq \|\mathfrak{A}_n(t_n)\mathfrak{U}_{n-1,l}(t_l)\|^{\xi}\|\mathfrak{U}_{n-1,l}(t_l)\|^{1-\xi}\\ &\leq C\|(\mathfrak{A}_n(t_n)-\mathfrak{A}_n(t_l))\,\mathfrak{U}_{n-1,l}(t_l)\|^{\xi} + \|\mathfrak{A}_n(t_l)\,\mathfrak{U}_{n-1,l}(t_l)\|^{\xi}. \quad (8.47)\end{aligned}$$

Since $\mathcal{T}$ is an exponentially balanced family of grids with basis $Q < |r(\infty)|^{-1}$, by Lemma 8.2 as in the proof of Theorem 8.1, it follows that

$$\|\mathfrak{A}_n(t_l)\,\mathfrak{U}_{n-1,l}(t_l)\| \leq C(t_{n+1}-t_l)^{-1}. \quad (8.48)$$

Using also Theorem 6.9 and Lemmas 8.2 and 8.3, it is not hard to see that

$$\|(\mathfrak{A}_n(t_n)-\mathfrak{A}_n(t_l))\,\mathfrak{U}_{n-1,l}(t_l)\| \leq C\left((t_{n+1}-t_l)^{\vartheta-1} + (t_{n+1}-t_l)^{-\gamma_1}\right). \quad (8.49)$$

Therefore, combining (8.47), (8.48), and (8.49) yields, for $t_l \leq t_n$,

$$\|\mathfrak{A}_n(t_n)^{\xi}\mathfrak{U}_{n-1,l}(t_l)\| \leq C(t_{n+1}-t_l)^{-\xi}.$$

[9] The fractional powers $\mathfrak{A}_n(t_n)^{\xi}$, $\xi \geq 0$, are defined and bounded on $\mathfrak{X}$ since $\sigma \geq 0$.

Further, by Theorem 6.1 it follows that the seminorm $\sum_{j=1}^{\nu} |v_j(-kA(t)) \cdot|_t$ forms a $\Lambda_1^k(\gamma_1)$-concordant pair with $k^{-1}\alpha(-kA(t))$. Hence applying Lemma 1.9 shows that with some $\mu_0 \geq 0$ and $\gamma_2 \in (\gamma_1, 1)$, for all $u \in \mathfrak{X}$,

$$\sum_{j=1}^{\nu} |v_j(-k_n A(t_n))\, u|_{t_n} \leq C\|(\mathfrak{A}_n(t_n) + \mu_0 I)^{\gamma_2} u\|. \tag{8.50}$$

Now, with the aid of (8.45), (8.49), (8.50), and the above argument, we obtain, with $\gamma_3 = \max\{1-\vartheta, \gamma_2\}$, for $t_l \leq t_n$,

$$\begin{aligned}\|(\mathfrak{A}_n(t_l) - \mathfrak{A}_n)\,\mathfrak{U}_{n-1,l}(t_l)\| &\leq \|\mathfrak{B}_n\mathfrak{U}_{n-1,l}(t_l)\| + \|\mathfrak{A}_n(t_n) - \mathfrak{A}_n(t_l))\,\mathfrak{U}_{n-1,l}(t_l)\| \\ &\leq C\left(k_n^{\vartheta} + (t_n - t_l)^{\vartheta}\right)(t_{n+1} - t_l)^{-1} + C(t_{n+1} - t_l)^{-\gamma_2} \\ &\leq C(t_{n+1} - t_l)^{-\gamma_3}.\end{aligned}$$

Using this inequality together with (8.48) for an estimation on the basis of (8.43), with both sides multiplied from the left by $\mathfrak{A}_n(t_n)^{\xi}$, the reader will have no difficulty in seeing that with $z_{n,l} = \|\mathfrak{A}_n(t_n)^{\xi}\mathfrak{U}_{n-1,l}\|$, for $t_l < t_n$,

$$z_{n,l} \leq C(t_{n+1} - t_l)^{-\xi} + C\sum_{q=l}^{n-1} k_q z_{n,q+1}(t_{q+1} - t_l)^{-\gamma_3}.$$

Noting that $\gamma_3 < 1$, observing that (8.36) holds, by the fact that the family $\mathcal{T}$ is exponentially balanced with basis Q, and applying then Lemma 8.4, we find, for $t_l \leq t_n$,

$$\|\mathfrak{A}_n(t_n)^{\xi}\mathfrak{U}_{n-1,l}\| \leq C(t_{n+1} - t_l)^{-\xi}. \tag{8.51}$$

So the claim follows by using the representation (8.42), with both sides multiplied from the left by $\mathfrak{A}_n(t_n)^{\xi}$, and taking the estimates (8.46) and (8.51) into account. □

Now we consider the case where $\xi = 1$ is allowable as well.

Theorem 8.3. *Let the conditions of Theorem 8.2 be satisfied with the difference that the operator $A(t)$ fulfils hypothesis* **höl**$[\vartheta]$ *instead of* **höl**$^{\oplus}[\vartheta; |\cdot|_t]$. *Then the stability estimate (8.44) holds for any fixed $\xi \in [0,1]$.*

Proof. Clearly, it suffices to show the result for $\xi = 1$ only. To that end, we put, for $t_l \leq t_n$,

$$w_{n,l} = \|(t_{n+1} - t_l)\mathfrak{A}_n(t_n)\mathfrak{U}_{n-1,l}\|.$$

Let p be the smallest integer such that

$$l \leq p \leq n+1 \text{ and } t_{n+1} - t_p \leq \frac{1}{2}(t_{n+1} - t_l).$$

Observe that $p \geq l+1$. If $p = n+1$, we have $t_{n+1} - t_l < 2k_n$, which gives, applying Theorem 8.2 with $\xi = 0$, Lemma 1.1, and Theorem 6.2,

$$w_{n,l} \leq 2\|k_n \mathfrak{A}_n(t_n)\| \, \|\mathfrak{U}_{n-1,l}\| \leq C.$$

Assume further that $p \leq n$. We then use the identity, for $t_l \leq t_n$,

$$\begin{aligned}(t_{n+1} - t_l)\mathfrak{A}_n(t_n)\mathfrak{U}_{n-1,l} &= (t_{n+1} - t_l)\mathfrak{A}_n(t_n)\mathfrak{U}_{n-1,p-1}(t_n)\mathfrak{U}_{p-2,l} \\ &+ (t_{n+1} - t_l) \sum_{q=p-1}^{n-1} k_q \mathfrak{A}_n(t_n)\mathfrak{U}_{n-1,q+1}(t_n)\left(\mathfrak{A}_q(t_n) - \mathfrak{A}_q\right)\mathfrak{U}_{q-1,l}. \end{aligned} \tag{8.52}$$

Under the accepted Hölder-continuity type hypothesis, instead of (8.45) it holds, for all $u \in \mathfrak{X}$,

$$\|\mathfrak{B}_n u\| \leq C k_n^{\vartheta}(\|\mathfrak{A}_n(t_n)u\| + \|u\|). \tag{8.53}$$

With this estimate in mind and applying Theorems 6.9 and 8.2 (the latter one with $\xi = 0$), we find, for $t_l \leq t_q \leq t_n$,

$$\begin{aligned}\|(\mathfrak{A}_q(t_n) - \mathfrak{A}_q)\mathfrak{U}_{q-1,l}\| &\leq \|\mathfrak{B}_q\mathfrak{U}_{q-1,l}\| + \|(\mathfrak{A}_q(t_n)) - \mathfrak{A}_q(t_q))\mathfrak{U}_{q-1,l}\| \\ &\leq C\left(k_n^{\vartheta} + (t_n - t_q)^{\vartheta}\right)\left(\|\mathfrak{A}_q(t_q)\mathfrak{U}_{q-1,l}\| + 1\right) \\ &\leq C\left(t_{n+1} - t_q\right)^{\vartheta}\left((t_{q+1} - t_l)^{-1} w_{q,l} + 1\right).\end{aligned}$$

Using this estimate, Theorem 8.2 with $\xi = 0$, and the argument employed in its proof for an estimation on the basis of (8.52), we now get, for $t_l \leq t_n$,

$$\begin{aligned} w_{n,l} &\leq C\frac{t_{n+1} - t_l}{t_{n+1} - t_{p-1}} + C(t_{n+1} - t_l) \\ &\times \sum_{q=p-1}^{n-1} k_q (t_{n+1} - t_{q+1})^{-1}(t_{n+1} - t_q)^{\vartheta}\left((t_{q+1} - t_l)^{-1} w_{q,l} + 1\right). \end{aligned} \tag{8.54}$$

The fact that $\mathfrak{T}$ is an exponentially balanced family of grids with basis Q implies (8.36). Therefore it holds

$$(t_{n+1} - t_{q+1})^{-1} \leq C(t_{n+1} - t_q)^{-1} \quad \text{for } t_q \leq t_n. \tag{8.55}$$

Also, by the definition of p we have

$$(t_{n+1} - t_{p-1})^{-1} \leq 2(t_{n+1} - t_l)^{-1} \text{ and } (t_p - t_l)^{-1} \leq 2(t_{n+1} - t_l)^{-1}. \tag{8.56}$$

By (8.55) and (8.56) it then follows from (8.54) that for $t_l \le t_n$,

$$w_{n,l} \le C + C \sum_{q=p-1}^{n-1} k_q (t_{n+1} - t_q)^{-(1-\vartheta)} w_{q,l},$$

and further, since $p \ge l + 1$,

$$w_{n,l} \le C + C \sum_{q=l}^{n-1} k_q (t_{n+1} - t_q)^{-(1-\vartheta)} w_{q,l}.$$

Finally, using Lemma 8.4 and recalling that $1 - \vartheta < 1$, the last estimate yields that $w_{n,l} \le C$ which shows the claim, with the aid of (8.42) and (8.46). □

Remark 8.2. *Let the conditions of Theorem 8.1 or of Theorem 8.2 be satisfied for some* $\varphi \in (0, \pi/2)$, $\sigma \ge 0$, *and* $\xi \in [0, 1]$,[10] *let a seminorm* $]\![\cdot]\!\![_t$ *form a* $(\xi|1|\varphi, \sigma)$*-concordant pair with the operator* $A(t)$, *and let* $\omega(z)$ *be a rational function which is* (φ)*-regular and satisfies the condition* $\deg[\omega] \le -\xi$. *Then the stability estimate (8.44) holds with* $]\![\omega(-k_n A(t_n))\, Y_n]\!\![_{t_n}$ *substituted for* $\|\mathfrak{A}_n(t_n)^{\xi} Y_n\|$.

This assertion can be established with the aid of the above argument, using, if $\xi \in (0, 1]$, the fact that for all $u \in \operatorname{Dom} A(t)$,

$$]\![u]\!\![_t \le C \|(A(t) + I)\, u\|^{\xi} \|u\|^{1-\xi}.$$

Note that for $\xi = 1$ the last inequality is a trivial consequence of the $(\xi|1|\varphi, \sigma)$-concordance of the pair $\big(]\![\cdot]\!\![_t; A(t)\big)$, while for $0 < \xi < 1$ it can be shown in a manner similar to that used for proving the convexity inequality (cf., e.g., Krein [96, The proof of Theorem 5.2]).

It turns out that if $A(t)$ is under hypothesis **lip**, in the case $\xi = 0$ it is not necessary to impose any restrictions on the family of grids.

Theorem 8.4. *Let the operator* $A(t)$ *and the RK method satisfy the conditions of Theorem 8.3 with the difference that* $A(t)$ *now fulfils hypothesis* **lip**[11] *and* $\mathfrak{T}$ *is an arbitrary family of grids. Then the stability estimate (8.44) holds with* $\xi = 0$.

The proof of the theorem will be based on the following.

---

[10] The case $\xi = 1$ is included under the conditions of Theorem 8.2 only.

[11] That is, one formally takes $\vartheta = 1$.

Lemma 8.5. *Under the conditions of Theorem 8.4, for all $\mho_k \in \mathcal{T}$, $t_n, t_l \in \mho_k$, and $t, s \in [0, T]$,*

$$\|\mathfrak{U}_{n,l}(t) - \mathfrak{U}_{n,l}(s)\| \le C|t - s|.$$

Proof. Given $t_n, t_l \in \mho_k$ such that $t_l \le t_n$, let $(h_j)_{j=1}^{n-l+1}$ be the ordered sequence for $(k_{j+l-1})_{j=1}^{n-l+1}$. With the aid of the identity (8.7) we then have

$$\mathfrak{U}_{n,l}(t) - \mathfrak{U}_{n,l}(s) = \sum_{j=1}^{n-l+1} r(\infty)^{n-l+1-j}(G_j(t) - G_j(s)), \tag{8.57}$$

where $G_j(t)$ is given by (8.8). Let further the contour $\Gamma^{(j)}$ be just the same as in the proof of Lemma 8.1. By the operator calculus formula (A.10), after a deformation of the path of integration, we have, for $j = 1, \dots, n - l + 1$,

$$\begin{aligned} G_j(t) - G_j(s) = \frac{1}{2\pi i} \int_{\Gamma^{(j)}} &(r(-h_j z) - r(\infty)) \\ &\times \prod_{q=1}^{j-1} r(-h_q z)((zI - A(t))^{-1} - (zI - A(s))^{-1})\, dz. \end{aligned} \tag{8.58}$$

By the accepted Lipschitz-continuity type hypothesis and the resolvent identity (6.45) with $k = 1$, we find, for $z \in \Gamma^{(j)}$,

$$\begin{aligned} &\|(zI - A(t))^{-1} - (zI - A(s))^{-1}\| \\ &\qquad \le \|(zI - A(t))^{-1}\| \, \|(A(t) - A(s))\,(zI - A(s))^{-1}\| \\ &\qquad \le C|t - s|\,(|z|^{-1} + |z|^{-2}). \end{aligned}$$

With this estimate in mind, the norm of the integral on the right-hand side of (8.58) can be bounded by the amount

$$C|t-s| \int_{\Gamma^{(j)}} |r(-h_j z) - r(\infty)| \prod_{q=1}^{j-1} |r(-h_q z)|\, |z|^{-1} |dz| + C|t-s| \int_{\Gamma^{(j)}} |z|^{-2} |dz|.$$

Clearly, the second integral in this expression is bounded by C and the same can be shown for the first integral, following the reasoning employed in the proof of Lemma 8.1. Therefore $\|G_j(t) - G_j(s)\| \le C|t - s|$. Using this fact for an estimation on the basis of (8.57) shows the claim. □

Proof of Theorem 8.4. Using the identity, for $t_l \leq t_n$,

$$\mathfrak{U}_n(t_n)\cdots\mathfrak{U}_l(t_l) = \mathfrak{U}_{n,l}(t_l) + \sum_{q=l}^{n-1}(\mathfrak{U}_{n,q+1}(t_{q+1}) - \mathfrak{U}_{n,q+1}(t_q))\mathfrak{U}_q(t_q)\cdots\mathfrak{U}_l(t_l),$$

and applying Lemmas 8.1 and 8.5, a simple calculation shows that

$$\|\mathfrak{U}_n(t_n)\cdots\mathfrak{U}_l(t_l)\| \leq C + C\sum_{q=l}^{n-1} k_q\|\mathfrak{U}_q(t_q)\cdots\mathfrak{U}_l(t_l)\|.$$

Therefore, by Lemma 8.4,

$$\|\mathfrak{U}_n(t_n)\cdots\mathfrak{U}_l(t_l)\| \leq C. \tag{8.59}$$

Furthermore, by virtue of (8.53) with $\vartheta = 1$ and Lemma 1.1 it follows that $\|\mathfrak{B}_n\| \leq C$. Using this fact together with (8.59) for an estimation on the basis of the identity, for $t_l \leq t_n$,

$$\mathfrak{U}_{n-1,l} = \mathfrak{U}_{n-1}(t_{n-1})\cdots\mathfrak{U}_l(t_l) - \sum_{q=l}^{n-1} k_q\mathfrak{U}_{n-1}(t_{n-1})\cdots\mathfrak{U}_{q+1}(t_{q+1})\mathfrak{B}_q\mathfrak{U}_{q-1,l},$$

and applying next Lemma 8.4 once more leads us to the desired result, with the aid of the representation (8.42). □

The following result is also of interest for practical applications.

Theorem 8.5. *Let the operator $A(t)$ and the RK method be under the conditions of Theorem 6.13 for some $\varphi \in (0,\pi/2)$, $\sigma \geq 0$, $\gamma_\oplus, \gamma, \gamma_1, \gamma_*, \vartheta_\bullet, \vartheta_{*\bullet} \in [0,1)$ and $\vartheta_\oplus, \vartheta, \vartheta_1 \in (0,1]$ such that $\gamma + \gamma_* = 1$, $\gamma_1 + \gamma_* < 1$, $\vartheta_\bullet + \vartheta_{*\bullet} < \vartheta$. Let further $\mathfrak{T}$ be an exponentially balanced family of grids with basis $Q > 1$ satisfying condition (8.28). Finally, suppose that a seminorm $]\![\cdot]\![_t$ forms a $(\xi|1|\varphi,\sigma)$-concordant pair with $A(t)$ for some $\xi \in [0,1]$ and that $\omega(z)$ is a rational function which is (φ)-regular and satisfies the condition $\deg[\omega] \leq -\xi$. Then, if $\xi < 1 - \gamma_* + \min\{\vartheta,\vartheta_1\}$, with some $K > 0$ sufficiently small, for all grids $\mho_k \in \mathfrak{T}$ such that $\max_{t_n\in\breve{\mho}_k} k_n \leq K$, the discrete solution Y_n satisfies the stability estimate (8.44) both as it is and with $]\![\omega(-k_nA(t_n))\,Y_n]\![_{t_n}$ substituted for $\|\mathfrak{A}_n^\xi Y_n\|$. Furthermore, if in addition $A(t)$ satisfies hypothesis* $\mathbf{lip}(\mathcal{A}(t;\cdot,\cdot);|\cdot|_{\mathfrak{m};t},\mathfrak{m}=\emptyset,*\emptyset)$, *then (8.44) holds with $\xi = 0$ even if $\mathfrak{T}$ is an arbitrary family of grids.*

This result can be shown in a manner similar to that used in the proofs of Theorems 8.2, 8.3, and 8.4, with the aid of Theorem 6.7 and Lemma 8.3, using as well some reasonings employed in the proof of Theorem 6.13. Also, note that if $A(t)$ is under hypothesis $\mathbf{lip}(\mathcal{A}(t;\cdot,\cdot);|\cdot|_{\mathfrak{m};t}, \mathfrak{m}=\emptyset, *\emptyset)$, the assertion of Lemma 8.5 is still valid, which allows one to use the argument applied in the proof of Theorem 8.4.

The reader will have no difficulty in stating an analogue to Theorem 6.14 in the case of non-uniform grids.

8.3 Error estimates

In this section we present two types of error estimates, namely, *a priori* and *a posteriori* estimates. Our reasons are that both types of estimates are of great importance for practical applications in the case of nonuniform time partitions. In particular, *a posteriori* error analysis is widely used as a basis for applying adaptive stepsize control.

Starting by showing certain *a priori* error estimates, we remark that the analysis of convergence in Chapter 7, carried out in the case of uniform time partitions, is based on a step-by-step estimation of the local error and on subsequent summing up the contribution to the total error coming from all steps of the numerical process. Clearly, error estimates of RK methods on non-uniform grids are derived in the same way provided that one has already shown the necessary stability estimates. Below we give some analogues of the convergence results of Chapter 7, dealing with families of non-uniform grids. Actually, we show, as an example, error estimates of the standard form, that is, in the norm $\|\cdot\|$. Note nevertheless that weighted error estimates can be derived as well by analogy with what is done in Chapter 7.

We first show an analogue of Theorem 7.4 for non-uniform grids, considering, for simplicity, only the case $m = 0$.

Theorem 8.6. *Let $T = \infty$, let the operator $A(t)$ and the RK method be under the conditions of Theorem 7.4, assuming that $\sigma \geq 0$ in addition, and let $\mathcal{T}$ be an arbitrary family of grids. Then instead of (7.21) it holds*

$$
\begin{aligned}
\|Y_n - y(t_n)\| \;&\leq\; C\sum_{j=0}^{n-1} k_j^{p+1}\left(\|y^0_{(p+1)}\| + \|f^{(p)}(0)\| + \sum_{l=q}^{p-1}\|A^{p-l}f^{(l)}(t_n)\|\right)\\
&\quad + C\sum_{j=0}^{n-1} k_j^{p+1}\int_0^{t_{j+1}} \|f^{(p+1)}(s)\|\,ds,
\end{aligned}
$$

for all $\mho_k \in \mathfrak{T}$ *and* $t_n \in \mho_k$.

The proof of this result follows that of Theorem 7.4, using Lemma 8.1 as well.

Next we state a convergence result in the non-autonomous case.

Theorem 8.7. *Let* $T < \infty$, *let the operator* $A(t)$ *and the RK method be under the conditions of one of the above Theorems 7.7, 7.8, or 7.9, assuming that* $\sigma > 0$ *in addition, and let* $\mathfrak{T}$ *be an arbitrary family of grids. Then, with* $K > 0$ *sufficiently small, for all* $\mho_k \in \mathfrak{T}$ *with* $\max_{t_n \in \breve{\mho}_k} k_n \le K$ *and for all* $t_n \in \mho_k$, *it holds, instead of (7.42),*

$$\|Y_n - y(t_n)\| \le Ck_0^{q+2}\|A(0)y^{(q+1)}(0)\| + Ck_n^{q+2}\|A(t_n)y^{(q+1)}(t_n)\|$$
$$+ C\sum_{j=0}^{n-1} k_j^{q+2} \sum_{l=q+1}^{q+2} \int_{t_j}^{t_{j+1}} \|A(s)\, y^{(l)}(s)\|\, ds + C\sum_{j=0}^{n-1} k_j^p \int_{t_j}^{t_{j+1}} \|y^{(p+1)}(s)\|\, ds,$$

and, instead of (7.51),

$$\|Y_n - y(t_n)\| \le Ck_0^{q+1}\|y^{(q+1)}(0)\| + Ck_n^{q+1}\|y^{(q+1)}(t_n)\|$$
$$+ C\sum_{j=0}^{n-1}\left(k_j^{q+1} \sum_{l=q+1}^{q+2} \int_{t_j}^{t_{j+1}} \|y^{(l)}(s)\|\, ds + k_j^p \int_{t_j}^{t_{j+1}} \|y^{(p+1)}(s)\|\, ds \right).$$

This assertion follows by analogy with the proofs of Theorems 7.7, 7.8, and 7.9, using Theorems 8.4 and 8.5.

Further, turning to giving an *a posteriori* error estimate on nonuniform grids, we consider such estimates, for simplicity, only in the case where the backward Euler method is applied to problem (5.1) with $A(t) \equiv A$ independent of t. Noting that under our below conditions on A the operator $(I + kA)^{-1}$ is uniformly bounded with respect to $k > 0$, the extended solution by the backward Euler method applied to (5.1) is represented by the recurrence formula, with $\mathfrak{U}_n = (I + k_n A)^{-1}$,

$$Y_{n+1} = \mathfrak{U}_n Y_n + k_n \mathfrak{U}_n f(t_{n+1}) \quad \text{for } t_n \in \breve{\mho}_k, \qquad Y_0 = y^0. \tag{8.60}$$

The next assertion gives an *a posteriori* error estimate for this method.

Theorem 8.8. *Let* $T = \infty$, *let* $A \in \mathcal{S}(\varphi, \sigma)$ *with some* $\varphi \in (0, \pi/2)$ *and* $\sigma > 0$, *and let* $\mathfrak{T}$ *be a family of grids with the property:* $k_n \le Ck_{n+1}$ *whenever* $t_n, t_{n+1} \in \breve{\mho}_k$, *for any* $\mho_k \in \mathfrak{T}$. *Then, if* $f(t)$ *is strongly continuously differentiable on* $[0, \infty)$, *for* Y_n *specified by (8.60) and for* $y(t)$ *the extended* *solution of problem (5.1) in the considered case* $A(t) \equiv A$, *it holds, for all* $\mho_k \in \mathfrak{T}$ *and* $t_n \in \mho_k$ *with* $n \geq 1$,

$$\begin{aligned}\|Y_n - y(t_n)\| &\leq C\left(1 + \log\frac{t_n}{k_{n-1}}\right)\max_{0\leq j\leq n-1}\|y_{j+1} - Y_j\| \\ &\quad + C\sum_{j=0}^{n-1} k_j \int_{t_j}^{t_{j+1}} \|f'(s)\|\,ds.\end{aligned}$$

Proof. Under the accepted conditions we have, for any $\mho_k \in \mathfrak{T}$ and $t_n \in \breve{\mho}_k$,

$$y(t_{n+1}) = e^{-k_n A} y(t_n) + k_n \int_{t_n}^{t_{n+1}} e^{-(t_{n+1}-s)A} f(s)\,ds.$$

Subtracting this equality from the equation in (8.60), we obtain, with $E_n = Y_n - y(t_n)$, for $n \geq 0$,

$$E_{n+1} = e^{-k_n A} E_n + R_n, \tag{8.61}$$

where the residual R_n is given by

$$\begin{aligned}R_n &= Y_{n+1} - e^{-k_n A} Y_n - \int_{t_n}^{t_{n+1}} e^{-(t_{n+1}-s)A} f(s)\,ds \\ &= \left(\mathfrak{U}_n - e^{-k_n A}\right) Y_n + k_n \mathfrak{U}_n f(t_{n+1}) - \int_{t_n}^{t_{n+1}} e^{-(t_{n+1}-s)A} f(s)\,ds.\end{aligned}$$

Next, observing that A^{-1} is bounded, by $\sigma > 0$, it follows that

$$\begin{aligned}\int_{t_n}^{t_{n+1}} e^{-(t_{n+1}-s)A} f(s)\,ds &= \int_{t_n}^{t_{n+1}} e^{-(t_{n+1}-s)A} (f(s) - f(t_{n+1})\,ds \\ &\quad + (I - e^{-k_n A})\,A^{-1} f(t_{n+1}).\end{aligned}$$

Also, since the equation in (8.60) can be equivalently rewritten in the form

$$\partial Y_n + A\mathfrak{U}_n Y_n = \mathfrak{U}_n f(t_{n+1}), \tag{8.62}$$

we get, using the fact that all the involved operators are pairwise permutable,

$$\left(\mathfrak{U}_n - e^{-k_n A}\right) Y_n = \left(I - (I + k_n A)\,e^{-k_n A}\right)\left(\mathfrak{U}_n A^{-1} f(t_{n+1}) - A^{-1}\partial Y_n\right),$$

whence

$$\begin{aligned}&\left(\mathfrak{U}_n - e^{-k_n A}\right) Y_n + k_n \mathfrak{U}_n f(t_{n+1}) - (I - e^{-k_n A}) A^{-1} f(t_{n+1}) \\ &\qquad = \left((I + k_n A)\,e^{-k_n A} - I\right) A^{-1}\partial Y_n.\end{aligned}$$

Therefore, with the aid of these reasonings, the above representation for R_n is rewritten as follows

$$R_n = ((I + k_n A)\, e^{-k_n A} - I)\, A^{-1} \partial Y_n + \int_{t_n}^{t_{n+1}} e^{-(t_{n+1}-s)A} (f(t_{n+1} - f(s))\, ds. \tag{8.63}$$

Further, since the functions $\omega_l(z) = z^{-l}((1-z)e^z - 1)$, $l = 1, 2$, are bounded in a neighbourhood of the origin, applying Lemma 5.4 gives, for all $k > 0$,

$$\|\omega_1(-kA)\| + \|\omega_2(-kA)\| \le C.$$

This estimate implies, for $l = 1, 2$,

$$\|((I + kA)\, e^{-kA} - I)\, u\| \le Ck^l \|A^l u\| \quad \text{for all } k > 0 \text{ and } u \in \operatorname{Dom} A^l.$$

Using the last estimate next shows that

$$\|((I + k_{n-1}A)\, e^{-k_{n-1}A} - I)\, A^{-1} \partial Y_{n-1}\| \le C k_{n-1} \|\partial Y_{n-1}\| = C\|Y_n - Y_{n-1}\|$$

and, taking as well (7.11) for $\eta = 1$ into account, for $t_{j+1} < t_n$,

$$\begin{aligned} \|((I + k_j A)\, e^{-k_j A} - I)\, A^{-1} e^{-(t_n - t_{j+1})A} \partial Y_j\| &\le C k_j^2 \|A e^{-(t_n - t_{j+1})A} \partial Y_j\| \\ &\le C k_j (t_n - t_{j+1})^{-1} \|Y_{j+1} - Y_j\|. \end{aligned}$$

Also, by (7.11) with $\eta = 0$, we find, for $t_{j+1} < t_n$,

$$\begin{aligned} &\|e^{-(t_n - t_{j+1})A} \int_{t_j}^{t_{j+1}} e^{-(t_{j+1}-s)A} (f(t_{j+1}) - f(s))\, ds\| \\ &\qquad \le C \int_{t_j}^{t_{j+1}} \|f(t_{j+1}) - f(s)\|\, ds \le C k_j \int_{t_j}^{t_{j+1}} \|f'(s)\|\, ds. \end{aligned}$$

Therefore combining the last three estimates and taking (8.63) into account shows that

$$\begin{aligned} \|e^{-(t_n - t_{j+1})A} R_j\| &\le C k_j \int_{t_j}^{t_{j+1}} \|f'(s)\|\, ds \\ &+ \begin{cases} C\|Y_n - Y_{n-1}\| & \text{if} \quad t_j = t_{n-1}, \\ C k_j (t_n - t_{j+1})^{-1} \|Y_{j+1} - Y_j\| & \text{if} \quad 0 \le t_j \le t_{n-1}. \end{cases} \end{aligned} \tag{8.64}$$

Now, since $E_0 = 0$, similarly to (7.6) it follows from (8.61) that

$$E_n = \sum_{j=0}^{n-1} e^{-(t_n - t_{j+1})A} R_j.$$

Using this formula as well as (8.64), a simple estimation yields

$$\begin{aligned}\|E_n\| &\leq C\left(1+\sum_{l=0}^{n-2} k_l(t_n-t_{l+1})^{-1}\right)\max_{0\leq j\leq n-1}\|Y_{j+1}-Y_j\| \\ &\quad +\sum_{j=0}^{n-1} k_j\int_{t_j}^{t_{j+1}}\|f'(s)\|\,ds.\end{aligned}$$

To conclude the proof, it suffices to note that since $k_n \leq Ck_{n+1}$, we have, for $n \geq 1$, that

$$\sum_{l=0}^{n-2} k_l(t_n-t_{l+1})^{-1} \leq C\sum_{l=0}^{n-2} k_{l+1}(t_n-t_{l+1})^{-1} \leq C\left(1+\log\frac{t_n}{k_{n-1}}\right).$$

□

It is also seen from the above proof that the second term on the right-hand side of the *a posteriori* estimate can be replaced by the expression $C\sum_{j=0}^{n-1}\int_{t_j}^{t_{j+1}}\|f(t_{j+1})-f(s)\|\,ds$.

When applying the *a posteriori* estimate of Theorem 8.8, one has to be certain of the behaviour of the quantity $\|Y_n - Y_{n-1}\|$. By (8.62) it follows that

$$Y_n - Y_{n-1} = -k_{n-1}\mathfrak{A}_{n-1}Y_{n-1} + k_{n-1}\mathfrak{U}_{n-1}f(t_n),$$

where $\mathfrak{A}_{n-1} = A\mathfrak{U}_{n-1}$. Therefore, assuming in addition that problem (5.1) possesses an extended solution of first order and that the family of grids used is exponentially balanced with basis $Q \geq 1$, after a slight modification of the proof of Theorem 8.1, we obtain, for all $\mathfrak{V}_k \in \mathfrak{T}$ and $t_n \in \mathfrak{V}_k$ such that $n \geq 1$,

$$\begin{aligned}\|Y_n - Y_{n-1}\| &\leq k_{n-1}\|\mathfrak{A}_{n-1}Y_{n-1}\| + Ck_{n-1}\|f(t_n)\| \\ &\leq Ck_{n-1}\left(\|Ay^0\| + \sum_{j=1}^{n-1} k_j(t_n-t_j)^{-1}\|f(t_j)\| + \|f(t_n)\|\right) \\ &\leq Ck_{n-1}\left(\|Ay^0\| + \left(1+\log\frac{t_n}{k_{n-1}}\right)\max_{1\leq j\leq n}\|f(t_j)\|\right),\end{aligned}$$

which shows that the above *a posteriori* estimate can be used for constructing adaptive algorithms.

8.4 Comments and bibliographical remarks

Section 8.1: Apparently, standard and weighted stability estimates for variable stepsize approximations to autonomous parabolic equations in a Banach space setting are first obtained in our work [44], dealing with the class S_I and quasi-uniform partitions. Palencia [127] further shows that the stability in the standard sense is still preserved for such partitions when employing methods of the classes S_{II} and S_{III} as well. The case of arbitrary time partitions is discussed in our papers [30, 31, 36], in [30, 36] for methods of class S_I and in [31] for methods of the classes S_{II} and S_{III}. Note that the stability estimate in [31] contains an additional factor $\log\big(\max_{t_n\in\breve{\mho}_k} k_n/\min_{t_n\in\breve{\mho}_k} k_n\big)$. Theorem 8.1 is shown first in our paper [30],[12] and it is repeated later by Palencia [128] who uses different techniques. We further mention a paper of Yan [166] who shows some stability estimates working with families of increasing quasi-quasiuniform grids (see Section 8.1). It appears, however, using the above observations concerning locally balanced families of grids and families of quasi-quasiuniform grids, that the stability results of [166] can easily be obtained as well from our earlier work [36], where exponentially balanced families of grids are in use. Also, it is worth noting that only the operator $\mathfrak{U}_{n,0}$ is under estimation in [166] while the paper [36] deals with the estimation of $\mathfrak{U}_{n,l}$, $0 \le l \le n$, which is essential when handling inhomogeneous equations. Apart from everything else, Saito [138] derives a weighted stability estimate in the homogeneous case, dealing with arbitrary grids but requiring instead that the operator in question is bounded and its norm is co-ordinated with the quantity $\max_{t_n\in\breve{\mho}_k} k_n$. The stability of variable stepsize rational approximations of arbitrary C_0 semigroups has been carried out in [42].

Section 8.2: The analysis of stability of variable stepsize Runge-Kutta approximations in the non-commutative case and in a Banach space setting seems to be first given in [44], for quasi-uniform partitions. More general time partitions are studied in [30, 36]. The statements and proofs of Theorems 8.2 and 8.3 are very close to those presented in [36]. The case where the operator in question is Lipschitz-continuous is considered first in González & Palencia [74]. Theorem 8.4 is due to these authors, but we give a different proof to their result. The above Theorems 8.2, 8.4, and 8.3 use, however, a Hölder/Lipschitz-continuity type condition on $A(t)$, which does not cover some important practical situations. For this reason we present Theorem

[12]We follow here the proof of this result, which is given in [30] and also reproduced in [36].

8.5 which then is useful and it is stated here for the first time.
Section 8.3: An *a priori* error analysis is carried out in González & Ostermann [71] under the hypothesis of quasi-uniformity and some additional restrictions on the used family of grids. In Section 8.3, when dealing with *a priori* error estimates, we consider more general time partitions, and it is seen that these results are simply compiled from those achieved in Chapter 7 and in Sections 8.1 and 8.2. Also, we mention a paper of Saito [138] who presents smooth and non-smooth error estimates in the homogeneous case, assuming no restrictions on the stepsize but acting under some conditions on the operator and method in question. Certain *a posteriori* error estimates for parabolic equations have been studied, for example, in Akrivis, Makridakis, and Nochetto [8], Eriksson, Johnson, and Larsson [68], Johnson, Nie, and Thomée [88], and Nochetto, Savaré, and Verdi [120], as a rule, in Hilbert space settings. The argument employed in the proof of Theorem 8.8 is close to that demonstrated in [8]. Note that for practical applications of Theorem 8.8 we have to control the size of the constant C in its assertion. In principle, it is possible, with certain provisos, but it is easier done in the Hilbert case.

Part III

OTHER DISCRETIZATION METHODS

In this part of the book we examine other methods which are widely used in practice for discretization of abstract parabolic problems. As such we consider the θ-method, certain operator splitting schemes, and linear multistep methods. The methods selected here for analysis are united by the purpose of demonstrating wide possibilities of the approach presented in Part I.

Unfortunately through our limited opportunities, many interesting discretization methods are left out of consideration. Nevertheless, we would like to comment briefly on some of them.

First of all we mention the discontinuous Galerkin method which has been examined in the context of parabolic equations in Jamet [87], Eriksson, Johnson, and Larsson [68], Eriksson, Johnson, and Thomée [69], and Thomée [162]. The virtue of this method is the suitability for dealing with variable time stepsizes and with solutions of low regularity. We expect that the discontinuous Galerkin method, developed exclusively by using Hilbert space techniques, can be studied in the near future within Banach space settings as well. At least, this possibility seems to be realistic.

Also, in the context of parabolic equations using variable stepsize multistep methods is important. Unfortunately, many stability results shown on those lines are not directly adapted for applications connected with stiff problems and, in particular, with parabolic problems in themselves (see, e.g., Crouzeix & Lisbona [63], Gear & Tu [70], Grigorieff [76], and Guglielmi & Zennaro [79]). At the same time we point out the work of Becker [46], Calvo & Grigorieff [54], and Grigorieff [77], where the time discretizations of parabolic problems by the variable 2-step and 3-step BDF methods are examined from the viewpoint of stability and convergence.

In order to concentrate on the most essential features of the methods considered below, we deal only with stability and carry out our investigation working not with the most general case. As above, we let Ω_k be a uniform grid in the segment $[0,T]$ with stepsize $k > 0$. In what follows we keep the notation introduced in Part II and which is connected with the use of uniform grids.

Chapter 9

The θ-Method

In this chapter we discuss the θ-method which seems to be the simplest in realization among all discretization methods. In principle, the θ-method can be thought of as a particular case of the linear multistep method whose analysis is carried out in Chapter 11. In the autonomous case $A(t) \equiv A$, the θ-method can be examined within the study of Runge-Kutta methods expounded in Part II. However, in the case with a constant operator there are some interesting features of the θ-method which are studied in the literature, as a rule, within Hilbert space settings (see, e.g., Samarskii [139][1]). Our main aim in the present chapter is to use similar developments in the Banach case.

Throughout the chapter the operator $A(t)$ is assumed to be independent of t. At the same time it is always assumed that the operator in question depends on a (scalar or vector) parameter $h \in \mathcal{H}$. So we write A_h instead of $A(t)$. Since we have as an aim certain applications, we assume A_h to be a bounded operator on $\mathfrak{X}$ for any fixed $h \in \mathcal{H}$.

The θ-method is usually specified by using the following discrete problem, with $\theta = \theta(k, h) \in \mathbb{R}$,[2]

$$\begin{aligned} &\partial Y_n + A_h((1-\theta)Y_n + \theta Y_{n+1}) \\ &\qquad = (1-\theta)f(t_n) + \theta f(t_{n+1}) \quad \text{for } t_n \in \breve{\Omega}_k, \qquad Y_0 = \mathfrak{R} y^0. \quad (9.1) \end{aligned}$$

The term $(1-\theta)f(t_n) + \theta f(t_{n+1})$ can be replaced by $f((1-\theta)t_n + \theta t_{n+1})$ or in some other way as well. For such modifications our analysis below

[1] In [139] this method is called the weighted difference scheme.

[2] Unlike the general use, we introduce an operator $\mathfrak{R} \in \mathcal{B}(\mathfrak{X})$ to correct the initial condition. The operator $\mathfrak{R}$ may depend on k and h as well.

changes in minor details, so subsequently we look only at problem (9.1). It is seen from (9.1) that the θ-method may be thought of as a 2-stage RK method[3] whose Butcher array is

$$\left(\begin{array}{c|cc} 0 & 0 & 0 \\ 1 & 1-\theta & \theta \\ \hline & 1-\theta & \theta \end{array}\right).$$

Clearly, in the case where θ does not depend on k and h, this method is covered by the theory developed in Part II. At the same time it is well known (see, e.g., Samarskii [139]) that the order of accuracy for the θ-method is raised at the expense of a suitable choice of θ as a function of k and h. Therefore we must admit that θ depends on k and h, which is in the case below.

9.1 Stability results

In the considerations below we require that at least $\Theta \geq \theta = \theta(k,h) \geq 0$ and $A_h \in \mathcal{S}(\varphi,\sigma)$, for any $k > 0$ and $h \in \mathcal{H}$, with some $\Theta \geq 0$, $\varphi \in (0,\pi/2)$, and $\sigma \in \mathbb{R}$. This assumption yields in particular that the operator $(I + k\theta A_h)^{-1}$ is defined and bounded on $\mathfrak{X}$ for any fixed $k \in (0,K]$ and $h \in \mathcal{H}$. Therefore problem (9.1) is equivalently represented in the form (1.1) with[4]

$$\begin{aligned} \mathfrak{A}_n \equiv \mathfrak{A} &= A_h(I + k\theta A_h)^{-1}, \\ \mathfrak{Q}_n \equiv \mathfrak{Q} &= (I + k\theta A_h)^{-1}, \\ F_n &= (1-\theta)f(t_n) + \theta f(t_{n+1}), \end{aligned} \tag{9.2}$$

which further implies that (9.1) possesses a unique solution. In what follows we use this fact without explicitly mentioning it on each occasion.

Now let, given $\varphi \in (0,\pi/2)$ and $\tau \geq 0$, the set $\mathcal{Z}(\varphi,\tau)$ be defined by (see Figure 9.1)

$$\mathcal{Z}(\varphi,\tau) = \mathcal{D}\left(-\frac{i\tau e^{i\varphi}}{2\sin\varphi}; \frac{\tau}{2\sin\varphi}\right) \bigcap \mathcal{D}\left(\frac{i\tau e^{i\varphi}}{2\sin\varphi}; \frac{\tau}{2\sin\varphi}\right).$$

In this chapter together with hypothesis $\mathcal{S}(\varphi,\sigma)$ we use certain stronger hypotheses. As such, we introduce, given numbers $\varphi \in (0,\pi/2)$ and $R \geq 0$ and a positive-valued function $\psi(h)$, the following.

---

[3]It may be thought of as a 1-stage RK method if the term $(1-\theta)f(t_n) + \theta f(t_{n+1})$ in (9.1) is replaced by $f(t_{n+1})$.

[4]We recall that the operator A_h is bounded.

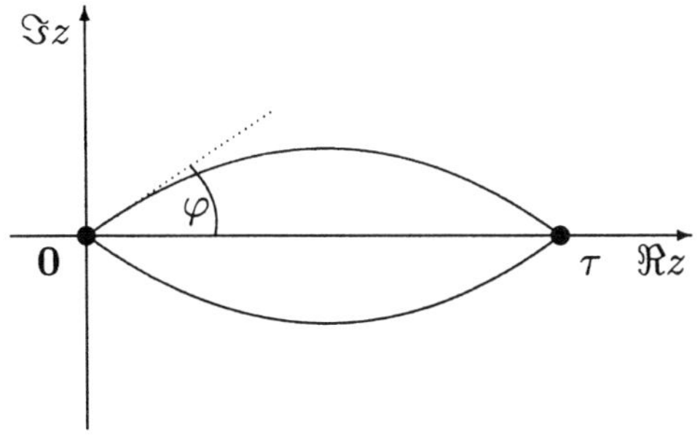

Figure 9.1: $\mathcal{Z}(\varphi, \tau)$

$\mathcal{S}_1(\varphi, R; \psi(\cdot))$: For all $h \in \mathcal{H}$ and $\lambda \notin \operatorname{Int} \mathcal{Z}(\varphi, \psi(h))$, $|\lambda| \geq R$,

$$\|(\lambda I - A_h)^{-1}\| \leq C|\lambda|^{-1},$$

and

$\mathcal{S}_2(\varphi, R; \psi(\cdot))$: For all $h \in \mathcal{H}$ and $\lambda \notin \operatorname{Int} \mathcal{Z}(\varphi, \psi(h))$, $|\lambda| \geq R$,

$$\|(\lambda I - A_h)^{-1}\| \leq C\left(|\lambda|^{-1} + |\lambda - \psi(h)|^{-1}\right).$$

We are now in a position to state our results on stability for the θ-method.

Theorem 9.1. *Suppose that $\Theta \geq \theta = \theta(k, h) \geq 1/2$ and $A_h \in \mathcal{S}(\varphi, \sigma)$ for all $k > 0$ and $h \in \mathcal{H}$, with some $\Theta \geq 1/2$, $\varphi \in (0, \pi/2)$, and $\sigma \in \mathbb{R}$. Let also the operator $\mathfrak{R}$ satisfy (6.48). Then, if $T < \infty$, the solution Y_n of (9.1) satisfies the following stability estimate, with $K > 0$ sufficiently small, for all $k \in (0, K]$, $h \in \mathcal{H}$, and $t_n \in \Omega_k$,*

$$\|Y_n\| \leq C\left(\|y^0\| + k\sum_{l=0}^{n}\|f(t_l)\|\right). \tag{9.3}$$

If $\sigma \geq 0$ in addition, the above holds for $T = \Theta = \infty$ and one can then take $K = \infty$.

Proof. For the sake of simplicity we show the claim in the particular case $\sigma \geq 0$ only. In the general case $\sigma \in \mathbb{R}$, the below argument is slightly modified with the aid of reasonings similar to those used in Part II.[5]

Having therefore assumed that $\sigma \geq 0$, we recall that for all $k > 0$ and $h \in \mathcal{H}$, the operator $(I + k\theta A_h)^{-1}$ is defined and bounded on $\mathfrak{X}$. Moreover, it obviously holds

$$\|(I + k\theta A_h)^{-1}\| \leq C. \tag{9.4}$$

[5] Cf., e.g., the proof of Theorem 6.1

Further, using the identity

$$\begin{aligned}(\lambda I - k\mathfrak{A})^{-1} &= \left(\lambda I - kA_h(I + k\theta A_h)^{-1}\right)^{-1} \\ &= \frac{\theta}{\theta\lambda - 1}I - (\theta\lambda - 1)^{-2}\left(\frac{\lambda}{\theta\lambda - 1}I + kA_h\right)^{-1}\end{aligned}$$

leads us to the estimate, for all $k > 0$ and $h \in \mathcal{H}$,

$$\|(\lambda I - k\mathfrak{A})^{-1}\| \leq |\lambda - \theta^{-1}|^{-1} + |1 - \theta\lambda|^{-2}\left\|\left(\frac{\lambda}{1 - \theta\lambda}I - kA_h\right)^{-1}\right\|. \tag{9.5}$$

Clearly, the mapping $\alpha(-z) = z(1 + \theta z)^{-1}$ takes the sector Σ_φ to the set $\mathcal{Z}(\varphi, \theta^{-1})$. Therefore, since $A_h \in \mathcal{S}(\varphi, \sigma)$ and $\lambda/(1 - \theta\lambda) \notin \operatorname{Int}\Sigma_\varphi$ for $\lambda \notin \operatorname{Int}\mathcal{Z}(\varphi, \theta^{-1})$, it follows from (9.5) that for all $k > 0$ and $\lambda \notin \operatorname{Int}\mathcal{Z}(\varphi, \theta^{-1})$,

$$\|(\lambda I - k\mathfrak{A})^{-1}\| \leq |\lambda - \theta^{-1}|^{-1} + C|1 - \theta\lambda|^{-1}|\lambda|^{-1},$$

which further implies, since $\theta \geq 1/2$, that for all $k > 0$ and $\lambda \notin \operatorname{Int}\mathcal{Z}(\varphi, 2)$,

$$\|(\lambda I - k\mathfrak{A})^{-1}\| \leq C\left(|\lambda|^{-1} + |\lambda - 2|^{-1}\right). \tag{9.6}$$

Observe that there exists a set Υ of Λ_2-configuration, with singular points $\zeta_1 = 0$ and $\zeta_2 = 2$, such that $\mathcal{Z}(\varphi, 2) \subseteq \Upsilon$. The desired result thus follows from Theorem 2.17 applied with $\xi = 0$ to problem (1.1), (9.2), taking as well the estimates (9.4) and (9.6) into account. □

Theorem 9.2. *Suppose that $\Theta \geq \theta = \theta(k, h) \geq \theta_0$ and $A_h \in \mathcal{S}(\varphi, \sigma)$ for all $k > 0$ and $h \in \mathcal{H}$, with some $\Theta \geq \theta_0 > 1/2$, $\varphi \in (0, \pi/2)$, and $\sigma \in \mathbb{R}$. Let also (6.48) hold. Then, if $T < \infty$, the solution Y_n of (9.1) satisfies the stability estimate, with some $K > 0$ sufficiently small, with any fixed $\mu_0 \geq 0$ sufficiently large, and with any fixed $\xi \geq 0$, for all $k \in (0, K]$, $h \in \mathcal{H}$, and $t_n \in \Omega_k$,*

$$\|(\mathfrak{A} + \mu_0 I)^{\xi} Y_n\| \leq C\left(t_{n+1}^{-\xi}\|y^0\| + k\sum_{l=0}^{n} t_{n-l+1}^{-\xi}\|f(t_l)\|\right). \tag{9.7}$$

If $\sigma \geq 0$ in addition, the above holds for $T = \Theta = \infty$ provided that $\mu_0 = 0$, and one can then take $K = \infty$.

The proof is similar to that of Theorem 9.1, applying as well Theorem 2.17 and using the fact that the pair $\left(\|(\mathfrak{A} + \mu_0 I)^{\xi} \cdot \|; \mathfrak{A}\right)$ is $\Lambda_1^k(\xi)$-concordant ($\Lambda_1(\xi)$-concordant if $\sigma \geq 0$ and $\mu_0 = 0$).

Now we show that the above results can be improved if we suppose that $A_h \in \mathcal{S}_1(\varphi, R; \psi(\cdot))$ or $A_h \in \mathcal{S}_2(\varphi, R; \psi(\cdot))$, instead of the condition $A_h \in \mathcal{S}(\varphi, \sigma)$ previously used.

Theorem 9.3. *Suppose that $A_h \in \mathcal{S}_2(\varphi, R, \psi(\cdot))$ for all $h \in \mathcal{H}$ and that $k > 0$ and $h \in \mathcal{H}$ are concordant in the sense that*

$$\theta = \theta(k,h) \geq \frac{1}{2} - (k\psi(h))^{-1} \geq 0, \tag{9.8}$$

with some $\varphi \in (0, \pi/2)$, $R \geq 0$, and $\Theta \geq 0$ and with some positive-valued function $\psi(h)$. Also, let (6.48) hold. Then, if $T < \infty$, the solution Y_n of (9.1) satisfies the stability estimate (9.3) under the additional condition $k \leq K$, with $K > 0$ sufficiently small. If $R = 0$ in addition, then the above holds for $T = \Theta = \infty$ and one can take $K = \infty$.

Proof. We show the result for $R = 0$ only, remarking that the below proof is slightly modified in the general case $R \geq 0$ (cf. the remark at the beginning of the proof of Theorem 9.1).

Having therefore assumed that $R = 0$, we observe that by spectral theory (see, e.g, Hille & Phillips [84]) the spectrum of the operator $k\mathfrak{A}$ belongs to the set $\operatorname{Int} \mathcal{Z}(\varphi, 2) \cup \{0\} \cup \{2\}$. Next we introduce the operators

$$\widetilde{A}_h = \psi(h)^{-1} A_h = (k\psi(h))^{-1} k A_h$$

and

$$B_h = \widetilde{A}_h (I - \widetilde{A}_h)^{-1}.$$

Our further argument will be based on making use of the representation

$$\begin{aligned}(\lambda I - B_h)^{-1} &= (I - \widetilde{A}_h)\left(\lambda I - (\lambda+1)\widetilde{A}_h\right)^{-1} \\ &= (\lambda+1)^{-1}(I - \widetilde{A}_h)\left(\frac{\lambda}{\lambda+1} I - \widetilde{A}_h\right)^{-1}.\end{aligned} \tag{9.9}$$

By the accepted conditions, the spectrum of the operator $\widetilde{A}_h$ belongs to the set $\widetilde{\mathcal{Z}} = \operatorname{Int} \mathcal{Z}(\varphi, 1) \cup \{0\} \cup \{1\}$ and, for all $\lambda \notin \widetilde{\mathcal{Z}}$,

$$\|(\lambda I - \widetilde{A}_h)^{-1}\| \leq C\left(|\lambda|^{-1} + |\lambda - 1|^{-1}\right),$$

which in particular implies

$$\|\widetilde{A}_h\| \leq C.$$

It thus follows from (9.9), using as well the last two estimates, that whenever $\lambda/(\lambda+1) \notin \operatorname{Int} \mathcal{Z}(\varphi, 1)$,

$$\|(\lambda I - B_h)^{-1}\| \leq C|\lambda+1|^{-1} \|(\frac{\lambda}{\lambda+1} I - \widetilde{A}_h)^{-1}\| \leq C|\lambda|^{-1}.$$

Since the conditions $\lambda/(\lambda+1) \notin \operatorname{Int} \mathcal{Z}(\varphi, 1)$ and $\lambda \notin \operatorname{Int} \Sigma_\varphi$ are equivalent, it follows from the last estimate that for all $\lambda \notin \operatorname{Int} \Sigma_\varphi$,

$$\|(\lambda I - B_h)^{-1}\| \leq C|\lambda|^{-1}.$$

In other words, we have $B_h \in \mathcal{S}(\varphi, 0)$ and, thus, $k^{-1}B_h \in \mathcal{S}(\varphi, 0)$. With this fact in mind, it remains to apply Theorem 9.1 with $k^{-1}B_h$ in place A_h and with $\theta + (k\psi(h))^{-1}$ in place of θ since $k\mathfrak{A} = B_h\left(I + (\theta + (k\psi(h))^{-1})B_h\right)^{-1}$ and $\theta + (k\psi(h))^{-1} \geq \frac{1}{2}$. □

Theorem 9.4. *Let $A_h \in \mathcal{S}_1(\varphi, R; \psi(\cdot))$ for all $h \in \mathcal{H}$ and let $k > 0$ and $h \in \mathcal{H}$ be subject to condition (9.8), with some $\varphi \in (0, \pi/2)$, $R \geq 0$, and $\Theta \geq 0$ and with some positive-valued function $\psi(h)$. Let also the operator $\mathfrak{R}$ be given by*

$$\mathfrak{R} = I - \frac{1}{2}k\mathfrak{A} = I - \frac{1}{2}kA_h(I + \theta k A_h)^{-1}.$$

Then, if $T < \infty$ and $\xi \in [0, 1]$, the solution Y_n of (9.1) satisfies the stability estimate (9.7) under the additional condition $k \leq K$, with $K > 0$ sufficiently small. If $R = 0$ and $\mu_0 = 0$ in addition, then the above holds for $T = \Theta = \infty$ and one can take $K = \infty$.

Proof. The operator $I - \frac{1}{2}k\mathfrak{A}$ can be represented as follows

$$I - \frac{1}{2}k\mathfrak{A} = (I + (\theta - \frac{1}{2})kA_h)\,(I + \theta k A_h)^{-1}. \tag{9.10}$$

Further, by the accepted conditions it holds $\left(k(\frac{1}{2} - \theta)\right)^{-1} \notin \operatorname{Int} \mathcal{Z}(\varphi, \psi(h))$, which yields since $A_h \in \mathcal{S}_1(\varphi, R; \psi(\cdot))$ that with $K > 0$ sufficiently small, for all $k \in (0, K]$ and $h \in \mathcal{H}$,

$$\|(I + (\theta - \frac{1}{2})kA_h)^{-1}\| \leq C. \tag{9.11}$$

Note that one can take $K = \Theta = \infty$ if $R = 0$. It now follows from (9.10) and (9.11) that[6]

$$\|(I + \theta k A_h)^{-1}(I - \frac{1}{2}k\mathfrak{A})^{-1}\| \leq C. \tag{9.12}$$

It is seen in the proof of Theorem 9.3 that the operator $\mathfrak{A}$ satisfies the Λ_2^k-condition, in particular, the Λ_2-condition if $R = 0$, with the singular points

[6]By (9.10) and (9.11) the operator $(I - \frac{1}{2}k\mathfrak{A})^{-1}$ is bounded on $\mathfrak{X}$ for any fixed $h \in \mathcal{H}$ since the operator A_h has this property.

$\zeta_1 = 0$ and $\zeta_2 = 2$. Taking then (9.12) into account, noting that the operator $I - \frac{1}{2}k\mathfrak{A}$ takes the form (1.39) with $J = 2$, $\zeta_2 = 2$, $\eta = 1$, and $\mathfrak{A}$ in place of $\mathfrak{L}_n$, and using the representation (1.4) with both sides multiplied from the left by $(\mathfrak{A} + \mu_0 I)^{\xi}$ (where $\mu_0 = 0$ if $R = 0$), the claim follows by applying Theorem 2.3 together with Remark 2.2. □

9.2 Comments and bibliographical remarks

Section 9.1: The above Theorem 9.3 actually arises from our work [23, 24] where a slightly different terminology is used. Also, it follows from [23, 24] that the hypotheses $\mathcal{S}_1(\varphi, R; \psi(\cdot))$ and $\mathcal{S}_2(\varphi, R; \psi(\cdot))$ can be verified for the standard difference approximations to elliptic differential operators. It turns out, with the aid of this fact, that the results of Section 9.1 can be helpful for showing stability in the L_p norms, $1 \leq p \leq \infty$, of certain higher order difference schemes applied to parabolic PDE problems in rectangle domains. Similar results have previously been obtained mainly in the particular case $p = 2$ (see, e.g., Samarskii [139]) while the case $p = \infty$ has been examined only for problems with constant coefficients in infinite domains (see Serdyukova [142, 144]).

Chapter 10

Methods with Splitting Operator

Here we consider approximations to problem (5.1), which employ certain constructions with splitting operator. In fact, in the general case where the operator in question is time-dependent, we examine only one operator splitting method based on the backward Euler discretization, which is accurate to the first order. However, completing this chapter, in the special case $A(t) \equiv A$ and $f(t) \equiv 0$, we discuss splitting constructions which may lead to second order solvers. Note that our below assertions are obtained on the basis of the general theory of operator splitting methods developed in Chapter 3. This is possible only under the condition $T < \infty$ which is therefore assumed throughout.

In this chapter the main distinctive feature is that the operator $A(t)$ in (5.1) is split by $A(t) = A_1(t) + A_2(t)$ for $t \in [0, T]$, where $A_j(t)$, $j = 1, 2$, are closed linear operators acting in $\mathfrak{X}$, for which $\operatorname{Dom} A_j(t) \supseteq \operatorname{Dom} A(t)$ for any $t \in [0, T]$.

So, given a uniform grid Ω_k in the interval $[0, T]$, we consider the following splitting version of the backward Euler method[1]

$$\begin{aligned} Y_{n+\frac{1}{2}} &= (I + kA_1(t_{n+1}))^{-1} Y_n, \\ Y_{n+1} &= (I + kA_2(t_{n+1}))^{-1} Y_{n+\frac{1}{2}} \\ &\quad + k(I + kA_2(t_{n+1}))^{-1} f(t_{n+1}) \quad \text{for } t_n \in \breve{\Omega}_k, \\ Y_0 &= y^0. \end{aligned}$$

[1]In this chapter we only deal with the situation when $\mathfrak{R} = I$.

Eliminating $Y_{n+\frac{1}{2}}$, we rewrite this discrete problem as a single step method, that is,

$$\begin{aligned} Y_{n+\frac{1}{2}} &= (I+kA_2(t_{n+1}))^{-1}(I+kA_1(t_{n+1}))^{-1}Y_n \\ &\quad +k(I+kA_2(t_{n+1}))^{-1}f(t_{n+1}) \quad \text{for } t_n \in \breve{\Omega}_k, \\ Y_0 &= y^0. \end{aligned} \tag{10.1}$$

Now we intend to apply the theory of Chapter 3 in order to examine the question of stability of problem (10.1). One of the most important points is then to show that the conditions (3.2) and (3.3) can be verified.[2] In our subsequent consideration we use the following hypotheses which are stated for some $\mu_1 \geq 0$ and $\kappa \in [0,1)$.

com1$[\kappa]$: For any $j = 1,2$ and $t \in [0,T]$ the operator $A_j(t)$ has dense domain[3] and it holds

$$\mathcal{Q}_j(t) := (A_{3-j}(t)+\mu_1 I)^{-1}\,\mathrm{Dom}\,A_j(t) \subseteq \mathrm{Dom}\,A_j(t). \tag{10.2}$$

Furthermore, for all $j=1,2$, $t\in[0,T]$, $u \in \mathcal{Q}_j(t)$, and $\chi \in \mathrm{Dom}\,A^*_{3-j}(t)$,

$$\begin{aligned} &|\langle \chi, A_j(t)\,A_{3-j}(t)u\rangle - \langle A^*_{3-j}(t)\chi, A_j(t)u\rangle| \\ &\qquad \leq C\|(A_{3-j}(t)+\mu_1 I)^*\chi\|_*^{\kappa}\|\chi\|_*^{1-\kappa}\|(A_j(t)+\mu_1 I)\,u\| \\ &\qquad\quad + C\|(A_{3-j}(t)+\mu_1 I)^*\chi\|_*\|(A_j(t)+\mu_1 I)\,u\|^{\kappa}\|u\|^{1-\kappa}. \end{aligned}$$

com2$[\kappa]$: For any $j = 1,2$ and $t \in [0,T]$ the operator $A_j(t)$ has dense domain and satisfies (10.2) as well as

$$\mathcal{Q}^*_j(t) := (A^*_{3-j}(t)+\mu_1 I)^{-1}\,\mathrm{Dom}\,A^*_j(t) \subseteq \mathrm{Dom}\,A^*_j(t).$$

Furthermore, for all $j=1,2$, $t\in[0,T]$, $u \in \mathcal{Q}_j(t)$, and $\chi \in \mathcal{Q}^*_j(t)$,

$$\begin{aligned} &|\langle \chi, A_j(t)\,A_{3-j}(t)u\rangle - \langle A^*_{3-j}(t)\chi, A_j(t)u\rangle| \\ &\qquad \leq C\sum_{l=1}^{2}\|(A_{3-l}(t)+\mu_1 I)^*\chi\|_*^{\kappa}\|\chi\|_*^{1-\kappa}\|(A_l(t)+\mu_1 I)\,u\|. \end{aligned}$$

Lemma 10.1. *Suppose $A_1(t), A_2(t) \in \mathcal{S}(\varphi,\sigma)$ and hypothesis* **com1**$[\kappa]$ *is satisfied, for some $\varphi \in (0,\pi/2)$, $\sigma \in \mathbb{R}$, and $\kappa \in [0,1)$. Then condition (3.2) is fulfilled with $\mathfrak{A}_{j;n} = A_j(t_{n+1})\,(I+kA_j(t_{n+1}))^{-1}$, $j=1,2$. If in addition hypothesis* **com2**$[\kappa]$ *is satisfied, then (3.3) holds with the same $\mathfrak{A}_{j;n}$.*

---

[2] It is obvious that problem (10.1) takes the form (1.1), (3.1).

[3] This fact implies in particular that each of the operators $A_j(t)$, $j = 1,2$, has an adjoint.

Proof. We show the claim in the case where $A_j(t) = A_j$, $j = 1, 2$, are independent of t for the reason that, on the one hand, the arguments do not alter[4] and, on the other hand, notation becomes shorter in the autonomous case. In particular, we write $\mathfrak{A}_j$ and $\mathcal{Q}_j$ instead of $\mathfrak{A}_{j;n}$ and $\mathcal{Q}_j(t)$, respectively.[5]

Next observe that, by Lemma 5.3, the operators $\mathfrak{A}_j$, $j = 1, 2$, are defined and bounded on $\mathfrak{X}$ for any fixed $k \in (0, K]$, with $K > 0$ sufficiently small. Moreover, by Theorem 6.2 they both satisfy the Λ_1^k-condition.

Denoting now $\varrho_j = (I + kA_j)^{-1}$, $j = 1, 2$, by hypothesis **com1**$[\kappa]$ we have $\varrho_2 u \in \mathcal{Q}_1$ for any $u \in \mathrm{Dom}\, A_1$. Then, for all $u \in \mathrm{Dom}\, A_1$ and $\chi \in \mathfrak{X}^*$,

$$\begin{aligned} \langle \chi, \mathrm{Com}\,(A_1|\varrho_2)u\rangle &= \langle (I - \varrho_2^*)\chi, A_1\varrho_2 u\rangle - \langle \varrho_2^*\chi, A_1(I - \varrho_2)u\rangle \\ &= k\langle A_2^*\varrho_2^*\chi, A_1\varrho_2 u\rangle - k\langle \varrho_2^*\chi, A_1 A_2 \varrho_2 u\rangle, \end{aligned}$$

thus, with any $\mu_1 \geq 0$,

$$\begin{aligned} \langle \chi, (A_1 + \mu_1 I)\, \varrho_2 u\rangle &= \langle \chi, \varrho_2\, (A_1 + \mu_1 I)\, u\rangle \\ &\quad + k\langle A_2^*\varrho_2^*\chi, A_1\varrho_2 u\rangle - k\langle \varrho_2^*\chi, A_1 A_2\varrho_2 u\rangle \\ &=: \langle \chi, \varrho_2\, (A_1 + \mu_1 I)\, u\rangle + \Psi. \end{aligned} \tag{10.3}$$

By hypothesis **com1**$[\kappa]$ we further obtain, for all $u \in \mathrm{Dom}\, A_1$ and $\chi \in \mathfrak{X}^*$,

$$\begin{aligned} |\Psi| &\leq Ck\|(A_2 + \mu_1 I)^*\varrho_2^*\chi\|_*^\kappa \|\chi\|_*^{1-\kappa}\|(A_1 + \mu_1 I)\,\varrho_2 u\| \\ &\quad + Ck\|(A_2 + \mu_1 I)^*\varrho_2^*\chi\|_*\|(A_1 + \mu_1 I)\,\varrho_2 u\|^\kappa \|u\|^{1-\kappa} \\ &\leq C\|\chi\|_*(k^{1-\kappa}\|(A_1 + \mu_1 I)\,\varrho_2 u\| + \|(A_1 + \mu_1 I)\,\varrho_2 u\|^\kappa\|u\|^{1-\kappa}). \end{aligned}$$

Clearly since $A_1 \in \mathcal{S}(\varphi, \sigma)$, the last estimate is in force with any $\mu_2 > 0$ sufficiently large in place of μ_1 while no constant C depends on μ_2. Therefore we get instead

$$\begin{aligned} |\Psi| &\leq C\left(k^{1-\kappa} + \|(A_1 + \mu_2 I)^{-1}\|^{1-\kappa}\right)\|(A_1 + \mu_2 I)\,\varrho_2 u\|\,\|\chi\|_* \\ &\leq C\left(k^{1-\kappa} + \mu_2^{\kappa-1}\right)\|(A_1 + \mu_2 I)\,\varrho_2 u\|\,\|\chi\|_*. \end{aligned}$$

Applying this inequality in an estimation on the basis of (10.3), we find that for all $k \in (0, K]$ and $u \in \mathfrak{X}$,

$$\|(A_1 + \mu_2 I)\,\varrho_2 u\| \leq C\|\varrho_2(A_1 + \mu_2 I)\,u\| \leq C\|(A_1 + \mu_2 I)\,u\| \tag{10.4}$$

[4] In fact, all the estimates retain their shapes, becoming uniform with respect to t_n.
[5] We recall that $\mathcal{Q}(t)$ is specified by (10.2).

if $K > 0$ is sufficiently small and $\mu_2 > 0$ is sufficiently large. Next, with the aid of the following identity, for all $u \in \mathfrak{X}$ and $\chi \in \mathfrak{X}^*$,

$$\begin{aligned}\langle (A_2+\mu_2 I)^* \varrho_2^* \varrho_1^* \chi, u\rangle &= \langle \chi, \varrho_1 (A_2 + \mu_2 I)\varrho_2 u\rangle \\ &= k^{-1}\langle \chi, \varrho_1 (I - \varrho_2)\, u\rangle + \mu_2 \langle \chi, \varrho_1 \varrho_2 u\rangle \\ &= \langle \chi, A_2 \varrho_2 \varrho_1 u\rangle + \mu_2 \langle \chi, \varrho_1 \varrho_2 u\rangle - k\langle \varrho_2^* \varrho_1^* \chi, A_1 A_2 \varrho_2 \varrho_1 u\rangle \\ &\quad + k\langle A_2^* \varrho_2^* \varrho_1^* \chi, A_1 \varrho_2 \varrho_1 u\rangle,\end{aligned}$$

employing hypothesis **com1**$[\kappa]$, and taking (10.4) into account leads, as above, to the following estimate

$$\|(A_2 + \mu_2 I)^* \varrho_2^* \varrho_1^* \chi\|_* \le C\|(A_2 + \mu_2 I)^* \varrho_2^* \chi\|_*, \tag{10.5}$$

with any fixed $\mu_2 > 0$ sufficiently large, for all $\chi \in \mathfrak{X}^*$. Furthermore, it is not hard to see that for all $u \in \mathfrak{X}$ and $\chi \in \mathfrak{X}^*$,

$$\begin{aligned}&\langle \chi, \mathrm{Com}\ (\mathfrak{A}_1|\mathfrak{A}_2)\, u\rangle = k^{-2}\langle \chi, \mathrm{Com}\ ((I - \varrho_1), (I - \varrho_2))\, u\rangle \\ &= k^{-2}\langle \chi, \mathrm{Com}\ (\varrho_1|\varrho_2)\, u\rangle = \langle \varrho_2^* \varrho_1^* \chi, A_1 A_2 \varrho_2 \varrho_1 u\rangle - \langle A_2^* \varrho_2^* \varrho_1^* \chi, A_1 \varrho_2 \varrho_1 u\rangle,\end{aligned}$$

so that we find, with the aid of hypothesis **com1**$[\kappa]$ and of the above argument, with any fixed $\mu_2 > 0$ sufficiently large, for all $k \in (0, K]$, $u \in \mathfrak{X}$, and $\chi \in \mathfrak{X}^*$,

$$\begin{aligned}\big|\langle \chi, \mathrm{Com}\ (\mathfrak{A}_1|\mathfrak{A}_2)\, u\rangle\big| &\le C\|(A_2 + \mu_2 I)^* \varrho_2^* \varrho_1^* \chi\|_*^\kappa \|\chi\|_*^{1-\kappa} \|(A_1 + \mu_2 I)\, \varrho_2 \varrho_1 u\| \\ &\quad + C\|(A_2 + \mu_2 I)^* \varrho_2^* \varrho_1^* \chi\|_* \|(A_1 + \mu_2 I)\, \varrho_2 \varrho_1 u\|^\kappa \|u\|^{1-\kappa}. \end{aligned} \tag{10.6}$$

Therefore combining (10.4), (10.5), and (10.6) gives

$$\begin{aligned}\big|\langle \chi, \mathrm{Com}\ (\mathfrak{A}_1|\mathfrak{A}_2)\, u\rangle\big| &\le C\|(A_2 + \mu_2 I)^* \varrho_2^* \chi\|_*^\kappa \|\chi\|_*^{1-\kappa} \|(A_1 + \mu_2 I)\, \varrho_1 u\| \\ &\quad + C\|(A_2 + \mu_2 I)^* \varrho_2^* \chi\|_* \|(A_1 + \mu_2 I)\, \varrho_1 u\|^\kappa \|u\|^{1-\kappa} \\ &\le C\left(\|\mathfrak{A}_2^* \chi\|_* + \|\chi\|_*\right)^\kappa \|\chi\|_*^{1-\kappa} \left(\|\mathfrak{A}_1 u\| + \|u\|\right) \\ &\quad + C\left(\|\mathfrak{A}_2^* \chi\|_* + \|\chi\|_*\right)\left(\|\mathfrak{A}_1 u\| + \|u\|\right)^\kappa \|u\|^{1-\kappa}.\end{aligned}$$

So since both $\mathfrak{A}_1$ and $\mathfrak{A}_2$ satisfy the Λ_1^k-condition, (3.2) with $j = 2$ follows from the last estimate in view of Lemma 1.5, and (3.2) with $j = 1$ is established by symmetry. If in addition hypothesis **com2**$[\kappa]$ is satisfied, it can be shown in a similar way that (3.3) holds with $\mathfrak{A}_j$ substituted for $\mathfrak{A}_{j;n}$. □

Now we state a stability result for the splitting method (10.1).

Theorem 10.1. *Suppose $A_1(t), A_2(t) \in \mathcal{S}(\varphi, \sigma)$, seminorms $|\cdot|_{1;t}$ and $|\cdot|_{2;t}$ form $(\gamma_1|\varphi,\sigma)$-concordant pairs with $A_1(t)$ and $A_2(t)$, respectively, each of $A_j(t)$ satisfies the respective hypothesis* $\mathbf{h\ddot{o}l}^{\oplus}[\vartheta; |\cdot|_{j;t}]$, *$j = 1, 2$, and the hypotheses* $\mathbf{com1}[\kappa]$ *and* $\mathbf{com2}[\kappa]$ *are fulfilled, for some $\varphi \in (0, \pi/2)$, $\sigma \in \mathbb{R}$, $\vartheta \in (0, 1]$, and $\gamma_1, \kappa \in [0, 1)$. Then, if $K > 0$ is sufficiently small, for the solution Y_n of problem (10.1), we have, with any fixed $\mu_0 \geq 0$ sufficiently large and with any fixed $\xi \in [0, 1)$, for all $k \in (0, K]$ and $t_n \in \Omega_k$, that*

$$\begin{aligned} \|(\mathfrak{A}_n + \mu_0 I)^{\xi} Y_n\| &+ \sum_{j=1}^{2} \|(\mathfrak{A}_{j;n} + \mu_0 I)^{\xi} Y_n\| \\ &\leq C t_{n+1}^{-\xi} \|y^0\| + Ck \sum_{l=1}^{n} t_{n-l+1}^{-\xi} \|f(t_l)\|, \end{aligned} \tag{10.7}$$

where the operators $\mathfrak{A}_{j;n}$, $j = 1, 2$, are defined as above and $\mathfrak{A}_n$ is specified by (3.1). In the case where $A_1(t)$ and $A_2(t)$ do not depend on t, the above holds for any fixed $\xi \in [0, 1]$ while the first term on the left-hand side of (10.7) is estimated as stated for any fixed $\xi \geq 0$, even if we do not assume hypothesis $\mathbf{com2}[\kappa]$.

Proof. We consider the general case where $A_j(t)$, $j = 1, 2$, generally depend on t. Since each of the operators $\mathfrak{A}_{j;n}$, $j = 1, 2$, satisfies the Λ_1^k-condition because of Theorem 6.2, using the fact that each of the operators $A_j(t)$, $j = 1, 2$, satisfies the respective hypothesis $\mathbf{h\ddot{o}l}^{\oplus}[\vartheta; |\cdot|_{j;t}]$ and applying Theorem 6.9, it follows that the Hölder-continuity type condition (3.40) holds. Since the operator $(I + kA_2(t))^{-1}$ is, by Lemma 5.3, uniformly bounded with respect to k and t, it thus remains to apply Theorem 3.4 with $\mathfrak{B}_{j;n} = 0$, $j = 1, 2$, taking Lemma 10.1 into account.

As concerns the more precise wording in the particular case where $A_j(t)$, $j = 1, 2$, are both independent of t, the claim follows by Theorem 3.3, by using Lemma 10.1 again. □

Note that Theorem 10.1 remains valid for splitting constructions generated by arbitrary RK methods of type $V_I(\varphi)$, but if the operators $A_1(t)$ and $A_2(t)$ are not permutable, approximations of that type would be only first order and of little practical use. At the same time, using the fact that the stability function of the backward Euler method vanishes at ∞ allows us easily to obtain (10.7) with $A_j(t_{n+1})$ substituted for $\mathfrak{A}_{j;n}$.

Now, in the particular case where $A(t) \equiv A$ and $f(t) \equiv 0$, we examine an operator splitting method which leads to second order solvers. In order to construct such a method, let the operator A be decomposed as above, with $A_j(t) \equiv A_j$, $j = 1, 2$, and let $r(z)$ be the stability function of some RK method. We consider the following procedure

$$Y_{n+1} = r(-\frac{k}{2}A_1)\, r(-\frac{k}{2}A_2)^2 r(-\frac{k}{2}A_1)\, Y_n \quad \text{for } t_n \in \breve{\Omega}_k, \qquad Y_0 = y^0. \quad (10.8)$$

The next result shows the stability of the splitting method (10.8).

Theorem 10.2. *Let $A_1(t), A_2(t) \in \mathcal{S}(\varphi, \sigma)$, let hypothesis* **com1**$[\kappa]$ *be satisfied, and let $r(z)$ be the stability function of an $A_I(\varphi)$-stable RK method, for some $\varphi \in (0, \pi/2)$, $\sigma \in \mathbb{R}$, and $\kappa \in [0, 1)$. Then, for Y_n the solution of problem (10.8) we have, with some $K > 0$ sufficiently small, with any fixed $\mu_0 \geq 0$ sufficiently large, and with any fixed $\xi \geq 0$, for all $k \in (0, K]$ and $t_n \in \Omega_k$, that*

$$\|(\mathfrak{A} + \mu_0 I)^\xi Y_n\| \leq C t_{n+1}^{-\xi} \|y^0\|,$$

where $\mathfrak{A} = \mathfrak{A}_1 + \mathfrak{A}_2 - k\mathfrak{A}_2\mathfrak{A}_1$, $\mathfrak{A}_j = k^{-1}\alpha(-kA_j)$, $j = 1, 2$, and, as above, $\alpha(z) = 1 - r(z)$. If in addition hypothesis **com2**$[\kappa]$ *is satisfied, then the quantity $\sum_{j=1}^2 \|(\mathfrak{A}_{j;n} + \mu_0 I)^\xi Y_n\|$ admits the same estimate for any fixed $\xi \in [0, 1]$.*

Proof. By virtue of Theorem 6.2 each of the operators $\mathfrak{A}_1, \mathfrak{A}_2$ satisfies the Λ_1^k-condition. Therefore the claim will follow by Theorem 3.3 if we have proved that the hypotheses **com1**$[\kappa]$ and **com2**$[\kappa]$ imply the conditions (3.2) and (3.3), respectively, with $\mathfrak{A}_{j;n} = \mathfrak{A}_j$, $j = 1, 2$. The last fact is however established in a manner similar to that used in the proof of Lemma 10.1. In fact, by the condition $\deg[\rho] = -1$, where, as above, $\rho(z) = z^{-1}\alpha(z)$, we have, with some $b_{lq} \in \mathbb{C}$, $l = 1, \ldots, l_0$, $q = 1, \ldots, q_l$,

$$\rho(-z) = \sum_{l=1}^{l_0} \sum_{q=1}^{q_l} b_{lq}(z - z_l)^{-q}, \qquad (10.9)$$

where z_l, $l = 1, \ldots, l_0$, are the poles of $\rho(-z)$ and q_l denotes the multiplicity of a pole z_l. By the $A_I(\varphi)$-stability it follows that $z_l \notin \Sigma_\varphi$, and the operators $(kA_j - z_l I)^{-1}$, $j = 1, 2$, $l = 1, \ldots, l_j$, are defined and bounded on $\mathfrak{X}$ for $k \in (0, K]$ and $t \in [0, T]$, with some $K > 0$ sufficiently small. Using therefore (10.9) leads to the representation

$$\rho(-kA_j) = \sum_{l=1}^{l_0} \sum_{q=1}^{q_l} b_{lq}(kA_j - z_l I)^{-q}, \quad j = 1, 2.$$

Now, with the aid of this expansion and of permutation techniques similar to those employed in the proof of Lemma 10.1, it is readily seen that this lemma remains valid with the newly defined $\mathfrak{A}_j$, $j = 1, 2$. $\square$

In conclusion, we remark that some related results are presented in Ostermann [122] under essentially stronger conditions on the splitting components.

Chapter 11

Linear Multistep Methods

In this chapter we consider one more field for application of the theory developed in Part I. More precisely, here we examine linear multistep methods applied to the Cauchy problem (5.1). We restrict ourselves to dealing with such methods for the sake of simplicity and for the reason that they are most important for practice. In principle, we could study even general multistep methods, which would lead, however, to a more involved analysis. It is remarkable that the investigation of linear multistep methods, in comparison with the one of Runge-Kutta methods, possesses its own distinctive features. Also, it is worth noting that the stability characteristics of the most popular linear multistep methods, among those considered below, are well adapted for their application to parabolic problems.

11.1 Linear multistep discretization

Let there be given numbers $m \in \mathbb{N}$ and $a_j, b_j \in \mathbb{C}$, $j = 0, \ldots, m$. Let further $\Omega_k = \{t_n : t_n = nk \text{ for } n = 0, 1, \ldots, N\}$ be a uniform grid in the segment $[0, T]$, with stepsize k,[1] and let $\Omega_k^{(m)} = \{t_n : t_n = nk \text{ for } n = 0, 1, \ldots, N - m + 1\}$ and $\breve{\Omega}_k^{(m)} = \{t_n : t_n = nk \text{ for } n = 0, 1, \ldots, N - m\}$.

Now we associate to problem (5.1) the following discrete problem

$$\sum_{l=0}^{m} \big(a_l Y_{n+m-l} + k b_l (A(t_{n+m-l})\, Y_{n+m-l} - f(t_{n+m-l}))\big) = 0 \quad \text{for } t_n \in \breve{\Omega}_k^{(m)},$$

$$Y_0 = y^0,\ Y_1 = y^1,\ \ldots,\ Y_{m-1} = y^{m-1}, \quad (11.1)$$

[1] As above, k and N are connected by $T = Nk$, with the standard proviso in the case $T = \infty$.

with some $y^0, y^1, \dots, y^{m-1} \in \mathfrak{X}$. Hereinafter we assume that the initial values $y^0, y^1, \dots, y^{m-1}$ are given. Problem (11.1) is just the linear multistep method applied formally to (5.1). However this form of writing is typical for application to ODEs. Unfortunately, in our case the fact that the domain of $A(t)$ does not generally coincide with $\mathfrak{X}$ implies that some terms of the form $A(t_{n+m-l})\, y(t_{n+m-l})$ in (11.1) may be undefined. We therefore have to specify what we mean by a solution of problem (11.1).

First of all note that, as above, our main assumption in this chapter is that $A(t) \in \mathcal{S}(\varphi, \sigma)$ with some $\varphi \in (0, \pi/2)$ and $\sigma \in \mathbb{R}$. If in addition to this condition we assume as well the following restriction[2]

$$a_0 b_0 \neq 0 \text{ and } |\arg(a_0/b_0)| < \pi - \varphi, \tag{11.2}$$

then, with $K > 0$ sufficiently small,[3] the operator $(a_0 I + k b_0 A(t))^{-1}$ is bounded on $\mathfrak{X}$ for $k \in (0, K]$ and $t \in [0, T]$ and, if the entry in (11.1) makes sense, we proceed instead from the following recursion procedure

$$\begin{aligned} Y_{n+m} = &- (a_0 I + k b_0 A(t_{n+m}))^{-1} \sum_{l=1}^{m} (a_l I + k b_l A(t_{n+m-l}))\, Y_{n+m-l} \\ &+ k (a_0 I + k b_0 A(t_{n+m}))^{-1} \sum_{l=0}^{m} b_l f(t_{n+m-l}) \quad \text{for } t_n \in \breve{\Omega}_k^{(m)}, \end{aligned} \tag{11.3}$$

with $Y_0, Y_1, \dots, Y_{m-1}$ specified the same as in (11.1). Before taking further steps we now have to distinguish between the cases where $A(t)$ does not depend and where it does depend on t.

In the particular case where $A(t) \equiv A$ is independent of t,[4] observe that under the above assumptions the operators $(a_l I + k b_l A)\,(a_0 I + k b_0 A)^{-1}$, $l = 1, \dots, m$, are defined and bounded on $\mathfrak{X}$. We therefore come to the following definition.

Definition 11.1. *In the autonomous case, the discrete function Y_n, $t_n \in \Omega_k$, as defined by the recursion*

$$\begin{aligned} Y_{n+m} = &- \sum_{l=1}^{m} (a_l I + k b_l A)\,(a_0 I + k b_0 A)^{-1} Y_{n+m-l} \\ &+ k (a_0 I + k b_0 A)^{-1} \sum_{l=0}^{m} b_l f(t_{n+m-l}) \quad \textit{for } t_n \in \breve{\Omega}_k^{(m)}, \end{aligned}$$

[2] Which is present in all below assertions

[3] With $K = \infty$ if $\sigma \geq 0$

[4] But Dom A is not necessarily dense

with $Y_0, Y_1, \dots, Y_{m-1}$ specified just the same as in (11.1), is called the C-extended solution *of the discrete problem (11.1) (or of (11.3)).*

In the general case, we assume in addition that $\mathrm{Cl}(\mathrm{Dom}\, A(t)) = \mathfrak{X}$ for any $t \in [0,T]$ and that the operator $(\lambda I - A(t))^{-1} A(s)$ extends to a linear bounded operator on the whole $\mathfrak{X}$ for any fixed $\lambda \notin \left(\mathrm{Int}\, \Sigma_\varphi \cup \{0\}\right) + \sigma$ and $t, s \in [0,T]$.[5] However, unlike Part II, we do not require that $\mathrm{Dom}\, A(t)$ is necessarily independent of t.

Definition 11.2. *In the non-autonomous case, under the accepted restrictions, the* V-extended solution Y_n, $t_n \in \Omega_k$ *of problem (11.1) (or of (11.3)) is defined by the recursion*

$$Y_{n+m} = -\sum_{l=1}^{m} (a_0 I + k b_0 A(t_{n+m}))^{-1} (a_l I + k b_l A(t_{n+m-l}))\, Y_{n+m-l}$$
$$+ k(a_0 I + k b_0 A(t_{n+m}))^{-1} \sum_{l=0}^{m} b_l f(t_{n+m-l}) \quad \textit{for } t_n \in \breve{\Omega}_k^{(m)},$$

with $Y_0, Y_1, \dots, Y_{m-1}$ specified just the same as in (11.1).

In what follows we employ again the product space $\mathfrak{X}^{(m)}$ which is defined the same as in Section 5.3, with m in place of ν. As above, the norm in $\mathfrak{X}^{(m)}$ is given by

$$\|(u_1, \dots, u_m)^T\|_{(m)} = \max_{1 \le j \le m} \|u_j\| \quad \text{for any } (u_1, \dots, u_m)^T \in \mathfrak{X}^{(m)}.$$

The symbol $\|\cdot\|_{(m)}$ is used to denote as well the induced operator norm in $\mathfrak{X}^{(m)}$. Next, by $\mathfrak{X}^{(m)*}$ we denote the dual space for $\mathfrak{X}^{(m)}$. For our subsequent needs recall that any member of $\mathfrak{X}^{(m)*}$ is, in fact, a vector of the form $(\chi_1, \dots, \chi_m)^T$ with $\chi_j \in \mathfrak{X}^*$, $j = 1, \dots, m$. The norm in $\mathfrak{X}^{(m)*}$ is given by

$$\|(\chi_1, \dots, \chi_m)^T\|_{(m)*} = \sum_{j=1}^{m} \|\chi_j\|_*,$$

and the duality pairing between $\mathfrak{X}^{(m)*}$ and $\mathfrak{X}^{(m)}$ is expressed as follows

$$\langle (\chi_1, \dots, \chi_m)^T, (u_1, \dots, u_m)^T \rangle_{(m)} = \sum_{j=1}^{m} \langle \chi_j, u_j \rangle. \tag{11.4}$$

[5] The last fact is observed if, for instance, $\mathrm{Dom}\, A^*(t)$ is independent of t.

Also, by $\mathbf{I}$ we denote the identity operator both in $\mathfrak{X}^{(m)}$ and in $\mathfrak{X}^{(m)*}$.

Now we show that if Y_n is the V-extended or C-extended solution of problem (11.3), the vector

$$\mathbf{Y}_n = (Y_n, \ldots, Y_{n+m-1})^T \tag{11.5}$$

solves, in fact, a discrete problem of the form (1.1) in the space $\mathfrak{X}^{(m)}$. To that end, we put

$$\mathbf{y}^0 = (y^0, \ldots, y^{m-1})^T, \tag{11.6}$$

$$B_{j;n} = a_{m-j}I + kb_{m-j}A(t_{n+j}), \quad j = 0, \ldots, m,$$

and

$$\mathbf{F}_n = (\underbrace{0, \ldots, 0, F_n}_{m})^T, \quad \text{where } F_n = B_{m;n}^{-1} \sum_{l=0}^{m} b_l f(t_{n+m-l}). \tag{11.7}$$

Also, we introduce the operator $\mathfrak{A}_n : \mathfrak{X}^{(m)} \to \mathfrak{X}^{(m)}$ by

$$\mathfrak{A}_n = k^{-1} \begin{pmatrix} I & -I & 0 & \cdots & 0 \\ 0 & I & -I & \cdots & 0 \\ \cdot & \cdot & \cdot & \cdots & \cdot \\ B_{m;n}^{-1}B_{0;n} & B_{m;n}^{-1}B_{1;n} & \cdot & \cdots & I + B_{m;n}^{-1}B_{m-1;n} \end{pmatrix}, \tag{11.8}$$

if $A(t)$ generally depends on t, and by

$$\mathfrak{A}_n = \mathfrak{A} = k^{-1} \begin{pmatrix} I & -I & 0 & \cdots & 0 \\ 0 & I & -I & \cdots & 0 \\ \cdot & \cdot & \cdot & \cdots & \cdot \\ \alpha_m(-kA) & \alpha_{m-1}(-kA) & \cdot & \cdots & I + \alpha_1(-kA) \end{pmatrix}, \tag{11.9}$$

with $\alpha_j(z) = (a_j - b_j z)/(a_0 - b_0 z)$, $j = 1, \ldots, m$, if $A(t) \equiv A$ is independent of t. It is then easily seen under the accepted conditions[6] that if Y_n is the V-extended or C-extended solution of (11.3), the vector $\mathbf{Y}_n$ solves the problem

$$\mathbf{Y}_{n+1} = (\mathbf{I} - k\mathfrak{A}_n)\,\mathbf{Y}_n + k\mathbf{F}_n \quad \text{for } t_n \in \breve{\Omega}_k^{(m)}, \qquad \mathbf{Y}_0 = \mathbf{y}^0,$$

or, what is actually the same,

$$\partial \mathbf{Y}_n + \mathfrak{A}_n \mathbf{Y}_n = \mathbf{F}_n \quad \text{for } t_n \in \breve{\Omega}_k^{(m)}, \qquad \mathbf{Y}_0 = \mathbf{y}^0, \tag{11.10}$$

[6] Which are different in the autonomous and non-autonomous cases

which takes the form (1.1).[7] We therefore attempt to apply the general results of Part I.

Throughout this chapter it is assumed, without explicit mention on each occasion, that it holds

$$\sum_{l=0}^{m} a_l = 0 \text{ and } \sum_{l=0}^{m} b_l = 1, \tag{11.11}$$

since this condition is satisfied by all methods being of practical value.

For our below needs we now give a series of further definitions.

Definition 11.3. *With $\alpha_j(z)$, $j = 1, \ldots, m$, just the same as above, the matrix*

$$\mathcal{A}(z) = \begin{pmatrix} 1 & -1 & 0 & \cdots & 0 \\ 0 & 1 & -1 & \cdots & 0 \\ \cdot & \cdot & \cdot & \cdots & \cdot \\ \alpha_m(z) & \alpha_{m-1}(z) & \cdot & \cdots & 1+\alpha_1(z) \end{pmatrix}$$

is called the amplification matrix *of the linear multistep method.*

Definition 11.4. *The linear multistep method is called of* class M_I *if, with $\mathcal{A}(z)$ the amplification matrix of the method, all the roots of the equation*

$$\det(\lambda \mathrm{i} - \mathcal{A}(0)) = 0, \tag{11.12}$$

except the single simple root $\lambda_1(0) = 0$, and all the roots of the equation

$$\det(\lambda \mathrm{i} - \mathcal{A}(\infty)) = 0 \tag{11.13}$$

lie in the open disk $\mathrm{Int}\,\mathcal{D}(1;1)$.[8]

Definition 11.5. *The linear multistep method is called of* type $LM(\varphi)$[9] *for some $\varphi \in (0, \pi/2)$ if*
(i) condition (11.2) holds;
and
(ii) all the roots of the equation (solved for λ)

$$\det(\lambda \mathrm{i} - \mathcal{A}(-z)) = 0 \tag{11.14}$$

belong to the closed disk $\mathcal{D}(1;1)$ whenever $|\arg z| \leq \varphi$.

---

[7] In the particular case $\mathfrak{Q}_n = \mathfrak{R}_n = I$

[8] This definition is easily modified in order to be applied to the general multistep methods. That is why we use M in the abbreviation.

[9] This abbreviation is relevant when dealing with linear multistep methods.

Definition 11.6. *The linear multistep method is called of* type $LM_I(\varphi)$ *if it is both of type* $LM(\varphi)$ *and of class* M_I.[10]

In this chapter, in the general case where the operator $A(t)$ depends on t, we employ certain Hölder-continuity type conditions on $A(t)$. In this connection we need to modify slightly the concept of testing functional introduced in Section 5.1.[11] Therefore the following definition will be valid hereinafter.

Definition 11.7. *Let* $\mathrm{Cl}(\mathrm{Dom}\, A(t)) = \mathfrak{X}$ *and let* $\mathcal{P}_*$ *and* $\mathcal{P}$ *be linear manifolds (independent of* t*) such that* $\mathrm{Dom}\, A^*(t) \subseteq \mathcal{P}_* \subseteq \mathfrak{X}^*$ *and* $\mathrm{Dom}\, A(t) \subseteq \mathcal{P} \subseteq \mathfrak{X}$ *for* $t \in [0,T]$. *A bilinear functional* $\mathcal{A}(t;\cdot,\cdot)$ *with* $\mathrm{Dom}\, \mathcal{A}(t;\cdot,\cdot) = \mathcal{P}_* \times \mathcal{P}$ *is then called a* testing functional *for* $A(t)$ *if, for any* $t \in [0,T]$,

$$\langle \chi, A(t)u \rangle = \mathcal{A}(t;\chi,u) \quad \textit{whenever } \chi \in \mathcal{P}_* \textit{ and } u \in \mathrm{Dom}\, A(t),$$

and

$$\langle A^*(t)\chi, u \rangle = \mathcal{A}(t;\chi,u) \quad \textit{whenever } \chi \in \mathrm{Dom}\, A^*(t) \textit{ and } u \in \mathcal{P}.$$

Apart from everything else, one more hypothesis of Hölder-continuity is in use below. More precisely, we introduce, with some $\mu_1 \geq 0$ and $\vartheta \in [0,1]$, the following.

$\mathbf{h\ddot{o}l}^*_{(k)}[\vartheta]$: The domain of $A(t)$ is dense in $\mathfrak{X}$ for any $t \in [0,T]$, the domain of $A^*(t)$ is independent of t, and it holds, for all $k \in (0,K]$, $t, s \in [0,T]$ such that $|t-s| \leq k$, and $\chi \in \mathfrak{X}^*$,

$$\|(A^*(t) - A^*(s))\,\chi\|_* \leq Ck^{\vartheta}\|(A^*(t) + \mu_1 I)\,\chi\|_*.$$

As a matter of fact we deal with linear multistep methods of type $LM_I(\varphi)$ only and if the dependence on t is generally present, with operators $A(t)$ satisfying the hypotheses $\mathbf{h\ddot{o}l}^*_{(k)}[\vartheta]$ and $\mathbf{h\ddot{o}l}(\vartheta;\mathcal{A}(t;\cdot,\cdot);|\cdot|_{\mathfrak{m};t}, \mathfrak{m} = \emptyset, *\emptyset)$[12] when showing below the assertions on stability with the help of the general theory developed in Part I. To create premises for application of this theory, in the next section we give statements of preparatory character.

---

[10] We do not give a complete classification of the linear multistep methods owing to the fact that our consideration in the present chapter is of confined character.

[11] The point is that the analysis in Part II is carried out under the assumption that $\mathrm{Dom}\, A(t)$ is independent of t, which is reasonable when dealing with RK methods. For the aims of the present chapter it is natural to assume instead that $\mathrm{Cl}(\mathrm{Dom}\, A(t)) = \mathfrak{X}$ for $t \in [0,T]$. It turns out that we then have to modify the concept of testing functional, previously defined, in order to make our subsequent argument correct.

[12] With the newly defined concept of testing functional the latter hypothesis is stated just the same as in Section 5.1.

11.2 Auxiliary results

We begin by showing a certain resolvent estimate.

Theorem 11.1. *Suppose that the linear multistep method is of type $LM_I(\varphi)$ and that $A(t) \in \mathcal{S}(\varphi, \sigma)$, for some $\varphi \in (0, \pi/2)$ and $\sigma \in \mathbb{R}$. Then there exist numbers $K > 0$, $\delta \geq 0$ and a set Υ of Λ_1-configuration such that for all $k \in (0, K]$, $t \in [0, T]$, and $\lambda \notin \operatorname{Int}(\Upsilon \cup \mathcal{D}(\delta k))$,*

$$\|(\lambda \mathbf{I} - \mathcal{A}(-kA(t)))^{-1}\|_{(m)} \leq C|\lambda|^{-1},$$

where $\mathcal{A}(z)$ is the amplification matrix. If $\sigma \geq 0$ in addition, we then take $K = \infty$ and $\delta = 0$.

Proof. First of all note that the operator $\mathcal{A}(-kA(t))$ is defined and bounded on $\mathfrak{X}^{(m)}$ since, by the accepted conditions, the operators $\alpha_j(-kA(t))$, $j = 1, \ldots, m$, are defined and bounded on $\mathfrak{X}$ for $k \in (0, K]$ and $t \in [0, T]$, owing to Lemma 5.3.

Further, letting $\varphi_1 \in (0, \varphi)$ and $R \geq 0$ be just the same as in the assertion of Lemma A.1 stated for $A(t)$ in place of $\mathfrak{L}$,[13] we select some $d \in (0, |a_0|/|2b_0|)$.[14] Moreover, for our further needs we choose $d > 0$ to be sufficiently small so that for $z \in \mathcal{D}(d)$ all the roots $\lambda_j(-z)$ of equation (11.14), except the single simple root $\lambda_1(-z)$, are bounded away from 0. The last is feasible by Reed & Simon [134, Theorem XII.1] since 0 is a simple root of (11.12) and, hence, a simple root of (11.14) when $z = 0$. Also, we set $K \leq d/R$ if $\sigma < 0$ and $K = \infty$ if $\sigma \geq 0$.

By the assumptions it follows from perturbation theory (see [134, Section XII.1]) that there exists a set Υ' of Λ_1-configuration such that all the roots $\lambda_j(-z)$ of (11.14) belong to Υ' for $z \in \Sigma_{\varphi_1}$. Note also that there are no poles of $\alpha_j(-z)$, $j = 1, \ldots, m$, lying in Σ_{φ_1} since $|\arg(a_0/b_0)| < \pi - \varphi$. Next, it is evident that there exists a set Υ of Λ_1-configuration such that $\Upsilon' \setminus \{0\} \subset \operatorname{Int}\Upsilon$.

We now show how to estimate the quantity $\|(\lambda \mathbf{I} - \mathcal{A}(-kA(t)))^{-1}\|_{(m)}$ for $\lambda \notin \Upsilon$, $|\lambda| \leq \lambda_0$, with some $\lambda_0 > 0$ which will be specified later. Observe that condition (11.11) implies that

$$c|z| \leq |\lambda_1(-z)| \leq C|z| \quad \text{for all } z \in \mathcal{D}(d). \tag{11.15}$$

Using this inequality we select a fixed $d_0 > 0$ sufficiently small to have

$$d_0|\lambda_1(-z)| \leq \frac{1}{2}|z| \quad \text{for all } z \in \mathcal{D}(d), \tag{11.16}$$

[13]Observe that if $\sigma \geq 0$, we can take $R = 0$.

[14]Therefore the functions $\alpha_j(z)$, $j = 1, \ldots, m$, have no poles in $\mathcal{D}(d)$.

and set $\lambda_0 = d/d_0$. Let further Γ_λ be the contour coinciding with the boundary of the set $\mathcal{D}(d_0|\lambda|) \cup \Sigma_{\varphi_1}$. Applying then the operator calculus formula (A.10) componentwise, we obtain, with $\delta = d_0^{-1}R$, for all $k \in (0, K]$ and $\lambda \notin \mathrm{Int}(\Upsilon \cup \mathcal{D}(\delta k))$, $|\lambda| \le \lambda_0$,

$$(\lambda\mathbf{I} - \mathcal{A}(-kA(t)))^{-1} = (\lambda\mathfrak{i} - \mathcal{A}(\infty))^{-1}\mathbf{I} + \frac{1}{2\pi i}\int_{\Gamma_\lambda} \left((\lambda\mathfrak{i} - \mathcal{A}(-z))^{-1} - (\lambda\mathfrak{i} - \mathcal{A}(\infty))^{-1}\right) \otimes (zI - kA(t))^{-1}dz, \tag{11.17}$$

where, as above, $\mathfrak{i}$ stands for the identity $(m \times m)$-matrix. For our needs below we further denote, given an $(m \times m)$-matrix $\mathfrak{p} = (p_{jl})$,

$$\nparallel \mathfrak{p} \nparallel = \max_{1\le j,l\le m} |p_{jl}|.$$

In order to evaluate the integrand in (11.17), we first observe that

$$\nparallel (\lambda\mathfrak{i} - \mathcal{A}(-z))^{-1} - (\lambda\mathfrak{i} - \mathcal{A}(\infty))^{-1} \nparallel \le C \nparallel \mathcal{A}(\infty) - \mathcal{A}(-z) \nparallel \nparallel (\lambda\mathfrak{i} - \mathcal{A}(\infty))^{-1} \nparallel \nparallel (\lambda\mathfrak{i} - \mathcal{A}(-z))^{-1} \nparallel . \tag{11.18}$$

Since the eigenvalues of $\mathcal{A}(\infty)$ lie in the interior of Υ, it follows that

$$\nparallel (\lambda\mathfrak{i} - \mathcal{A}(\infty))^{-1} \nparallel \le C(1 + |\lambda|)^{-1} \quad \text{for } \lambda \notin \mathrm{Int}\,\Upsilon. \tag{11.19}$$

It is also readily seen that

$$\nparallel \mathcal{A}(-z) - \mathcal{A}(\infty) \nparallel \le C(1 + |z|)^{-1} \quad \text{for all } z \in \mathcal{D}(d) \cup \Sigma_{\varphi_1}. \tag{11.20}$$

Furthermore since $d_0\lambda_0 = d$, observing that the entries of the matrix $\mathcal{A}(-z)$ are bounded and that $\lambda_j(-z)$, $j = 2, \dots, m$, are bounded away from 0 for $z \in \mathcal{D}(d) \cup \Sigma_{\varphi_1}$ and using the formula for the entries of the inverse matrix (see, e.g., Horn & Johnson [85]), we obtain, with $\mathcal{A}_{jl}(\lambda; -z)$ the cofactors of the respective entries of $\lambda\mathfrak{i} - \mathcal{A}(-z)$, for all $\lambda \notin \mathrm{Int}\,\Upsilon$, $\lambda \in \mathcal{D}(\lambda_0)$, and $z \in \Gamma_\lambda$,

$$\nparallel (\lambda\mathfrak{i} - \mathcal{A}(-z))^{-1} \nparallel \le C \max_{1\le j,l\le m} |\mathcal{A}_{jl}(\lambda; -z)|\, |\det(\lambda\mathfrak{i} - \mathcal{A}(-z))|^{-1} \le C(1 + |\lambda|)^{m-1} \prod_{j=1}^{m} |\lambda - \lambda_j(-z)|^{-1} \le C|\lambda - \lambda_1(-z)|^{-1}. \tag{11.21}$$

Since $\lambda_1(-z) \in \Upsilon'$ for $z \in \Sigma_{\varphi_1}$, it holds

$$|\lambda - \lambda_1(-z)| \geq c(|\lambda| + |\lambda_1(-z)|) \quad \text{for } \lambda \notin \operatorname{Int}\Upsilon \text{ and } z \in \Sigma_{\varphi_1},$$

and it follows from (11.16) that

$$|\lambda - \lambda_1(-z)| \geq |\lambda| - |\lambda_1(-z)| \geq |\lambda| - \frac{|z|}{2d_0} \geq \frac{|\lambda|}{2} \quad \text{for } |z| \leq d_0|\lambda|.$$

Using these observations, instead of (11.21) we get, for all $\lambda \notin \operatorname{Int}\Upsilon$, $\lambda \in \mathcal{D}(\lambda_0)$, and $z \in \Gamma_\lambda$,

$$\nparallel (\lambda \mathrm{i} - \mathcal{A}(-z))^{-1} \nparallel \leq C\big(|\lambda| + |\lambda_1(-z)|\big)^{-1}. \tag{11.22}$$

Combining (11.18), (11.19), (11.20), (11.22) and taking (11.15) into account therefore yields, for all $\lambda \notin \operatorname{Int}\Upsilon$, $\lambda \in \mathcal{D}(\lambda_0)$, and $z \in \Gamma_\lambda$,

$$\begin{aligned} \nparallel (\lambda \mathrm{i} - \mathcal{A}(-z))^{-1} - (\lambda \mathrm{i} - \mathcal{A}(\infty))^{-1} \nparallel \ &\leq \ C(1+|z|)^{-1}\big(|\lambda| + |\lambda_1(-z)|\big)^{-1} \\ &\leq \ C(|\lambda| + |z|)^{-1}. \end{aligned} \tag{11.23}$$

Also, similarly to (5.7), we have, for $k \in (0, K]$, $\lambda \notin \operatorname{Int}\big(\Upsilon \cup \mathcal{D}(\delta k)\big)$, and $z \in \Gamma_\lambda$,

$$\|(zI - kA(t))^{-1}\| \leq C|z|^{-1}. \tag{11.24}$$

Using now (11.19), (11.23), and (11.24) for estimation on the basis of (11.17), we find, for all $k \in (0, K]$ and $\lambda \in \operatorname{Int}\big(\Upsilon \cup \mathcal{D}(\delta k)\big) \cap \mathcal{D}(\lambda_0)$,

$$\begin{aligned} \|(\lambda \mathbf{I} - \mathcal{A}(-kA(t)))^{-1}\|_{(m)} &\leq C \nparallel (\lambda \mathbf{I} - \mathcal{A}(-kA(t)))^{-1} \nparallel \\ &\leq C + C\int_{\Gamma_\lambda} (|\lambda| + |z|)^{-1}|z|^{-1}|dz| \leq C|\lambda|^{-1}. \end{aligned} \tag{11.25}$$

Finally, note that the estimate

$$\nparallel (\lambda \mathrm{i} - \mathcal{A}(-z))^{-1} - (\lambda \mathrm{i} - \mathcal{A}(\infty))^{-1} \nparallel \leq C(|z| + |\lambda|)^{-1}$$

obviously holds for $\lambda \notin \operatorname{Int}\big(\Upsilon \cup \mathcal{D}(\lambda_0)\big)$ and $z \in \mathcal{D}(d) \cup \Sigma_{\varphi_1}$ as well. Using this fact and (11.19) for an estimation on the basis of (11.17), with Γ_{λ_0} in place of Γ_λ, gives, for $\lambda \notin \operatorname{Int}\big(\Upsilon \cup \mathcal{D}(\lambda_0)\big)$,

$$\|(\lambda \mathbf{I} - \mathcal{A}(-kA(t)))^{-1}\|_{(m)} \leq C|\lambda|^{-1}.$$

Combined with (11.25), this inequality shows the claim. □

Remark 11.1. *The assertion of Theorem 11.1 yields that, under the assumed conditions, the operator* $\mathfrak{A}(t_n) = k^{-1}A(-kA(t_n)) \in \mathcal{B}(\mathfrak{X}^{(m)})$ *satisfies the* Λ_1^k*-condition, in particular, the* Λ_1*-condition if* $\sigma \geq 0$*, in the space* $\mathfrak{X}^{(m)}$.

Theorem 11.2. *Suppose that* $A(t) \in \mathcal{S}(\varphi, \sigma)$, $\mathrm{Cl}(\mathrm{Dom}\, A(t)) = \mathfrak{X}$*, and the linear multistep method is of type* $LM_I(\varphi)$*, for some* $\varphi \in (0, \pi/2)$ *and* $\sigma \in \mathbb{R}$*. Let further seminorms* $|\cdot|_t$ *and* $|\cdot|_{*;t}$ *form respectively* $(\gamma|\varphi, \sigma)$*- and* $(*;\gamma|\varphi,\sigma)$*-concordant pairs with* $A(t)$ *for some* $\gamma \geq 0$ *and let a rational function* $\omega(z)$ *be* (φ)*-regular and subject to* $\deg[\omega] \leq -\lfloor\gamma\rfloor - 1$*. Then, with* $K > 0$ *sufficiently small, the seminorms*

$$|||(u_1, \ldots, u_m)^T|||_n = \max_{1 \leq j \leq m} |\omega(-kA(t_n))u_j|_{t_n}$$

and

$$|||(\chi_1, \ldots, \chi_m)^T|||_{*;n} = \sum_{j=1}^{m} |\omega(-kA(t_n))^* \chi_j|_{*;t_n}$$

are well defined for $k \in (0, K]$, $t_n \in \Omega_k^{(m)}$ *and form respectively* $\Lambda_1^k(\gamma)$*- and* $\Lambda_1^k(*;\gamma)$*-concordant* ($\Lambda_1(\gamma)$*- and* $\Lambda_1(*;\gamma)$*-concordant if* $\sigma \geq 0$) *pairs with* $\mathfrak{A}(t_n) = k^{-1}A(-kA(t_n))$*. In the case where* γ *is an integer, under the extra requirement that*

the pairs $\big(|\cdot|_t; A(t)\big)$ *and* $\big(|\cdot|_{*;t}; A(t)\big)$ *are respectively* $(\gamma|\gamma|\varphi,\sigma)$ – *and* $(*;\gamma|\gamma|\varphi,\sigma)$ – *concordant,*

if $\gamma \geq 1$*, and*

$$|u|_t \leq C\|u\|, \quad |\chi|_{*;t} \leq C\|\chi\|_* \qquad \text{for all } t \in [0,T],\ u \in \mathfrak{X},\ \text{and } \chi \in \mathfrak{X}^*,$$

if $\gamma = 0$*, it is sufficient to suppose that* $\deg[\omega] \leq -\lfloor\gamma\rfloor$ *instead of* $\leq -\lfloor\gamma\rfloor - 1$.

Furthermore, for any fixed $\zeta \in (0,1]$ *and* $\mu_1 \geq 0$ *sufficiently large, with some* $K > 0$ *sufficiently small, the seminorms*

$$]\![(u_1, \ldots, u_m)^T]\!]_n = k^{\zeta} \max_{1 \leq j \leq m} \left\|(A(t_n) + \mu_1 I)\,(a_0 I + kb_0 A(t_n))^{-1} u_j\right\|$$

and

$$]\![(\chi_1, \ldots, \chi_m)^T]\!]_{*;n} = k^{\zeta} \sum_{j=1}^{m} \left\|(A^*(t_n) + \mu_1 I)\,(a_0^* I + kb_0^* A^*(t_n))^{-1} \chi_j\right\|_*$$

are well defined for $k \in (0, K]$, $t_n \in \Omega_k^{(m)}$ *and form respectively* $\Lambda_1^k(1-\zeta)$*- and* $\Lambda_1^k(*;1-\zeta)$*-concordant* ($\Lambda_1(1-\zeta)$*- and* $\Lambda_1(*;1-\zeta)$*-concordant if* $\sigma \geq 0$) *pairs with* $\mathfrak{A}(t_n)$*. Apart from everything else, in the case* $\sigma \geq 0$ *the above holds with* $K = \infty$.

Proof. This result is established as in Theorem 11.1, applying as well the argument employed in the proof of Theorem 6.1. As concerns the seminorms $\|\!|\cdot\|\!|_n$ and $\|\!|\cdot\|\!|_{*;n}$, to show the claim we let K, δ, λ_0, Υ, and Γ_λ be the same as in the proof of Theorem 11.1[15] and, taking Γ_{λ_0} in place of Γ_λ if $|\lambda| > \lambda_0$, we use the operator calculus formula, for all $k \in (0, K]$, $t \in [0, T]$, and $\lambda \notin \operatorname{Int}(\Upsilon \cup \mathcal{D}(\delta k))$,

$$(\lambda\mathbf{I} - k\mathfrak{A}(t))^{-\gamma_0} \otimes \omega(-kA(t)) = (\lambda\mathbf{I} - k\mathfrak{A}(\infty))^{-\gamma_0} \otimes \omega(-kA(t))$$
$$+\frac{1}{2\pi i}\int_{\Gamma_\lambda} \omega(-z)\left((\lambda\mathrm{i} - \mathcal{A}(-z))^{-\gamma_0} - (\lambda\mathrm{i} - \mathcal{A}(\infty))^{-\gamma_0}\right)\otimes(zI - kA(t))^{-1}dz,$$

with $\gamma_0 = \lfloor\gamma\rfloor + 1$. Applying the same formula with $\gamma_0 = 1$ and with $(a_0 - b_0 z)^{-1}$ in place of $\omega(z)$ allows us to prove the desired estimates for the other two seminorms involved. □

Theorem 11.3. *Suppose that* $A(t) \in \mathcal{S}(\varphi, \sigma)$, $\mathrm{Cl}(\mathrm{Dom}\, A(t)) = \mathfrak{X}$ *for any* $t \in [0, T]$, *and the linear multistep method is of type* $LM_I(\varphi)$, *for some* $\varphi \in (0, \pi/2)$ *and* $\sigma \in \mathbb{R}$. *Let further* $\mathcal{A}(t; \cdot, \cdot)$ *be a testing functional for* $A(t)$, *let seminorms* $|\cdot|_{\mathfrak{m};t}$, $\mathfrak{m} = \emptyset, *\emptyset$, *form respectively* $(\gamma|\varphi, \sigma)$- *and* $(*; \gamma_*|\varphi, \sigma)$-*concordant pairs with* $A(t)$, *and let* $A(t)$ *satisfy* $\mathbf{h\ddot{o}l}(\vartheta; \mathcal{A}(t; \cdot, \cdot); |\cdot|_{\mathfrak{m};t}, \mathfrak{m} = \emptyset, *\emptyset)$, *for some* $\gamma, \gamma_* \in [0, 1)$ *and* $\vartheta \in [0, 1]$. *Then the operator* $\mathfrak{A}(t_n) = k^{-1}\mathcal{A}(-kA(t_n))$ *satisfies hypothesis* $\mathbf{H\ddot{O}L}(\vartheta; \|\!|\cdot\|\!|_{\mathfrak{m};n}, \mathfrak{m} = \emptyset, *\emptyset)$ *in the space* $\mathfrak{X}^{(m)}$, *where*

$$\begin{aligned} \|\!|(u_1, \ldots, u_m)^T\|\!|_{\emptyset;n} &= \max\nolimits_{1\le j\le m} \left|(a_0 I + kb_0 A(t_n))^{-1} u_j\right|_{\emptyset;t_n}, \\ \|\!|(\chi_1, \ldots, \chi_m)^T\|\!|_{*\emptyset;n} &= \textstyle\sum_{j=1}^m \left|(a_0^* I + kb_0^* A^*(t_n))^{-1}\chi_j\right|_{*\emptyset;t_n}. \end{aligned} \tag{11.26}$$

Proof. An easy calculation shows that

$$\mathfrak{A}(t) - \mathfrak{A}(s) = k^{-1}\begin{pmatrix} 0 & \cdots & 0 \\ \cdot & \cdots & \cdot \\ 0 & \cdots & 0 \\ V_1(t,s) & \cdots & V_m(t,s) \end{pmatrix},$$

where $V_j(t, s) = \alpha_{m-j}(-kA(t)) - \alpha_{m-j}(-kA(s))$, $j = 1, \ldots, m$. At the same time we have, denoting for short $c_j = a_{m-j}b_0 - b_{m-j}a_0$ and $\varrho(t) = (a_0 I + kb_0 A(t))^{-1}$,

$$V_j(t, s) = \frac{c_j}{b_0}\left(\varrho(t) - \varrho(s)\right), \quad j = 1, \ldots, m.$$

[15]Therefore $\delta = 0$ and $K = \infty$ if $\sigma \geq 0$.

With the aid of this representation, we find, for $j = 1, \ldots, m$ and for all $u \in \mathfrak{X}$ and $\chi \in \mathfrak{X}^*$,

$$\begin{aligned}\langle \chi, V_j(t,s)\, u\rangle &= \frac{c_j}{b_0}\big(\langle \chi, \varrho(t)u\rangle - \langle \varrho^*(s)\chi, u\rangle\big)\\ &= kc_j\big(\langle A^*(s)\varrho^*(s)\chi, \varrho(t)u\rangle - \langle \varrho^*(s)\chi, A(t)\varrho(t)u\rangle\big)\\ &= -kc_j\Delta\mathcal{A}(t,s;\varrho^*(s)\chi, \varrho(t)u),\end{aligned}$$

where, as above, $\Delta\mathcal{A}(t,s;\chi,u) = \mathcal{A}(t;\chi,u) - \mathcal{A}(s;\chi,u)$. Using the last equality and recalling that $A(t)$ satisfies $\mathbf{h\ddot{o}l}(\vartheta,\vartheta_1;\mathcal{A}(t;\cdot,\cdot);|\cdot|_{\mathfrak{m};t},\mathfrak{m}=\emptyset,*\emptyset)$, it follows that for $j = 1, \ldots, m$ and for all $t, s \in [0,T]$, $u \in \mathfrak{X}$, and $\chi \in \mathfrak{X}^*$,

$$k^{-1}|\langle \chi, V_j(t,s)u\rangle| \le C|t-s|^{\vartheta}|\varrho^*(s)\chi|_{*\emptyset;s}|\varrho(t)u|_{\emptyset;t}.$$

Owing to this estimate the claim now follows taking (11.4) into account. □

11.3 Stability

We are now in a position to state our results on stability. As above, we distinguish between the cases with a constant and a variable operator.

Theorem 11.4. *Suppose that $A(t) \equiv A$ is independent of t, $A \in \mathcal{S}(\varphi,\sigma)$, the linear multistep method is of type $LM_I(\varphi)$, and a seminorm $]\!\![\cdot]\!\![_t$ forms a $(\xi|\varphi,\sigma)$-concordant pair with A, for some $\varphi \in (0,\pi/2)$, $\sigma \in \mathbb{R}$, and $\xi \ge 0$. Let also $\omega(z)$ be a (φ)-regular rational function satisfying $\deg[\omega] \le -\lfloor\xi\rfloor$, where the last condition is replaced by $\deg[\omega] \le -\xi$ if $\xi \in \mathbb{N}$ and if the pair $(]\!\![\cdot]\!\![_t; A)$ is $(\xi|\xi|\varphi,\sigma)$-concordant or if $\xi = 0$ and if $]\!\![u]\!\![_t \le C\|u\|$ for all $t \in [0,T]$ and $u \in \mathfrak{X}$. Then, if $T < \infty$, the C-extended solution Y_n of (11.1) satisfies the estimate, with $K > 0$ sufficiently small, for all $k \in (0,K]$ and $t_n \in \Omega_k$,*

$$]\!\![\omega(-kA)\,Y_n]\!\![_{t_n} \le C\left(t_{n+1}^{-\xi}\max_{0\le j\le m-1}\|y^j\| + k\sum_{l=0}^{n} t_{n-l+1}^{-\xi}\|f(t_l)\|\right). \quad (11.27)$$

If $\sigma \ge 0$ in addition, the above holds for $T = K = \infty$.

Proof. It is pointed out in Section 11.1 that if Y_n is the C-extended solution of (11.1), then the vector $\mathbf{Y_n}$ specified by (11.5) fits (11.10) with $\mathbf{y}^0$, $\mathbf{F}_n$, and $\mathfrak{A}_n = \mathfrak{A}$ given by (11.6), (11.7), and (11.9), respectively. Also, the seminorm $]\!\![\!\![\cdot]\!\!]\!\!]_n =]\!\![\omega(-kA)\cdot]\!\![_{t_n}$ forms a $\Lambda_1^k(\xi)$-concordant pair with $\mathfrak{A}$, by Theorem 11.2.

It thus remains to apply Theorem 2.1 to problem (11.10), taking into account that

$$\|\mathbf{F}_n\|_{(m)} \le C \sum_{j=0}^{m} \|f_{n+m-j}\| \tag{11.28}$$

and

$$\|\mathbf{Y}_0\|_{(m)} \le C \max_{0 \le j \le m-1} \|y^j\|. \tag{11.29}$$

□

Concerning the case with a variable operator, we present the following.

Theorem 11.5. *Let $T < \infty$ and let the operator $A(t)$ and the linear multistep method be under the conditions of Theorem 11.3 for some $\varphi \in (0, \pi/2)$, $\sigma \in \mathbb{R}$, $\gamma, \gamma_* \in [0, 1)$, and $\vartheta \in (0, 1]$ such that $\gamma + \gamma_* < 1 + \vartheta$. Let further $A(t)$ satisfy hypothesis $\mathbf{h\ddot{o}l}^*_{(k)}[\vartheta_2]$ for some $\vartheta_2 \in (0, 1]$ subject to $\gamma < \vartheta + \vartheta_2$. Finally, suppose that a seminorm $]\![\cdot]\!\![_t$ forms a $(\xi|1|\varphi, \sigma)$-concordant pair with $A(t)$ for some $\xi \in [0, \min\{\xi_1, \vartheta_2\}]$, where $\xi_1 = 1 - \gamma_* + \vartheta$. Then the V-extended solution Y_n of (11.1) is defined and satisfies the stability estimate (11.27) with $A(t_n)$ substituted for A and with $(I + kA(t_n))^{-1}$ substituted for $\omega(-kA)$, where the factor $t_j^{-\xi}$ is replaced by $t_j^{-\xi} + \log(t_{j+1}/k)$ if $\xi = \vartheta_2 < \xi_1$ and by $t_j^{-\xi} + t_j^{-\gamma} \log(t_{j+1}/k)$ if $\xi = \xi_1 \le \vartheta_2$. Furthermore, (11.27) for $\xi = 0$ holds with $\|Y_n\|$ substituted for $]\![(I + kA(t_n))^{-1} Y_n]\!\![_{t_n}$.*

Proof. Since the operator $A(t)$ is under hypothesis $\mathbf{h\ddot{o}l}^*_{(k)}[\vartheta_2]$, it follows that $A(t)$ possesses an adjoint operator $A^*(t)$ and $\operatorname{Dom} A^*(t)$ is independent of t. This fact further implies that the operator $(\lambda I - A(t))^{-1} A(s)$ admits a unique extension to a linear bounded operator on the whole $\mathfrak{X}$ for any fixed $\lambda \notin (\operatorname{Int} \Sigma_\varphi \cup \{0\}) + \sigma$ and $t, s \in [0, T]$, and, thus, the V-extended solution Y_n of (11.1) is defined. As already mentioned in Section 11.1, the vector $\mathbf{Y}_n$ as specified by (11.5) then fits (11.10) with $\mathbf{y}^0$, $\mathbf{F}_n$, and $\mathfrak{A}_n$ as defined by (11.6), (11.7), and (11.8), respectively.

Next note that the operator $\mathfrak{A}_n$ is represented as $\mathfrak{A}_n = \mathfrak{A}(t_n) + \mathfrak{B}_n$, where $\mathfrak{A}(t) = k^{-1} A(-kA(t))$, as above, and the operator $\mathfrak{B}_n$ is given by the operator matrix

$$\mathfrak{B}_n = \begin{pmatrix} 0 & \cdots & 0 \\ \vdots & \cdots & \vdots \\ 0 & \cdots & 0 \\ W_{n,0} & \cdots & W_{n,m-1} \end{pmatrix},$$

with[16]

$$W_{n,j} = k^{-1}(a_0 I + kb_0 A(t_{n+m}))^{-1}(a_{m-j} I + kb_{m-j} A(t_{n+j})) \\ - k^{-1}\alpha_{m-j}(-kA(t_n)), \quad j = 0, \ldots, m-1.$$

It follows from Theorem 11.3 that the operator $\mathfrak{A}(t_n)$ satisfies hypothesis $\mathbf{HÖL}(\vartheta; \|\!| \cdot \|\!|_{\mathfrak{m};n}, \mathfrak{m} = \emptyset, *\emptyset)$ with the same ϑ as in the statement of the present theorem where the seminorms $\|\!| \cdot \|\!|_{\mathfrak{m};n}$, $\mathfrak{m} = \emptyset, *\emptyset$, are given by (11.26). By Theorem 11.2 these seminorms form respectively $\Lambda_1^k(\gamma)$- and $\Lambda_1^k(*;\gamma_*)$-concordant pairs with $\mathfrak{A}(t_n)$.

Further, applying the following identity, with c_j defined the same as in the proof of Theorem 11.3, for $j = 0, \ldots, m-1$,

$$W_{n,j} = b_{m-j}(a_0 I + kb_0 A(t_{n+m}))^{-1}(A(t_{n+j}) - A(t_{n+m})) \\ + \frac{c_j}{a_0}(a_0 I + kb_0 A(t_{n+m}))^{-1}(A(t_n) - A(t_{n+m}))\,(a_0 I + kb_0 A(t_n))^{-1},$$

where the factors $(a_0 I + kb_0 A(t_{n+m}))^{-1}(A(t) - A(t_{n+m}))$, $t = t_{n+j}, t_n$, extend to linear bounded operators on $\mathfrak{X}$, and using the fact that $A(t)$ satisfies hypothesis $\mathbf{höl}^*_{(k)}[\vartheta_2]$ we have, with some $\mu_1 \geq 0$, for all $\chi \in \mathfrak{X}^*$ and $u \in \mathfrak{X}$, that

$$|\langle \chi, W_{n,j} u\rangle| \leq Ck^{\vartheta_2}\|(A^*(t_n) + \mu_1 I)\,(a_0^* I + kb_0^* A^*(t_n))^{-1}\chi\|_*\|u\|. \quad (11.30)$$

Noting now that by (11.4) it holds, for any $(\chi_1, \ldots, \chi_m)^T \in \mathfrak{X}^{(m)*}$ and $(u_1, \ldots, u_m)^T \in \mathfrak{X}^{(m)}$,

$$\langle (\chi_1, \ldots, \chi_m)^T, \mathfrak{B}_n(u_1, \ldots, u_m)^T\rangle_{(m)} = \langle \chi_m, \sum_{j=1}^{m} W_{n,j-1} u_j\rangle,$$

with the aid of (11.30) we obtain

$$|\langle (\chi_1, \ldots, \chi_m)^T, \mathfrak{B}_n(u_1, \ldots, u_m)^T\rangle_{(m)}| \\ \leq Ck^{\vartheta_2}\|(A^*(t_n) + \mu_1 I)\,(a_0^* I + kb_0^* A^*(t_n))^{-1}\chi_m\|_*\|(u_1, \ldots, u_m)^T\|_{(m)} \\ \leq C\|\!|(\chi_1, \ldots, \chi_m)^T\|\!|_{*;\vartheta_2;n}\|(u_1, \ldots, u_m)^T\|_{(m)},$$

where the seminorm

$$\|\!|(\chi_1, \ldots, \chi_m)^T\|\!|_{*;\vartheta_2;n} = k^{\vartheta_2}\sum_{j=1}^{m}\|(A^*(t_n) + \mu_1 I)\,(a_0^* I + kb_0^* A^*(t_n))^{-1}\chi_j\|_*$$

[16] We remark that actually $W_{n,j} \in \mathcal{B}(\mathfrak{X})$, which follows from the above comment.

forms a $\Lambda_1^k(*; 1-\vartheta_2)$-concordant pair with $\mathfrak{A}(t_n)$, because of Theorem 11.2.

Now we use the above shown properties of the operators $\mathfrak{A}(t_n)$ and $\mathfrak{B}_n$ and note that the seminorm $]\![\cdot]\!]_n =]\![(I + kA(t_n))^{-1}\cdot]\!]_{t_n}$ forms a $\Lambda_1^k(\xi)$-concordant pair with $\mathfrak{A}(t_n)$, in virtue of Theorem 11.2,[17] while (11.28) and (11.29) are valid in the considered case. Thus the claim follows by Theorem 2.17 which is applied to problem (11.10) with $\mathfrak{A}(t_n)$ in place of $\mathfrak{A}_n$, with the same $\gamma, \gamma_*, \vartheta$ as in the statement of the present theorem and with $\gamma_1 = \gamma$, $\vartheta_1 = \vartheta$, $\gamma_0 = \vartheta_0 = 0$, $\gamma_{*0} = 1 - \vartheta_2$. □

As a possible application, we consider the family of m-step BDF methods, $m = 1, 2, \ldots$, for which

$$b_1 = b_2 = \ldots = b_m = 0, \quad b_0 = 1, \tag{11.31}$$

and

$$\sum_{l=0}^{m} a_l = 0, \quad \sum_{l=0}^{m} l a_l = -1, \quad \sum_{l=0}^{m} l^j a_l = 0, \; j = 2, \ldots, m. \tag{11.32}$$

Note that (11.31) and the first equality in (11.32) together imply condition (11.11). Further, by (11.31) it is immediate that equation (11.13) has only one root $\lambda(\infty) = 0$ (of multiplicity m). Also, it easily follows from the order conditions (11.32) that all the roots of (11.12), except the single simple root $\lambda_1(0) = 0$, belong to the open disk Int $D(1; 1)$. It is known (see, e.g., Hairer & Wanner [81]) that for the m-step BDF methods with $1 \le m \le 6$, condition (ii) in Definition 11.5 holds for $\varphi \in (0, \varphi_m]$, with $\varphi_m \in (0, \pi/2)$ depending on m. Therefore, for a given $1 \le m \le 6$, the m-step BDF method is of type $LM_I(\varphi)$ with some $\varphi \in (0, \pi/2)$.

In conclusion, we discuss a frequently used family of 2-step methods which is specified by

$$a_0 = 1 + \theta, \quad a_1 = -(1 + 2\theta), \quad a_2 = \theta, \quad b_0 = 1, \quad b_1 = b_2 = 0,$$

where θ is a fixed parameter. (For the sake of simplicity we assume θ to be real.) An elementary (but cumbersome) computation shows that for this method condition (ii) in Definition 11.5 is fulfilled for any fixed $\varphi \in (0, \pi/2)$ if $\theta \in [-1/2, 1/2]$ and for $\varphi \in (0, \varphi_\theta]$ if $\theta > 1/2$, where

$$\varphi_\theta = \arctan \frac{(1 + 4\theta)^{3/2}}{(1 + 2\theta)^{1/2}(2\theta - 1)^{3/2}}.$$

---

[17] In the case $\xi = 0$ we also can take $]\![\cdot]\!]_n = \|\cdot\|$, by Theorem 11.1.

Further, it is easily checked that equation (11.13) has a root $\lambda(\infty) = 1$ of multiplicity 2 while one of the roots of (11.12) is $\lambda_1(0) = 0$ and the other lies in the open disk $\operatorname{Int} D(1;1)$ if $\theta > -1/2$. Consequently, the considered method is of type $LM_I(\varphi)$ for any fixed $\varphi \in (0, \pi/2)$ if $\theta \in (-1/2, 1/2]$ and for any fixed $\varphi \in (0, \varphi_\theta]$ if $\theta > 1/2$.

11.4 Comments and bibliographical remarks

Apparently, multistep approximation in a Banach space setting has been examined for the first time in Crouzeix [60], in the case with constant operator but for general C_0 semigroups. At the same time an earlier paper of Le Roux [102],[18] although written in a Hilbert space setting, demonstrates techniques that are extended to the Banach case, dealing with sectorial operators. Note that both mentioned papers aim at showing convergence only but they do not allow us to settle the question of stability. The consideration of linear multistep methods in the present chapter covers the cases both with a time-independent and a time-dependent operator; in fact, our analysis here is close to the lines found in [28].[19] Later Palencia [129], considering the autonomous case only, shows the stability of general multistep approximations. For a special class of such approximations, some strong stability and nonsmooth error estimates are pointed out by Hansbo [83]. The case with a variable operator has also been considered recently in González & Palencia [74], but their results are applicable only if the operator $A(t)$ is Lipschitz-continuous while the situation in which $A(t)$ is properly Hölder-continuous is not covered in [74].

[18] Apart from single step methods it also treats linear multistep ones.

[19] We are now based on a different form of Hölder-continuity for $A(t)$.

Part IV

INTEGRO-DIFFERENTIAL EQUATIONS UNDER DISCRETIZATION

In this part of the book we present some results concerning stability of discretizations of abstract linear integro-differential equations. As the starting object, we consider the initial value problem

$$y_t + A(t)y = \int_0^t P(t,s)\, y(s)\, ds + f(t) \quad \text{for } t \in (0,T], \qquad y(0) = y^0, \quad (11.33)$$

where $P(t,s)$ is a linear (generally unbounded) operator in $\mathfrak{X}$ for $0 \leq s \leq t \leq T$ and all the other objects have the same meaning as in Part II. Throughout this part we suppose that for any $\mu_1 \geq 0$ sufficiently large and for any $t, s \in [0,T]$ such that $t \geq s$, the operator $(A(t) + \mu_1 I)^{-1} P(t,s)$ is extended to a linear bounded operator defined on the whole $\mathfrak{X}$. Further requirements on $A(t)$ and $P(t,s)$ are imposed in the subsequent statements.[20]

Problem (11.33) is more complicated in comparison with (5.1) since an extra term is involved in (11.33), which is of nonlocal nature and brings about additional difficulties. As concerns discretizations of problem (11.33), they are accordingly harder for analysis than those studied above for problem (5.1). In our opinion, this topic deserves to be the main subject for a separate book; nevertheless, in what follows, just to reveal further the possibilities of our approach, we discuss certain selected discretization schemes for problem (11.33). In Section 12.1 we show the stability of these discretizations on finite time intervals while Section 12.2 is devoted to discussing suitable sparse quadrature rules for an approximation of the memory term, which are of considerable interest in practical applications.

[20]Here we derive no error estimates so that no conditions that provide the unique solvability and high regularity properties of (11.33) are employed. The below requirements on $A(t)$ and $P(t,s)$ are merely used to show the stability of the discretizations that are considered.

Chapter 12

Integro-Differential Equations under Discretization

12.1 Stability of discretizations

In the present chapter we examine two concrete discretizations schemes of which the first is the backward Euler method and the second the Crank-Nicolson one.[1] Both of these methods, when applied to purely differential equations of the form (5.1), are analysed with the aid of certain assertions established in Part II.

Let, as above, Ω_k be a uniform grid in the segment $[0, T]$, with stepsize $k > 0$, and let $\breve{\Omega}_k$ retain the same meaning as before. Since our techniques are now based on making use of the results of Chapters 4 and 6, throughout this chapter it is assumed that $T < \infty$. The memory term in (11.33) is approximated by applying the quadrature formula, with some weights $\kappa_{n,l}$,

$$\Delta^n(\psi) = \sum_{l=0}^{n-1} \kappa_{n,l}\psi^l \approx \int_0^{t_n} \psi(s)\, ds, \quad \text{with } \psi^l = \psi(t_l), \tag{12.1}$$

where $\psi(t)$ can be thought of as a scalar or $(\mathfrak{X})$-valued function. Our main requirement below on the quadrature formulas used is that with some $p > 0$,

$$\Delta^n(1) = t_n,\ \kappa_{n,l} \geq 0, \text{ and } \sum_{l=0}^{j-1} |\kappa_{j,l} - \kappa_{n,l}| \leq Ck^p, \tag{12.2}$$

[1]The latter method is considered in the case $A(t) \equiv A$ only.

for all $t_n, t_j \in \Omega_k$ such that $0 < t_j \le t_n$. The backward Euler discretization of problem (11.33) is then formally introduced as

$$\partial Y_n + A(t_{n+1})Y_{n+1} = \Delta^{n+1}(P(t_{n+1},\cdot)Y) + f(t_{n+1}) \quad \text{for } t_n \in \breve{\Omega}_k, \quad Y_0 = y^0.$$

Since under the conditions subsequently stated the operator $I + kA(t_{n+1})$ has a bounded inverse, we take this discretization method in the following form, denoting $\mathfrak{U}_n = (I + kA(t_{n+1}))^{-1}$ and $\widehat{P}(t,s) = (I + kA(t))^{-1}P(t,s)$,

$$\begin{aligned} Y_{n+1} &= \mathfrak{U}_n Y_n + k\Delta^{n+1}(\widehat{P}(t_{n+1},\cdot)Y) + k\mathfrak{U}_n f(t_{n+1}) \\ &= \mathfrak{U}_n Y_n + k\sum_{l=0}^{n} \kappa_{n+1,l}\widehat{P}(t_{n+1},t_l)Y_l + k\mathfrak{U}_n f(t_{n+1}) \quad \text{for } t_n \in \breve{\Omega}_k, \\ Y_0 &= y^0, \end{aligned} \tag{12.3}$$

which can be equivalently rewritten in the form (4.2) with

$$\mathfrak{A}_n = A(t_n)\,(I + kA(t_n))^{-1}, \tag{12.4}$$

$$\mathfrak{B}_n = \mathfrak{A}_{n+1} - \mathfrak{A}_n, \tag{12.5}$$

$$F_n = \mathfrak{U}_n f(t_{n+1}), \tag{12.6}$$

and

$$\mathfrak{P}_{n,l} = \kappa_{n+1,l}\widehat{P}(t_{n+1},t_l). \tag{12.7}$$

In the following three theorems we show the stability of the backward Euler method (12.3) in different situations.

Theorem 12.1. *Let the operator $A(t)$ and a seminorm $|\cdot|_t$ satisfy the conditions of Theorem 6.12 for some $\varphi \in (0,\pi/2)$, $\sigma \in \mathbb{R}$, $\vartheta \in (0,1]$, $\gamma_1 \in [0,1)$, and let seminorms $]\![\cdot]\!]_t$ and $|\cdot|_{*;t}$ form respectively $(\xi|\varphi,\sigma)$- and $(*;\gamma_{*3}|\varphi,\sigma)$-concordant pairs with $A(t)$ for some $\xi \in [0,1)$ and $\gamma_{*3} \in [0, \min\{\vartheta, 1-\gamma_1, 1-\xi\})$. Also, suppose that the quadrature formula (12.1) satisfies condition (12.2) while the operator $\widehat{P}(t,s)$ is under the bound, for all $\chi \in \mathfrak{X}^*$ and $t, s \in [0,T]$ such that $s \le t$,*

$$\|\widehat{P}^*(t,s)\chi\|_* \le C|((I + kA(t))^{-1})^*\chi|_{*;t}. \tag{12.8}$$

Then the solution Y_n of the discrete problem (12.3) satisfies the stability estimate

$$]\![(I + kA(t_n))^{-1}Y_n]\!]_{t_n} \le C\left(t_{n+1}^{-\xi}\|y^0\| + k\sum_{j=1}^{n} t_{n-j+1}^{-\xi}\|f(t_j)\|\right), \tag{12.9}$$

for all $k \in (0,K]$ and $t_n \in \Omega_k$. Furthermore, in the case $\xi = 0$ the last inequality holds with $]\![Y_n]\!]_{t_n}$ substituted for $]\![(I + kA(t_n))^{-1}Y_n]\!]_{t_n}$.

Proof. We recall that problem (12.3) is equivalent to (4.2), (12.4)−−(12.7). Since, by Lemma 5.3, the operator $\mathfrak{U}_n$ is uniformly bounded, for F_n given by (12.6) we have

$$\|F_n\| \le C\|f(t_{n+1})\| \quad \text{for all } t_n \in \breve{\Omega}_k. \tag{12.10}$$

Now, observing that the backward Euler method is of type $V_I(\varphi)$ for any $\varphi \in (0, \pi/2)$, by Theorem 6.1 it follows that the seminorm

$$]\![\cdot]\![_n = \begin{cases}]\![(I + kA(t_n))^{-1}\cdot]\![_{t_n} & \text{if } \xi > 0, \\ ]\![\cdot]\![_{t_n} & \text{if } \xi = 0, \end{cases}$$

forms a $\Lambda_1^k(\xi)$-concordant pair with $\mathfrak{A}_n$ given by (12.4), and, assuming further that all the conditions of Theorem 4.1 which do not involve the seminorm $]\![\cdot]\![_n$ are satisfied for some suitable ϑ, γ_1, γ_0, γ_3, γ_{*3}, the claim directly follows from that theorem in view of the above reasoning.

Below we will show that the needed conditions of Theorem 4.1 are verified for $\gamma_0 = \max\{1 - \vartheta, \gamma_1\}$, $\gamma_3 = 0$, and for the same ϑ, γ_1, γ_{*3} as in the statement of the present theorem, which is sufficient to conclude the proof.

In the matter of the conditions on the operators $\mathfrak{A}_n$ and $\mathfrak{B}_n$ given by (12.4) and (12.5), respectively, their verification can be done by using the resoning employed in the proof of Theorem 6.12, choosing suitably $|\!|\!|\cdot|\!|\!|_n$ and $|\!|\!|\cdot|\!|\!|_{1;n}$.

Note further that in virtue of (12.8) the operator $\mathfrak{P}_{n,l}$ specified by (12.7) fulfils condition (4.14) with $\pi_{n,l} = C\kappa_{n+1,l}$, $|\!|\!|\cdot|\!|\!|_{*3;n} = |\mathfrak{U}_n^* \cdot|_{*;t_{n+1}}$, and $|\!|\!|\cdot|\!|\!|_{3;n} = \|\cdot\|$. Although by Theorem 6.1* the seminorm $|\!|\!|\cdot|\!|\!|_{*3;n}$ forms a $\Lambda_1^k(*;\gamma_{*3})$-concordant pair with $\mathfrak{A}_{n+1}$, not with $\mathfrak{A}_n$, it is easy to see, using the above techniques, that Theorem 2.11 remains in force even if the seminorm $]\![\cdot]\![_{*;n}$ in its statement is assumed to form a $\Lambda_1^k(*;\xi_*)$-concordant pair with the operator $\mathfrak{A}_{n+1}$. Hence Theorem 4.1, whose proof uses Theorem 2.11, is still applicable in the case where the pair $\left(|\!|\!|\cdot|\!|\!|_{*3;n}; \mathfrak{A}_{n+1}\right)$ is $\Lambda_1^k(*;\gamma_{*3})$-concordant. It is also obvious that the pair $\left(|\!|\!|\cdot|\!|\!|_{3;n}; \mathfrak{A}_n\right)$ is $\Lambda_1^k(0)$-concordant.

In view of the above observations it remains to establish that (4.15) and (4.16) hold with $\gamma_3 = 0$ and with $\kappa_{n+1,l}$ substituted for $\pi_{n,l}$. To that end, since $\gamma_{*3} < 1$, by a simple calculation, changing the order of summation and using (12.2) we have, for all $t_l \leq t_m \leq t_s \leq t_n$, that

$$k \sum_{j=m}^{s-1} \sum_{q=j}^{n-1} t_{n-q}^{-\gamma_{*3}} \kappa_{q+1,j} = k \sum_{q=m}^{n-1} t_{n-q}^{-\gamma_{*3}} \sum_{j=m}^{\min\{q,s-1\}} (\kappa_{q+1,j} - \kappa_{n,j})$$

$$+ k \sum_{q=m}^{n-1} t_{n-q}^{-\gamma_{*3}} \sum_{j=m}^{\min\{q,s-1\}} \kappa_{n,j} \leq Ck^p + C \sum_{j=m}^{s-1} \kappa_{n,j}$$

and

$$\begin{aligned} \sum_{j=m}^{s-1} \kappa_{n,j} &= \sum_{j=0}^{s-1} \kappa_{s,j} - \sum_{j=0}^{m-1} \kappa_{m,j} + \sum_{j=0}^{s-1} (\kappa_{n,j} - \kappa_{s,j}) + \sum_{j=0}^{m-1} (\kappa_{m,j} - \kappa_{n,j}) \\ &\leq t_{s-m} + Ck^p. \end{aligned}$$

Combining both these estimates, it follows that condition (4.15) is satisfied for $\gamma_3 = 0$, with $\kappa_{n+1,l}$ substituted for $\pi_{n,l}$. The same is similarly established in respect of (4.16).

So the proof is complete. □

Theorem 12.2. *Let the operator $A(t)$ and seminorms $|\cdot|_{\oplus;t}$ and $|\cdot|_{\mathfrak{m};t}$, $\mathfrak{m} = \emptyset, 1, *\emptyset$, satisfy the conditions of Theorem 6.13 for some $\varphi \in (0, \pi/2)$, $\sigma \in \mathbb{R}$, $\gamma_\oplus, \gamma, \gamma_1, \gamma_*, \vartheta_\bullet, \vartheta_{*\bullet} \in [0,1)$ and $\vartheta_\oplus, \vartheta, \vartheta_1 \in (0,1]$ such that $\gamma + \gamma_* = 1$, $\gamma_1 + \gamma_* < 1$, $\vartheta > \vartheta_\bullet + \vartheta_{*\bullet}$. Let also seminorms $]\!\![\cdot]\!\![_t$ and $|\cdot|_{*;t}$ form respectively $(\xi|\varphi,\sigma)$- and $(*;\gamma_{*3}|\varphi,\sigma)$-concordant pairs with $A(t)$ for some $\xi \in \big[0, 1 - \gamma_* + \min\{\vartheta, \vartheta_1\}\big)$ and $\gamma_{*3} \in \big[0, 1 + \min\{\vartheta - \gamma, \vartheta_1 - \gamma_1, -\xi\}\big)$. Suppose further that the conditions (12.8) and (12.2) hold for the operator $\widehat{P}(t,s)$ and for the quadrature formula (12.1), respectively. Then the solution Y_n of problem (12.3) satisfies the stability estimate (12.9) in which, if $\xi = 0$, the expression on the left-hand side can be replaced by $]\!\![Y_n]\!\![_{t_n}$.*

The claim is shown by checking the conditions of Theorem 4.2 with the aid of the argument employed in the proof of Theorem 6.13 since the backward Euler method is both of type $V_I(\varphi)$ and of type $V_I^*(\varphi)$ for any $\varphi \in (0, \pi/2)$.

Theorem 12.3. *Let $A(t) \equiv A \in \mathcal{S}(\varphi, \sigma)$ for some $\varphi \in (0, \pi/2)$ and let the requirement (12.2) hold for the quadrature formula (12.1). Also, suppose that for some $\mu_1 \geq 0$ sufficiently large and some $\zeta \in (0,1]$, the operator $\widetilde{P}(t,s) = (A + \mu_1 I)^{-1} P(t,s)$ satisfies the condition, for all $t, t_1, s \in [0,T]$ such that $s \leq \min\{t, t_1\}$,*

$$\|\widetilde{P}(t,s)\| + |t - t_1|^{-\zeta} \|(\widetilde{P}(t,s) - \widetilde{P}(t_1,s))\| \leq C. \tag{12.11}$$

Then, for Y_n the solution of problem (12.3) we have, for all $k \in (0, K]$ and $t_n \in \Omega_k$, that

$$\|Y_n\| \leq C \left(\|y^0\| + k \sum_{j=1}^{n} \|f(t_j)\| \right).$$

Proof. So far as the backward Euler method is $A_I(\varphi)$-stable for any $\varphi \in (0, \pi/2)$, by Theorem 6.2 it follows that the operator $\mathfrak{A} = A(I + kA)^{-1}$ satisfies the Λ_1^k-condition. Therefore, using the estimate (12.10) which is obviously valid in the considered case, the claim follows by Theorem 4.3 if we assume that condition (4.22) holds.

It thus remains to verify (4.22). Note that in view of the identity

$$(\mathfrak{A} + \mu_0 I)^{-1}(I + kA)^{-1} = ((1 + \mu_0 k)A + \mu_0 I)^{-1},$$

it suffices to show (4.22) with $\kappa_{n+1,l}\widetilde{P}(t_{n+1}, t_l)$ substituted for $\widetilde{\mathfrak{P}}_{n,l}$. By the obvious inequality

$$\begin{aligned} \|\kappa_{q+1,l}\widetilde{P}(t_{q+1}, t_l) - \kappa_{n,l}\widetilde{P}(t_n, t_l)\| \;\leq\; & |\kappa_{q+1,l} - \kappa_{n,l}| \, \|\widetilde{P}(t_{q+1}, t_l)\| \\ & + \kappa_{n,l}\|\widetilde{P}(t_{q+1}, t_l) - \widetilde{P}(t_n, t_l)\|, \end{aligned}$$

taking (12.11) into account now gives, for all $t_l \leq t_m \leq t_s \leq t_n$,

$$\begin{aligned} & \sum_{l=m}^{s-1} \left(k \sum_{q=l}^{n-1} t_{n-q}^{-1} \|\kappa_{q+1,l}\widetilde{P}(t_{q+1}, t_l) - \kappa_{n,l}\widetilde{P}(t_n, t_l)\| + \kappa_{n,l}\|\widetilde{P}(t_n, t_l)\| \right) \\ & \qquad \leq C \sum_{l=m}^{s-1} k \sum_{q=l}^{n-1} t_{n-q}^{-1} |\kappa_{q+1,l} - \kappa_{n,l}| + C \sum_{l=m}^{s-1} \kappa_{n,l} \left(k \sum_{q=l}^{n-1} t_{n-q}^{\zeta-1} + 1 \right). \end{aligned}$$

Finally, using actually the same reasoning as in the proof of Theorem 12.1 when proving (4.15) with $\gamma_3 = 0$ and with $\kappa_{n+1,l}$ substituted for $\pi_{n,l}$, we conclude that the expression on the right-hand side of the last inequality is bounded by $Ck^p \log(T/k) + Ct_{s-m}$. This completes the proof. □

Now we also discuss, for the sake of simplicity in the case $A(t) \equiv A$ only, the Crank-Nicolson discretization method whose formal application to problem (11.33) then yields, with $t_{n+1/2} = t_n + k/2$,

$$\begin{aligned} \partial Y_n + \frac{1}{2}A(Y_{n+1} + Y_n) &= \Delta^{n+1}(P(t_{n+1/2}, \cdot)Y) + f(t_{n+1/2}) \quad \text{for } t_n \in \breve{\Omega}_k, \\ Y_0 &= y^0. \end{aligned}$$

Since under the below restrictions the operator $I+\frac{k}{2}A$ has a bounded inverse, we take this method in the following form, with $\mathfrak{Q} = (I + \frac{k}{2}A)^{-1}$, $\mathfrak{A} = A\mathfrak{Q}$, $F_n = \mathfrak{Q}f(t_{n+1/2})$, and $\widehat{P}(t,s) = \mathfrak{Q}P(t,s)$,

$$\begin{aligned} Y_{n+1} &= (I - k\mathfrak{A})Y_n + k\Delta^{n+1}(\widehat{P}(t_{n+1/2},\cdot)Y) + kF_n \\ &= (I - k\mathfrak{A})Y_n + k\sum_{l=0}^{n} \kappa_{n+1,l}\widehat{P}(t_{n+1/2},t_l)Y_l + kF_n \quad \text{for } t_n \in \breve{\Omega}_k, \\ Y_0 &= y^0. \end{aligned} \tag{12.12}$$

Clearly, (12.12) can be equivalently rewritten in the form (4.1) with $\mathfrak{A}_n \equiv \mathfrak{A}$, $\mathfrak{P}_{n,l} = \kappa_{n+1,l}\widehat{P}(t_{n+1/2},t_l)$, and with the same F_n as here.

We present only one stability result for problem (12.12).

Theorem 12.4. *Let $A(t) \equiv A \in \mathcal{S}(\varphi,\sigma)$ for some $\varphi \in (0,\pi/2)$ and let the requirement (12.2) hold for the quadrature formula (12.1). Suppose further that seminorms $[\![\cdot]\!]_t$ and $|\cdot|_{*;t}$ form respectively $(\xi|\varphi,\sigma)$- and $(*;\gamma_{*3}|\varphi,\sigma)$-concordant pairs with A for some $\xi,\gamma_{*3} \in [0,1/2]$ subject to $\xi + \gamma_{*3} < 1$ and that the newly defined operator $\widehat{P}(t,s)$ satisfies condition (12.8). Then the solution Y_n of problem (12.12) is under the stability estimate (12.9), with $f(t_{j-1/2})$ substituted for $f(t_j)$. Furthermore, the expression on the left-hand side of this estimate can be replaced by $\|Y_n\|$ in the case where $\xi = 0$ and $[\![\cdot]\!] = \|\cdot\|$.*

Proof. The result follows by applying Theorem 4.2 whose conditions, in the case $\mathfrak{A}_n \equiv \mathfrak{A}$ now considered, are checked with the aid of Theorems 2.1 and 2.2. At the same time the fact that the last two theorems are applicable can be established by using Theorems 6.5 and 6.5* since the Crank-Nicolson method is $A_{III}(\varphi)$-stable for any $\varphi \in (0,\pi/2)$ and fulfils the condition $\deg[\alpha(\cdot) - \alpha(\infty)] = -1$. □

12.2 Quadrature rules

Here we give examples of quadrature rules that satisfy our above assumptions. The most obvious choice is the rectangle rule for which $\kappa_{n,l} = k$ for $t_l \le t_{n-1}$. A drawback of this method is that all already computed values of the solution are involved in further calculations so that all of them have to be stored for future use. Following the philosophy of Sloan & Thomée [147] and Zhang [168], we now turn to describing certain sparse quadrature rules for which the storage requirement is really reduced.

We begin with a quadrature formula based on the trapezoidal rule on intervals of length $\mathcal{O}\left(k^{1/2}\right)$, with a slight modification near t_n. More precisely, with $\nu = \lfloor k^{-1/2} \rfloor$ and $k_1 = \nu k$, we put $\bar{t}_l = lk_1$, and let l_n be the largest integer with $\bar{t}_{l_n} < t_n$. For the interval $[0, t_n]$ we then apply the composite trapezoidal rule with stepsize k_1 on $[0, \bar{t}_{l_n}]$, then the one-interval trapezoidal rule on $[\bar{t}_{l_n}, t_{n-1}]$, and finally the left side rectangle rule on $[t_{n-1}, t_n]$. We therefore come to the quadrature formula

$$\begin{aligned}\Delta^n(\psi) &= \frac{k_1}{2}\sum_{l=1}^{l_n}\left(\psi(\bar{t}_l) + \psi(\bar{t}_{l-1})\right) \\ &\quad + \frac{1}{2}\left(t_{n-1} - \bar{t}_{l_n}\right)\left(\psi(t_{n-1}) + \psi(\bar{t}_{l_n})\right) + k\psi(t_{n-1}).\end{aligned}$$

Since the rule is second order in k_1 over $[0, \bar{t}_{l_n}]$ and $[\bar{t}_{l_n}, t_{n-1}]$, and first order on $[t_{n-1}, t_n]$, it is first order in k on the whole. Observe that $\kappa_{n,l} \leq k^{1/2}$ for $t_l \leq t_{n-1}$, and it is easy to see that (12.2) holds for $p = 1/2$. The number of time levels that enter the computation is $\mathcal{O}\left(k^{-1/2}\right)$ for this rule, as compared with $\mathcal{O}\left(k^{-1}\right)$ for the standard rectangle rule.

Developing the idea of reducing the storage requirement, we set $\nu = \lfloor k^{-1/4} \rfloor$ and $k_2 = \nu^3 k = \mathcal{O}\left(k^{1/4}\right)$ and do the following. We first use Simpson's rule on as many intervals of length $2k_2$ that can be fitted into $[0, t_{n-1}]$, and then, on the remaining interval, which is of length at most $\mathcal{O}\left(k^{1/2}\right)$, the composite trapezoidal rule on as many intervals of length $k_1 = \nu^2 k = \mathcal{O}\left(k^{1/2}\right)$ as fit in, thus reaching $\bar{t}_{l_n}$, then the one-interval trapezoidal rule on the interval $[\bar{t}_{l_n}, t_{n-1}]$, and finally the left rectangle rule on $[t_{n-1}, t_n]$. Similarly to above, this quadrature rule is first order on the whole, while (12.2) holds for $p = 1/4$. Note that the number of time levels that need to be stored per unit time is now $\mathcal{O}\left(k_2^{-1}\right) + \mathcal{O}\left(k_2 k_1^{-1}\right) + 1 = \mathcal{O}\left(k^{-1/4}\right)$. Also, it is not hard to construct quadrature rules which further reduce the storage requirement.

We remark that our considerations in Section 12.1 admitting quadrature rules that reduce the storage requirement are related to Bakaev, Larsson, and Thomée [40], where condition (12.2) appears for the first time.

In conclusion, note that certain quadrature rules with $\mathcal{O}\left(k^{-1/2}\right)$ time levels that are needed for the computation are presented in [147] and [168], but the techniques used in these works do not allow one to deal with further reducing the storage requirement. At the same time, for instance, for the Crank-Nicolson method the results of [147] and [168] are applicable under weaker conditions on the kernel $P(t, s)$ than in the preceding section.

Appendix A

Functions of Linear Operators

Here we collect some facts of operator theory which are in use in the course of our main considerations. As therein, we let $\mathfrak{X}$ be a complex Banach space and accept all the attendant notation introduced in the book.

A.1 Analytic continuation of the resolvent

In this section, given a linear operator $\mathfrak{L}$, we denote by $\rho(\mathfrak{L})$ the resolvent set of $\mathfrak{L}$.

It is well known that if $\mathfrak{L}$ is a linear operator and if $\lambda \in \rho(\mathfrak{L})$, then $\rho(\mathfrak{L})$ contains as well some open disk centered at λ (cf. Hille & Phillips [84, Theorem 5.8.2]), which can be shown by analytic continuation. As a consequence of this fact, we obtain that $\rho(\mathfrak{L})$ is, in fact, an open set. For our purposes in the main parts of the book, we need, however, to have at our disposal some quantative estimates accompanying the procedure of analytic continuation.

Theorem A.1. *Let $\mathfrak{L}$ be a linear operator and let $\mu \in \rho(\mathfrak{L})$, with*

$$\|(\mu I - \mathfrak{L})^{-1}\| \leq M.$$

Then, for any $\lambda \in \mathbb{C}$ such that $|\lambda - \mu| = r < M$, there exists an operator $\mathfrak{S}(\mu, \lambda) \in \mathcal{B}(\mathfrak{X})$ which is permutable with the resolvent of $\mathfrak{L}$, obeys the equality

$$(\lambda I - \mathfrak{L})^{-1} = (\mu I - \mathfrak{L})^{-1}\mathfrak{S}(\mu, \lambda),$$

and satisfies the estimate

$$\|\mathfrak{S}(\mu, \lambda)\| \leq \frac{1}{1 - rM}.$$

As a consequence, it holds

$$\|(\lambda I - \mathfrak{L})^{-1}\| \leq \frac{M}{1 - rM}.$$

Proof. The claim follows when inserting

$$\frac{d^n}{d\mu^n}(\mu I - \mathfrak{L})^{-1} = (-1)^n n!(\mu I - \mathfrak{L})^{-(n+1)}$$

into the Taylor expansion of the resolvent

$$(\lambda I - \mathfrak{L})^{-1} = \sum_{n=0}^{\infty} \frac{(\lambda - \mu)^n}{n!} \frac{d^n}{d\mu^n}(\mu I - \mathfrak{L})^{-1},$$

with the aid of a standard argument (cf. Hille & Phillips [84, Section 5.8]). □

Now the last result is used for analytic continuation of the resolvent of a sectorial operator.

Theorem A.2. *Let $\mathfrak{L} \in \mathcal{S}(M; \varphi, 0)$ for some $\varphi \in (0, \pi/2)$ and $M \geq \frac{1}{\sin\varphi}$ and let φ_1 be an arbitrary number such that $\varphi - \arcsin\frac{1}{M} < \varphi_1 < \varphi$. Then for each $\lambda \in \Sigma_\varphi \setminus \operatorname{Int}\Sigma_{\varphi_1}$ there exists an operator $\mathfrak{G}(\lambda) \in \mathcal{B}(\mathfrak{X})$ which is permutable with the resolvent of $\mathfrak{L}$, obeys the equality*

$$(\lambda I - \mathfrak{L})^{-1} = \Big(\frac{\lambda}{\cos(\varphi - \varphi_1)} I - \mathfrak{L}\Big)^{-1} \mathfrak{G}(\lambda),$$

and satisfies the estimate

$$\|\mathfrak{G}(\lambda)\| \leq \frac{1}{1 - M\sin(\varphi - \varphi_1)}.$$

As a consequence, it holds, for all $\lambda \notin \operatorname{Int}\Sigma_{\varphi_1}$,

$$\|(\lambda I - \mathfrak{L})^{-1}\| \leq \frac{M}{1 - M\sin(\varphi - \varphi_1)}|\lambda|^{-1}.$$

Proof. Let $\lambda \in \mathbb{C}$ be arbitrary with $\arg\lambda = \varphi_1$. Then, for $\mu = \frac{|\lambda|}{\cos(\varphi - \varphi_1)} e^{i\varphi}$, we have $|\mu - \lambda| = |\mu|\sin(\varphi - \varphi_1)$. Using this result and noting that

$$\|(\mu I - \mathfrak{L})^{-1}\| \leq M|\mu|^{-1},$$

since $\mathfrak{L} \in \mathcal{S}(M;\varphi,0)$ and $\arg\mu = \varphi$, by Theorem A.1 we therefore find, with some $\mathfrak{S}(\lambda) \in \mathcal{B}(\mathfrak{X})$, for all λ with $\arg\lambda = \varphi_1$, that

$$(\lambda I - \mathfrak{L})^{-1} = (\mu I - \mathfrak{L})^{-1}\mathfrak{S}(\lambda),$$

where

$$\|\mathfrak{S}(\lambda)\| \le \frac{1}{1 - M|\mu|^{-1}|\mu - \lambda|} = \frac{1}{1 - M\sin(\varphi - \varphi_1)}.$$

Now, in view of the inequality $|\mu| \ge |\lambda|$, the claim follows at least for all λ with $\arg\lambda = \varphi_1$ and, by symmetry, for all λ with $\arg\lambda = \pm\varphi_1$. It is however clear, using the above reasons, that the same assertion remains valid for all $\lambda \notin \operatorname{Int}\Sigma_{\varphi_1}$ as well. □

Now we consider some applications of the above results.

Lemma A.1. *Suppose that $\mathfrak{L} \in \mathcal{S}(\varphi,\sigma)$ with some $\varphi \in (0,\pi/2)$ and $\sigma \in \mathbb{R}$. Then there exist a $\varphi_1 \in (0,\varphi)$ and an $R \ge 0$ such that for all $\lambda \notin \operatorname{Int}(\Sigma_{\varphi_1} \cup \mathcal{D}(R))$,*

$$\|(\lambda I - \mathfrak{L})^{-1}\| \le C|\lambda|^{-1}. \tag{A.1}$$

In the case $\sigma \ge 0$, (A.1) holds with $R = 0$.

Proof. By Theorem A.2 there exists a $\varphi_2 \in (0,\varphi)$ such that

$$\|(\lambda I - \mathfrak{L})^{-1}\| \le C|\lambda - \sigma|^{-1},$$

if $\lambda - \sigma \notin \operatorname{Int}\Sigma_{\varphi_2}$. Selecting further any fixed $\varphi_1 \in (\varphi_2,\varphi)$, we can take $R \ge 0$ sufficiently large to be confident that (A.1) holds for $\lambda \notin \operatorname{Int}\big(\Sigma_{\varphi_1} \cup \mathcal{D}(R)\big)$.

The more precise wording in the case $\sigma \ge 0$ is obvious. □

Lemma A.2. *Let $\mathfrak{L} \in \mathcal{S}(\varphi,\sigma)$ and let a seminorm $|\cdot|$ form a $(\xi|m|\varphi,\sigma)$-concordant pair with $\mathfrak{L}$, for some $\varphi \in (0,\pi/2)$, $\sigma \in \mathbb{R}$, $\xi \ge 0$, and $m \in \mathbb{N}$ such that $\xi \le m$. Then there exist a $\varphi_1 \in (0,\varphi)$ and an $R \ge 0$ such that for all $\lambda \notin \operatorname{Int}(\Sigma_{\varphi_1} \cup \mathcal{D}(R))$ and $u \in \mathfrak{X}$,*

$$|(\lambda I - \mathfrak{L})^{-m}u| \le C|\lambda|^{\xi - m}\|u\|.$$

In the case $\sigma \ge 0$ the above holds with $R = 0$.

Proof. The claim is established in a manner similar to that used in the proof of Lemma A.1. In fact, it suffices to mention that it follows from Theorem A.2 that with some $\varphi_2 \in (0,\varphi)$, for $\arg(\lambda - \sigma) = \varphi_2$, $\arg(\mu - \sigma) = \varphi$, and $|\lambda - \sigma| = |\mu - \sigma|\cos(\varphi - \varphi_2)$,

$$(\lambda I - \mathfrak{L})^{-m} = (\mu I - \mathfrak{L})^{-m}\mathfrak{S}(\lambda)^m,$$

where

$$\|\mathfrak{S}(\lambda)\| \leq C.$$

□

A.2 Functions of bounded and sectorial operators

We now discuss some means to define functions of linear operators; using such functions is an essential ingredient of our techniques in the present book. For our purposes we employ a simplified notion of contour which differs slightly from what is usually accepted in the theory of functions of a complex variable. More precisely, by a contour we mean a simple piecewise smooth curve in the complex plane, which either is of finite length and closed or passes through ∞. It is accepted in the book that a contour is always oriented counter-clockwise or downwards if it is finite or infinite, respectively, unless it is specified differently. Instead of contour sometimes we say on equal terms integration path.

Let first $\mathfrak{L} \in \mathcal{B}(\mathfrak{X})$ and let Ξ be a finite contour such that the open set Δ, of which Ξ is the boundary, contains the spectrum of $\mathfrak{L}$. Let further $f(\lambda)$ be a complex function that is holomorphic in a neighbourhood[1] of Δ. Then the operator $f(\mathfrak{L})$ given by

$$f(\mathfrak{L}) = \frac{1}{2\pi i} \int_{\Xi} f(\lambda)\,(\lambda I - \mathfrak{L})^{-1} d\lambda \tag{A.2}$$

is defined and bounded on $\mathfrak{X}$ (see, e.g., Hille & Phillips [84, Chapter V]). Furthermore, the following implications hold, for $\alpha_1, \alpha_2 \in \mathbb{C}$,

$$f(\lambda) \equiv 1 \Rightarrow f(\mathfrak{L}) = I, \tag{A.3}$$

$$f(\lambda) = \lambda \Rightarrow f(\mathfrak{L}) = \mathfrak{L}, \tag{A.4}$$

$$g(\lambda) = \alpha_1 f_1(\lambda) + \alpha_2 f_2(\lambda) \Rightarrow g(\mathfrak{L}) = \alpha_1 f_1(\mathfrak{L}) + \alpha_2 f_2(\mathfrak{L}), \tag{A.5}$$

$$g(\lambda) = f_1(\lambda) f_2(\lambda) \Rightarrow g(\mathfrak{L}) = f_1(\mathfrak{L}) f_2(\mathfrak{L}). \tag{A.6}$$

Apart from these properties, there exist some other premises allowing us to think of the operator $f(\mathfrak{L})$ given by (A.2) as a function of $\mathfrak{L}$. The above representation (A.2) is often called the Dunford operator calculus formula.

Next we consider possibilities of defining functions of closed unbounded operators. Actually, for our purposes it suffices to deal with sectorial operators.

[1]By a neighbourhood of a set $\mathcal{G} \subset \mathbb{C}$ we mean any open set containing $\mathrm{Cl}\,\mathcal{G}$.

Let now $\mathfrak{L} \in \mathcal{S}(\varphi, \sigma)$ for some $\varphi \in (0, \pi/2)$ and $\sigma \in \mathbb{R}$. Suppose further that a complex function $f(\lambda)$ satisfies the following conditions, for any fixed $\psi \in (0, \varphi)$, with some $\zeta_\psi > 0$ depending possibly on ψ,

$$f(\lambda) \text{ is holomorphic in a neighbourhood of } \Sigma_\psi + \sigma, \tag{A.7}$$

$$\text{there exists a finite limit } f(\infty) = \lim_{\substack{|\lambda| \to \infty, \\ \lambda \in \Sigma_\psi + \sigma}} f(\lambda), \tag{A.8}$$

$$|f(\lambda) - f(\infty)| \le C(1 + |\lambda|)^{-\zeta_\psi} \quad \text{for all } \lambda \in \Sigma_\psi + \sigma. \tag{A.9}$$

By Lemma A.1 there exist a $\varphi_1 \in (0, \varphi)$ and an $R \ge 0$ such that (A.1) holds. Then, letting, for any $\delta > 0$, Ξ_δ be the contour given by $\Xi_\delta = \partial\big(\Sigma_{\varphi_1} \cup \mathcal{D}(\delta)\big)$, owing to (A.1) and to the accepted conditions on $f(\lambda)$ the expression

$$f(\mathfrak{L}) = f(\infty)I + \frac{1}{2\pi i} \int_{\Xi_{R+\varepsilon}} \left(f(\lambda) - f(\infty)\right) (\lambda I - \mathfrak{L})^{-1} d\lambda \tag{A.10}$$

makes sense and defines a linear bounded operator on $\mathfrak{X}$ for any fixed $\varepsilon \ge 0$ subject to $R + \varepsilon > 0$. It is easy to show, using Cauchy's Theorem (see, e.g., Hille & Phillips [84, Theorem 3.11.1]), that the definition of $f(\mathfrak{L})$ does not depend on ε. Moreover, by Cauchy's Theorem there exist other deformations of the integration path in (A.10), which are tolerable as far as it is regulated by the accepted conditions on $\mathfrak{L}$ and $f(\lambda)$. Further, by standard means (cf. Taylor [160] or Hille & Phillips [84, Chapter V]) it is established that the properties (A.3), (A.5), and (A.6) are still in force while (A.4) is now absurd for the reason that formula (A.10) involves bounded functions only. In principle, following Taylor [160] one could show that formula (A.10) possesses all the features that are intrinsic in an operator calculus, dealing with the class of functions $f(\lambda)$ with the above stated properties (A.7) −−(A.9). Fortunately, there is no necessity in showing this fact because, for the aims of this book, it suffices to treat the algebra of functions which is constituted by $e^{-\lambda}$ and all suitable rational functions. At the same time, if $f(\lambda)$ is a rational function, using Cauchy's Theorem yields that formula (A.10) defines $f(\mathfrak{L})$ just the same as Taylor's operator calculus formula (see [84, Formula (5.11.2)]), while in the case $f(\lambda) = e^{-\lambda}$, applying (A.10) defines the operator $e^{-\mathfrak{L}}$ in conformity with semigroup theory (see, e.g., Reed & Simon [133, Section X.8]). Nevertheless, we remark, following [160], that if $f(\lambda)$ satisfies (A.7) −−(A.9) and if with some $m \in \mathbb{N}$,

$$\text{there exists a finite limit } g(\infty) = \lim_{\substack{|\lambda| \to \infty, \\ \lambda \in \Sigma_\psi + \sigma}} \left(f(\lambda)\, \lambda^m\right), \tag{A.11}$$

for any fixed $\psi \in (0, \varphi)$, it then holds, with $g(\lambda) = f(\lambda)\lambda^m$, for $u \in \operatorname{Dom} \mathfrak{L}^m$,

$$g(\mathfrak{L}) = f(\mathfrak{L})\, \mathfrak{L}^m, \tag{A.12}$$

where $f(\mathfrak{L})$ and $g(\mathfrak{L})$ are defined by (A.10).

The above reasoning can be extended to the case where $\mathfrak{L} \in \mathcal{S}(M;\varphi,\sigma)$, $M \geq 1$, $\varphi \in (0,\pi/2)$, $\sigma \in \mathbb{R}$ (see Remark 6.4). Since, according to our assumptions, the constant M varies considerably afterwards, we cannot now apply Lemma A.1 in order to obtain (A.10). At the same time, if we assume in addition to the conditions (A.7) —-(A.9) that $\mathfrak{L} \in \mathcal{S}(M;\varphi_1,\sigma)$ with some $\varphi_1 \in (0,\varphi)$ and with the same $\varphi \in (0,\pi/2)$ and $\sigma \in \mathbb{R}$ as above, then formula (A.10) remains valid with $R > \max\{-\sigma, 0\}$ even if the size of M varies considerably; however, the influence of M clearly appears in any estimates involving the norms of functions of the operator $\mathfrak{L}$. Note also that under the new assumption on $\mathfrak{L}$ the additional condition (A.11) implies, as above, (A.12).

In the next section we consider also other functions of sectorial operators.

A.3 Fractional powers of operators

Fractional powers are usually defined for sectorial operators, however, if the operator in question is unbounded, one then meets certain technical difficulties (see, e.g., Komatsu [93]). At the same time, for the aims of this book, it suffices to treat the positive fractional powers of bounded sectorial operators and certain negative fractional powers of sectorial operators possessing a bounded inverse, which facilitates our consideration.

Let first $\mathfrak{L} \in \mathcal{S}(\varphi,0) \cap \mathcal{B}(\mathfrak{X})$ for some $\varphi \in (0,\pi/2)$. In principle, there are several ways equivalent to each other that define the fractional powers $\mathfrak{L}^\xi$, $\xi \geq 0$. For the sake of being definite, we take the definition suggested by Balakrisnan [45] (cf. also Komatsu [93]), avoiding the difficulties that are proper for unbounded operators. More precisely, given $\xi \in (m-1,m)$ for $m \in \mathbb{N}$, the operator $\mathfrak{L}^\xi$ is introduced by

$$\mathfrak{L}^\xi = \frac{1}{2\pi i}\int_\Xi \lambda^{\xi-m}\mathfrak{L}^m(\lambda I - \mathfrak{L})^{-1}d\lambda, \tag{A.13}$$

where $\Xi = \partial\Sigma_\varphi$, while $\mathfrak{L}^\xi$ is defined in the standard way if ξ is an integer. It then follows from [45, 93] that[2]

$$\mathfrak{L}^\xi\mathfrak{L}^\eta = \mathfrak{L}^{\xi+\eta} \quad \text{for } \xi,\eta \geq 0. \tag{A.14}$$

It is significant that since $\mathfrak{L}$ is bounded, by Cauchy's Theorem the contour Ξ in (A.13) can be replaced by a suitable closed (that is, finite) contour,

[2]We take into account that $\mathfrak{L} \in \mathcal{B}(\mathfrak{X})$.

for instance, Ξ can be replaced by $\Xi_L = \partial\big(\Sigma_\varphi \cap \mathcal{D}(L)\big)$ with $L > 0$ sufficiently large.

Now let $\mathfrak{L} \in \mathcal{S}(\varphi, 0)$ for some $\varphi \in (0, \pi/2)$ and let $\mathfrak{L}^{-1} \in \mathcal{B}(\mathfrak{X})$. Then, given $\xi \in (0,1)$, the operator $\mathfrak{L}^{-\xi}$ is defined as follows (cf. Krasnosel'skii & Sobolevskii [95], Krein [96, Chapter I], or Komatsu [93])

$$\mathfrak{L}^{-\xi} = \frac{1}{2\pi i}\int_{\Xi} \lambda^{-\xi}(\lambda I - \mathfrak{L})^{-1} d\lambda, \tag{A.15}$$

where the contour Ξ is defined the same as above. Also, it is proved in a standard way (see, e.g., Krein [96, Chapter I, Theorem 5.1] [3]) that

$$\mathfrak{L}^{-\xi}\mathfrak{L}^{-\eta} = \mathfrak{L}^{-(\xi+\eta)} \qquad \text{for } \xi, \eta \geq 0 \text{ such that } \xi + \eta \leq 1. \tag{A.16}$$

In principle, $\mathfrak{L}^{-\xi}$ is defined for any $\xi \geq 0$ and the property (A.16) is then valid for any $\xi, \eta \geq 0$, but we do not deal with such an extension.

It is clear that the integral representations (A.13) and (A.15) are related to the operator calculus formulas (A.2) and (A.10).[4] Moreover, it appears that fractional powers are included in the algebra of functions of a bounded sectorial operator, as stated in the following result.

Theorem A.3. *Let $\mathfrak{L} \in \mathcal{S}(\varphi, 0) \cap \mathcal{B}(X)$ for $\varphi \in (0, \pi/2)$, let $L > 0$ be sufficiently large so that the spectrum of $\mathfrak{L}$ belongs to the set $\Delta_L = \operatorname{Int}\big(\Sigma_\varphi \cap \mathcal{D}(L)\big) \cup \{0\}$, and let $f(\lambda)$ be holomorphic in a neighbourhood of Δ_L. Then it holds, with $\Xi_L = \partial\Delta_L$, for any fixed $\xi > 0$,*[5]

$$\mathfrak{L}^{\xi} f(\mathfrak{L}) = \frac{1}{2\pi i}\int_{\Xi_L} \lambda^{\xi} f(\lambda)\,(\lambda I - \mathfrak{L})^{-1} d\lambda. \tag{A.17}$$

In fact, this is established with the aid of the representations (A.2) and (A.13) by using a standard argument (see, e.g., Krein [96, Chapter I, Section 5]).

We state also the so-called convexity inequality.

---

[3]It is assumed in [96] that the operator in question has dense domain. This assumption is not really needed for showing (A.16).

[4]The difference is that the integrand in (A.13) and (A.15) is unbounded (at 0), as a function of λ. It is easily seen that the corresponding integrals nevertheless converge under the accepted conditions.

[5]In principle, Ξ_L may be replaced by another closed contour, among those which are obtained by a deformation of Ξ_L within the bounds of possibility, as regulated by Cauchy's Theorem.

Theorem A.4. *Let* $\mathfrak{L} \in \mathcal{S}(\varphi, 0) \cap \mathcal{B}(\mathfrak{X})$ *for some* $\varphi \in (0, \pi/2)$. *Then, for any fixed* ξ *and* η *such that* $0 < \xi < \eta$,

$$\|\mathfrak{L}^{\xi} u\| \leq C \|\mathfrak{L}^{\eta} u\|^{\xi/\eta} \|u\|^{1-\xi/\eta} \quad \text{for all } u \in \mathfrak{X}. \tag{A.18}$$

If instead of $\mathfrak{L} \in \mathcal{B}(\mathfrak{X})$ *it is assumed that* $\mathfrak{L}^{-1} \in \mathcal{B}(\mathfrak{X})$, *the last estimate holds with* $\mathfrak{L}^{-\xi}$ *and* $\mathfrak{L}^{-\eta}$ *in place of* $\mathfrak{L}^{\xi}$ *and* $\mathfrak{L}^{\eta}$, *respectively, where* $\xi + \eta \leq 1$.

For a proof, see, e.g., Krein [96, Chapter I, Theorem 5.2].

A.4 Functions of ordered non-commutative operators

To meet our needs it suffices to discuss the subject for bounded operators only. Let therefore $\mathfrak{L}_1, \mathfrak{L}_2 \in \mathcal{B}(\mathfrak{X})$ and let Ξ_1 and Ξ_2 be closed contours surrounding the spectrum of $\mathfrak{L}_1$ and that of $\mathfrak{L}_2$, respectively. Let further $f(\lambda_2, \lambda_1)$ be a holomorphic function on a connected set in $\mathbb{C}^2$ containing $\Xi_2 \times \Xi_1$ in its interior. By the Dunford operator calculus formula (A.2), one can define the operator

$$f(\mathfrak{L}_2, \lambda) = \frac{1}{2\pi i} \int_{\Xi_2} f(\lambda_2, \lambda)\, (\lambda_2 I - \mathfrak{L}_2)^{-1} d\lambda_2.$$

It is readily ascertained that $f(\mathfrak{L}_2, \lambda)$ is a $(\mathcal{B}(\mathfrak{X}))$-valued holomorphic function of λ on a connected set containing Ξ_1 in its interior. This fact further yields that the expression

$$[\![f(\overset{2}{\mathfrak{L}}_2, \overset{1}{\mathfrak{L}}_1)]\!] = \frac{1}{2\pi i} \int_{\Xi_1} f(\mathfrak{L}_2, \lambda)\, (\lambda I - \mathfrak{L}_1)^{-1} d\lambda \tag{A.19}$$

makes sense and defines a linear bounded operator on $\mathfrak{X}$.

Definition A.1. *The operator* $[\![f(\overset{2}{\mathfrak{L}}_2, \overset{1}{\mathfrak{L}}_1)]\!]$ *given by (A.19) is called a* function of ordered operators.

In fact, the above notation arises from Maslov [111]: the indices over operators show the order in which they act while autonomous brackets $[\![\cdot]\!]$ are employed to restrict the region where this symbolism is valid.

It is easily checked that if, with $\mathcal{D}_j \subset \mathbb{C}$, $j = 1, 2$, some open disks centered at 0 and containing Ξ_j, $j = 1, 2$, respectively, $f(\lambda_2, \lambda_1)$ is holomorphic in the bidisk $\mathcal{D}_2 \times \mathcal{D}_1$ so that

$$f(\lambda_2, \lambda_1) = \sum_{j_1, j_2 = 0}^{\infty} a_{j_1, j_2} \lambda_1^{j_1} \lambda_2^{j_2},$$

then the expression

$$\sum_{j_1,j_2=0}^{\infty} a_{j_1,j_2} \mathfrak{L}_2^{j_2} \mathfrak{L}_1^{j_1}$$

defines a linear bounded operator on $\mathfrak{X}$ which coincides with that given by (A.19).

Appendix B

Linear Cauchy Problems in Banach Space

Here we give a brief report of introductive facts from the theory of linear abstract differential equations.

Let there be given a complex Banach space $\mathfrak{X}$, an interval $[0,T]$, a linear (generally, unbounded) operator $A(t) : \operatorname{Dom} A(t) \subseteq \mathfrak{X} \to \mathfrak{X}$ depending on $t \in [0,T]$, a function $f(t) : [0,T] \to \mathfrak{X}$, and a vector $y^0 \in \mathfrak{X}$. We consider the following Cauchy problem

$$y_t + A(t)\, y = f(t) \quad \text{for } 0 < t \leq T, \qquad y(0) = y^0. \tag{B.1}$$

By a classical solution of problem (B.1) we mean any function $y(t) : [0,T] \to \mathfrak{X}$ that is strongly continuously differentiable in $(0,T]$ and strongly continuous at 0, possesses the property $y(t) \in \operatorname{Dom} A(t)$ for $t \in (0,T]$ and fits the differential equation and the initial condition in (B.1). For brevity, we say everywhere in the book solution instead of classical solution.

In the parabolic case, which is under consideration in this book, the unique solvability of problem (B.1) is extensively examined in the literature. For the earliest results, we refer to Sobolevskii [148] and Tanabe [159], where the situation when $A(t)$ has dense, constant[1] domain is studied. In the case of variable or nondense domains, more or less recent results are contained in Amann [9] or Lunardi [108]. However, for showing convergence, higher regularity results are needed, about which not much is known, particularly for variable or nondense domains. Therefore, in the general case with a variable operator, in order to avoid quoting isolated results of higher reg-

[1]That is, independent of t

ularity, posed often under too restrictive conditions, we assume the unique solvability of problem (B.1) and as much regularity as needed.

However we act differently in the case where the operator $A(t)$ is independent of t[2] since in Chapter 7 we deal with error estimates in terms of the data. It is known that if $A \in \mathcal{S}(\varphi, \sigma)$ for some $\varphi \in (0, \pi/2)$ and $\sigma \in \mathbb{N}$, letting $\Xi = \partial(\Sigma_\varphi + \sigma)$, the integral

$$e^{-tA} = \frac{1}{2\pi i} \int_\Xi e^{-t\lambda}(\lambda I - A)^{-1} d\lambda \quad \text{for } t > 0, \tag{B.2}$$

makes sense and defines on $\mathfrak{X}$ a linear bounded operator e^{-tA} which possesses all the properties of holomorphic semigroups, except strong continuity at 0 unless A has a dense domain (see, e.g., Sinestrari [146]). In the autonomous case we employ the concept of extended solution and its development introduced further.

Definition B.1. *Under the assumptions that $A \in \mathcal{S}(\varphi, \sigma)$ for some $\varphi \in (0, \pi/2)$ and $\sigma \in \mathbb{N}$ and that $f(t)$ is strongly continuous for $t \in [0, T]$, the function*

$$y(t) = \begin{cases} e^{-tA} y^0 + \int_0^t e^{-(t-s)A} f(s)\, ds & \text{if } t > 0, \\ y^0 & \text{if } t = 0, \end{cases} \tag{B.3}$$

is called the extended solution *of problem (B.1).*[3]

Definition B.2. *Suppose that $A \in \mathcal{S}(\varphi, \sigma)$, $f(t)$ is p times strongly continuously differentiable for $t \in [0, T]$, and $y^0_{(j)} \in \operatorname{Dom} A$, $j = 0, \ldots, p-1$, for some $\varphi \in (0, \pi/2)$, $\sigma \in \mathbb{N}$, and $p \in \mathbb{N}$, where*

$$y^0_{(0)} = y^0 \text{ and } y^0_{(j)} = f^{(j-1)}(0) - A y^0_{(j-1)}, \; j = 1, \ldots, p-1. \tag{B.4}$$

Then the function $y(t)$ given by (B.3) is called the p-th order extended solution *of problem (B.1).*

It is not hard to see, using the properties of the integral in (B.2) (see, e.g., Sinestrari [146]), that the p-th order extended solution of (B.1) obeys the equality, with $y^0_{(p)} = f^{(p-1)}(0) - A y^0_{(p-1)}$,

$$y^{(p)}(t) = e^{-tA} y^0_{(p)} + \int_0^t e^{-(t-s)A} f^{(p)}(s)\, ds \quad \text{for } t > 0. \tag{B.5}$$

[2]In this case we write therefore A instead of $A(t)$.

[3]If $y^0 \in \operatorname{Cl}(\operatorname{Dom} A)$, the concept of extended solution coincides with that of mild solution (cf., e.g., Sinestrari [146]). Note however that the extended solution is defined under the accepted conditions even if $y^0 \notin \operatorname{Cl}(\operatorname{Dom} A)$.

Observe that it is then not necessary that $y^0_{(p)} \in \operatorname{Dom} A$. Also, it is worth noting that the conditions imposed on $f(t)$ and $y^0_{(j)}$, $j = 0, \ldots, p-1$, in Definition B.2 express, in fact, a certain kind of time regularity for $y(t)$. In particular, the extended solution of order ≥ 1 is simultaneously a solution of (B.1), as it follows from Sinestrari [146, Theorem 4.4].

Bibliography

[1] S. Abarbanel, D. Gottlieb, and M. H. Carpenter, *On the removal of boundary errors caused by Runge-Kutta integration of nonlinear partial differential equations,* SIAM J. Sci. Comput. **17**(1996), 777-782.

[2] Sh. Agmon, *On the eigenfunctions and on the eigenvalues of general elliptic boundary value problems,* Comm. Pure Appl. Math. **15**(1962), 119-147.

[3] S. Agmon, A. Douglis, and L. Nirenberg, *Estimates near the boundary for solutions of elliptic partial differential equations satisfying general boundary conditions. I, II,* Comm. Pure Appl. Math. **12**(1959), 623-727; **17**(1964), 35-92.

[4] G. Akrivis and M. Crouzeix, *Linearly implicit methods for nonlinear parabolic equations,* Math. Comp. **73**(2004), 613-635.

[5] G. Akrivis, M. Crouzeix, and Ch. Makridakis, *Implicit-explicit multstep methods for quasilinear parabolic equations,* Numer. Math. **82**(1999), 521-541.

[6] G. Akrivis, O. Karakashian, and F. Karakatsani, *Linearly implicit methods for nonlinear evolution equations,* Numer. Math. **94**(2003), 403-418.

[7] G. Akrivis and Ch. Makridakis, *Galerkin time-stepping methods for nonlinear parabolic equations,* M2AN **38**(2004), 261-289.

[8] G. Akrivis, Ch. Makridakis, and R. H. Nochetto, *A posteriori error estimates for the Crank-Nicolson for parabolic equations,* to appear in Math. Comp.

[9] H. Amann, *Linear and Quasilinear Parabolic Problems. Vol.I. Abstract Linear Theory,* Birkhäuser Verlag, Basel - Boston - Berlin, 1995.

[10] N. Yu. Bakaev, *Stability theory of difference schemes in arbitrary norms,* Dokl. Akad. Nauk SSSR **297**(1987), 275-279 (in Russian); English transl. in Soviet Math. Dokl. **36**(1988).

[11] N. Yu. Bakaev, *Stability theory of difference schemes for parabolic equations in arbitrary norms. Part 1,* Vopr. Atomn. Nauki i Tekhniki, ser. Metodi i Progr. Chisl. Reshen. Zadach Matem. Fiz. (1987), Iss. 1, 69-75 (in Russian).

[12] N. Yu. Bakaev, *Stability theory of difference schemes for parabolic equations in arbitrary norms. Part 2,* Vopr. Atomn. Nauki i Tekhniki, ser. Metodi i Progr. Chisl. Reshen. Zadach Matem. Fiz. (1987), Iss. 2, 29-34 (in Russian).

[13] N. Yu. Bakaev, *Stability theory of additive difference schemes in Banach norms,* Deposited in VINITI, No. 6044-88, 1988 (in Russian).

[14] N. Yu. Bakaev, *On the resolvent bound of the second order difference operator with aperiodic boundary conditions,* Probl. Sovr. Teor. Periodich. Dvizhen. **9**(1988), 73-84 (in Russian).

[15] N. Yu. Bakaev, *Stability theory of additive difference schemes in Banach spaces,* Dokl. Akad. Nauk SSSR **305**(1989), 18-21 (in Russian); English transl. in Soviet Math. Dokl. **39**(1989).

[16] N. Yu. Bakaev, *Stability estimates for a certain general discretization method,* Dokl. Akad. Nauk SSSR **309**(1989), 11-15 (in Russian); English transl. in Soviet Math. Dokl. **40**(1990).

[17] N. Yu. Bakaev, *Stability estimates of difference schemes for a differential equation with constant operator. I,* Partial Diff. Eqs., Novosibirsk, AN SSSR, Siberian Division, 1989, 3-14 (in Russian).

[18] N. Yu. Bakaev, *Stability estimates of difference schemes for a differential equation with constant operator.II,* Embed. Theorems Appls Probls Math. Phys., Novosibirsk, SO AN SSSR, 1989, 18-37 (in Russian).

[19] N. Yu. Bakaev, *Investigation of a certain class of difference schemes for an equation with variable operator,* Deposited in VINITI, No. 7324-89, Minsk, 1989 (in Russian).

[20] N. Yu. Bakaev, *On the well-posedness of difference problems with generalized input data,* Dokl. Akad. Nauk SSSR **314**(1990), 1293-1297 (in Russian); English transl. in Soviet Math. Dokl. **42**(1991), 624-629.

[21] N. Yu. Bakaev, *Stability estimates for some splitting difference schemes,* Probl. Sovr. Teor. Periodich. Dvizhen. **10**(1990), 56-58 (in Russian).

[22] N. Yu. Bakaev, *On the stability of Runge-Kutta methods for abstract linear equations,* Ukranian Math. J. 42(1990), 689-694 (in Russian).

[23] N. Yu. Bakaev, *Investigation of stability of some two-layer difference schemes,* Zhurn. Vychisl. Mat. i Mat. Fiz. **30**(1990), 151-156 (in Russian); English transl. in Comput. Maths Math. Phys.

[24] N. Yu. Bakaev, *On the stability of weighted difference schemes,* Ukranian Math. J. 42(1990), 1254-1258 (in Russian).

[25] N. Yu. Bakaev, *On the stability of some difference problems with smoothing operators,* Zhurn. Vychisl. Mat. i Mat. Fiz. **31**(1991), 75-85 (in Russian); English transl. in Comput. Maths Math. Phys.

[26] N. Yu. Bakaev, *Stability estimates of Runge-Kutta methods for differential equations with variable operator,* Ukranian Math. J. 43(1991), 280-283 (in Russian).

[27] N. Yu. Bakaev, *Conditional stability estimates of Runge-Kutta methods for an equation with variable operator,* Ukranian Math. J. 43(1991), 572-575 (in Russian).

[28] N. Yu. Bakaev, *Multistep difference method for parabolic equations,* Zhurn. Vychisl. Mat. i Mat. Fiz. **32**(1992), 261-276 (in Russian); English transl. in Comput. Maths Math. Phys. **32**(1992), 216-228.

[29] N. Yu. Bakaev, *To the theory of difference operators in L_{ph} spaces,* Vopr. Atomn. Nauki i Tekhniki, ser. Matem. Modelir. Fiz. Prots. (1992), Iss. 1, 18-20 (in Russian).

[30] N. Yu. Bakaev, *Some problems of well-posedness of difference schemes on non-uniform grids,* Zhurn. Vychisl. Mat. i Mat. Fiz. **33**(1993), 561-577 (in Russian); English transl. in Comput. Maths Math. Phys. **33** (1993), 511-524.

[31] N. Yu. Bakaev, *On the bounds of approximations of holomorphic semigroups,* BIT **35**(1995), 605-608.

[32] N. Yu. Bakaev, A priori *estimates for some classes of difference schemes,* Canad. J. Math. **48**(1996), 244-257.

[33] N. Yu. Bakaev, *The upper bounds for powers of linear operators and some applications to the stability analysis of difference problems,* J. Difference Eqs. Appl. **4**(1998), 343-364.

[34] N. Yu. Bakaev, *Analysis of linear discrete parabolic problems. Part 1. Well-posedness of a Cauchy problem for the evolution equation in discrete time,* Departamento de Matemática Aplicada y Computación, Facultad de Ciencias, Universidad de Valladolid, Valladolid, Spain, Report 1997/1.

[35] N. Yu. Bakaev, *Analysis of linear discrete parabolic problems. Part 2. Well-posedness of single step time-discrete approximations to abstract parabolic equations,* Departamento de Matemática Aplicada y Computación, Facultad de Ciencias, Universidad de Valladolid, Valladolid, Spain, Report 1998/6.

[36] N. Yu. Bakaev, *On variable stepsize Runge-Kutta approximations of a Cauchy problem for the evolution equation,* BIT **38**(1998), 462-485.

[37] N. Yu. Bakaev, *Analysis of discretizations of parabolic problems in pairs of spaces,* BIT **40**(2000), 1-23.

[38] N. Yu. Bakaev, *Maximum norm resolvent estimates for elliptic finite element operators,* BIT **41**(2001), 215-239.

[39] N. Yu. Bakaev, *Resolvent estimates of elliptic differential and finite-element operators in pairs of function spaces,* Intern. J. Math. Math. Sciences **2004:5**(2004), 217-238.

[40] N. Yu. Bakaev, S. Larsson, and V. Thomée, *Backward Euler type methods for parabolic integro-differential equations in Banach space,* RAIRO Modél. Math. Anal. Numér. **32**(1998), 85-99.

[41] N. Yu. Bakaev, S. Larsson, and V. Thomée, *Long time behavior of backward difference type methods for parabolic equations with memory in Banach space,* East-West J. Numer. Math. **6**(1998), 185-206.

[42] N. Bakaev and A. Ostermann, *Long-term stability of variable stepsize approximations of semigroups,* Math. Comp. **71**(2002), 1545-1567.

[43] N. Yu. Bakaev, V. Thomée, and L. B. Wahlbin, *Maximum-norm estimates for resolvents of finite element operators,* Math. Comp. **72**(2003), 1597-1610.

[44] N. Yu. Bakaev and V. V. Vyazankin, *The stability of difference schemes approximating parabolic differential equations on nonuniform grids,* VANT, ser. Mat. Modelir. Fiz. Prots. **2**(1991), 20-24 (in Russian).

[45] A. V. Balakrishnan, *Fractional powers of closed operators and the semigroups generated by them,* Pacific J. Math. **10**(1960), 419-437.

[46] J. Becker, *A second order backward difference method with variable steps for a parabolic problem,* BIT **38**(1998), 644-662.

[47] W.-J. Beyn and B. M. Garay, *Estimates of variable stepsize Runge-Kutta methods for sectorial evolution equations with nonsmooth data,* Appl. Numer. Math. **41**(2002), 369-400.

[48] N. Borovykh, D. Drissi, and M. N. Spijker, *A note about Ritt's condition, related resolvent conditions and power bounded operators,* Numer. Funct. Anal. and Optimiz. **21**(2000), 425-438.

[49] J. H. Bramble, A. H. Schatz, V. Thomée, and L. B. Wahlbin, *Some convergence estimates for semidiscrete Galerkin type approximations for parabolic equations,* SIAM J. Numer. Anal. **14**(1977), 218-241.

[50] Ph. Brenner, M. Crouzeix, and V. Thomée, *Single step methods for inhomogeneous linear differential equations in Banach space,* RAIRO Anal. Numér. **16**(1982), 5-26.

[51] Ph. Brenner and V. Thomée, *Stability and convergence rates in L_p for certain difference schemes,* Math. Scand. **27**(1970), 5-23.

[52] Ph. Brenner and V. Thomée, *On rational approximations of semigroups,* SIAM J. Numer. Anal. **16**(1979), 683-694.

[53] J. C. Butcher, *The Numerical Analysis of Ordinary Differential Equations: Runge-Kutta and General Linear Methods,* J. Wiley & Sons, Chichester, 1987.

[54] M. Calvo and R. D. Grigorieff, *Time discretization of parabolic problems with the variable 3-step BDF,* BIT **42**(2002), 689-701.

[55] M. P. Calvo and C. Palencia, *Avoiding the order reduction of Runge-Kutta methods for linear initial boundary value problems,* Math. Comp. **71**(2002), 1529-1543.

[56] S. Campanato, *Generation of analytic semigroups by elliptic operators of second order in Hölder spaces,* Ann. Sc. Norm. Sup. Pisa **8**(1981), 495-512.

[57] M. H. Carpenter, D. Gottlieb, S. Abarbanel, and W. S. Don, *The theoretical accuracy of Runge-Kutta time discretizations for the initial boundary value problem: a study of the boundary error,* SIAM J. Sci. Comput. **16**(1995), 1241-1252.

[58] F. Colombo and V. Vespri, *Generation of analytic semigroups in* $W^{k,p}(\Omega)$ *and* $C^k(\overline{\Omega})$, Diff. Integr. Equations **9**(1996), 421-436.

[59] M. Crouzeix, *Sur l'approximation des équations différentielles opérationnelles linéaires par des méthodes de Runge-Kutta,* Thése, Université Paris VI, 1975.

[60] M. Crouzeix, *On multistep approximation of semigroups in Banach spaces,* J. Comput. Appl. Math. **20**(1987), 25-35.

[61] M. Crouzeix, S. Larsson, S. Piskarev, and V. Thomée, *The stability of rational approximations of analytic semigroups,* BIT **33**(1993), 74-84.

[62] M. Crouzeix, S. Larsson, and V. Thomée, *Resolvent estimates for elliptic finite element operators in one dimension,* Math. Comp. **63**(1994), 121-140.

[63] M. Crouzeix and F. J. Lisbona, *The convergence of variable-stepsize, variable-formula, multistep methods,* SIAM J. Numer. Anal. **21**(1984), 512-534.

[64] M. Crouzeix and P. A. Raviart, *Approximation des Problèmes d'Évolution,* Lecture notes (unpublished), Rennes 1980.

[65] K. Dekker and J. G. Verwer, *Stability of Runge-Kutta Methods for Stiff Nonlinear Differntial Equations,* North Holland, Amsterdam-New York-Oxford, 1984.

[66] J. Douglas and J. E. Gunn, *Two high-order correct difference analogues for the equation on multi-dimensional heat flow,* Math. Comp. **17**(1963), 71-80.

[67] O. El-Fallah and T. Ransford, *Extremal growth of powers of operators satisfying resolvent conditions of Kreiss-Ritt type,* J. Function. Anal. **196**(2002), 135-154.

[68] K. Eriksson, C. Johnson, and S. Larsson, *Adaptive finite element methods for parabolic problems. VI. Analytic semigroups,* Dep. of Math., Chalmers University of Technology and Göteborg University, Preprint No. 1996-32, 1996.

[69] K. Eriksson, C. Johnson, and V. Thomée, *Time discretization of parabolic problems by the discontinuous Galerkin method,* RAIRO Modél. Math. Anal. Numér. **19**(1985), 611-643.

[70] C. W. Gear and K. W. Tu, *The effect of variable mesh size on the stability of multistep methods,* SIAM J. Numer. Anal. **11**(1974), 1025-1043.

[71] C. González and A. Ostermann, *Optimal convergence results for Runge-Kutta discretizations of linear nonautonomous parabolic problems,* BIT **39**(1999), 79-95.

[72] C. González, A. Ostermann, C. Palencia, and M. Thalhammer, *Backward Euler discretization of fully nonlinear parabolic problems,* Math. Comp. **71**(2002), 125-145.

[73] C. González and C. Palencia, *Stability of Runge-Kutta methods for abstract time-dependent parabolic problems: the Hölder case,* Math. Comp. **68**(1999), 73-89.

[74] C. González and C. Palencia, *Stability of time-stepping methods for abstract time-dependent parabolic problems,* SIAM J. Numer. Anal. **35**(1998), 979-989.

[75] C. González and C. Palencia, *Stability of Runge-Kutta methods for quasilinear parabolic problems,* Math. Comp. **69**(2000), 609-628.

[76] R. D. Grigorieff, *Stability of multistep-methods on variable grids,* Numer. Math. **42**(1983), 359-377.

[77] R. D. Grigorieff, *Time discretization of semigroups by the variable two-step BDF method,* in: Numerical Treatment of Differential Equations, K. Strehmel, ed., (NUMDIFF-5, Halle 1989), Teubner, Leipzig, 1991, 204-216.

[78] N. N. Gudovich and A. A. Terteryan, *On the stability of block difference schemes for parabolic equations,* Dokl. Akad. Nauk SSSR **297**(1987), 531-534 (in Russian); English transl. in Soviet Math. Dokl.

[79] N. Guglielmi and M. Zennaro, *On the zero-stability of variable stepsize multistep methods: the spectral radius approach,* Numer. Math. **88**(2001), 445-458.

[80] E. Hairer, S. P. Nørsett, and G. Wanner, *Solving Ordinary Differential Equations I. Nonstiff Problems,* Springer-Verlag, Berlin, 2nd ed., 1993.

[81] E. Hairer and G. Wanner, *Solving Ordinary Differential Equations II. Stiff and Differential-Algebraic Problems,* Springer-Verlag, Berlin, 1991.

[82] A. Hansbo, *Nonsmooth data error estimates for damped single step methods for parabolic equations in Banach space,* Calcolo **36**(1999), 75-101.

[83] A. Hansbo, *Strong stability and nonsmooth data error estimates for discretizations of linear parabolic problems,* BIT **42**(2002), 351-379.

[84] E. Hille and R. S. Phillips, *Functional analysis and semi-groups,* Providence, AMS, Colloquium Publs, vol. XXXI, 1957.

[85] R. A. Horn and C. R. Johnson, *Matrix Analysis,* Cambridge University Press, Cambridge, 1985.

[86] W. Hundsdorfer and J. G. Verwer, *Numerical Solution of Time-Dependent Advection-Diffusion-Reacrion Equations,* Springer-Verlag, Berlin, 2003.

[87] P. Jamet, *Galerkin-type approximations which are discontinuous in time for parabolic equations in a variable domain,* SIAM J. Numer. Anal. **15**(1978), 912-928.

[88] C. Johnson, Y.-Y. Nie, and V. Thomée, *An a posteriori error estimate and adaptive timestep control for a backward Euler discretization of a parabolic problem,* SIAM J. Numer. Anal. **27**(1990), 277-291.

[89] N. Kalton, S. Montgomery-Smith, K. Oleszkiewicz, and Yu. Tomilov, *Power-bounded operators and related norm estimates,* J. London Math. Soc. **70**(2004), 463-478.

[90] O. A. Karakashian, *On Runge-Kutta methods for parabolic equations with time-dependent coefficients,* Math. Comp. **47**(1986), 77-101.

[91] T. Kato, *Perturbation Theory for Linear Operators,* 2nd ed., Berlin, Springer-Verlag, 1976.

[92] S. L. Keeling, *Galerkin/Runge-Kutta discretizations for parabolic equations with time-dependent coefficients,* Math. Comp. **52**(1989), 561-586.

[93] H. Komatsu, *Fractional powers of operators,* Pacific J. Math. **19**(1966), 285-346.

[94] H. Komatsu, *Fractional powers of operators, VI. Interpolation of non-negative operators and imbedding theorems,* J. Fac. Sci. Univ. Tokyo, Sec. I A, Math. **19**(1972), 1-63.

[95] M. A. Krasnosel'skii and P. E. Sobolevskii, *Fractional powers of operators acting in Banach spaces,* Dokl. Akad. Nauk SSSR **129**(1959), 499-502 (in Russian).

[96] S. G. Krein, *Linear Differential Equations in Banach Space,* Nauka, Moscow, 1973 (in Russian).

[97] H. O. Kreiss, *Über die stabilitätsdefinition für differenzengleichungen die partielle differentialgleichungen approximieren,* BIT **2**(1962), 153-181.

[98] H. O. Kreiss, *On difference approximations of the dissipative type for hyperbolic differential equations,* Comm. Pure Appl. Math. **17**(1964), 335-353.

[99] S. Larsson, *Preservation of strong stability associated with analytic semigroups,* in: Collected Lectures on Preservation of Stability under Discretization, D.Estep and S.Tavener, eds., SIAM, 2002, 11-24.

[100] S. Larsson, V. Thomée, and L. B. Wahlbin, *Finite element methods for a strongly damped wave equation,* IMA J. Numer. Anal. **11**(1991), 115-142.

[101] P. D. Lax and B. Wendroff, *Difference schemes for hyperbolic equations with high order of accuracy,* Comm. Pure Appl. Math. **17**(1964), 381-398.

[102] M. N. Le Roux, *Semidiscretization in time for parabolic problems,* Math. Comp. **33**(1979), 919-931.

[103] R. J. Le Veque and L. N. Trefethen, *On the resolvent condition in the Kreiss matrix theorem,* BIT **24**(1984), 584-591.

[104] C. Lubich and O. Nevanlinna, *On resolvent conditions and stability estimates,* BIT **31**(1991), 293-313.

[105] Ch. Lubich and A. Ostermann, *Runge-Kutta methods for parabolic equations and convolution quadrature,* Math. Comp. **60**(1993), 105-131.

[106] Ch. Lubich and A. Ostermann, *Runge-Kutta approximation of quasi-linear parabolic equations,* Math. Comp. **64**(1995), 601-627.

[107] Ch. Lubich and A. Ostermann, *Runge-Kutta time discretization of reaction-diffusion and Navier Stokes equations: nonsmooth-data error estimates and applications to long-time behaviour,* Appl. Numer. Math. **22**(1996), 279-292.

[108] A. Lunardi, *Analytic Semigroups and Optimal Regularity in Parabolic Problems,* Birkhäuser, Basel - Boston - Berlin, 1995.

[109] M. Luskin and R. Rannacher, *On the smoothing property of the Crank-Nicolson scheme,* Appl. Anal. **14**(1982), 117-135.

[110] Yu. Lyubich, *Spectral localization, power boundedness and invariant subspaces under Ritt's type condition,* Studia Math. **134**(1999), 153-167.

[111] V. P. Maslov, *Operator Methods,* Moscow, Nauka, 1973 (in Russian).

[112] J. Miller and G. Strang, *Matrix theorems for partial differential and difference equations,* Math. Scand. **18**(1966), 113-123.

[113] K. W. Morton, *On a matrix theorem due to H.-O. Kreiss,* Comm. Pure Appl. Math. **17**(1964), 375-379.

[114] B. Nagy and J. Zemánek, *A resolvent condition implying power boundedness,* Studia Math. **134(2)**(1999), 143-151.

[115] E. Nakaguchi and A. Yagi, *Error estimates of implicit Runge-Kutta methods for quasilinear abstract equations of parabolic type,* Japan J. Math. (N.S.) **25**(1999), 181-226.

[116] O. Nevanlinna, *Converegence of Iterations of Linear Equations,* Birkhäuser Verlag, Basel - Boston - Berlin, 1993.

[117] O. Nevanlinna, *On the growth of the resolvent operators for power bounded operators,* in: Banach Center Publ. **38**, Inst. Math., Polish Acad. Sci., 1997, 247-264.

[118] O. Nevanlinna, *Resolvent conditions and powers of operators,* Studia Math. **145(2)**(2001), 113-134.

[119] J. A. Nitsche and M. F. Wheeler, *L_∞-boundedness of the finite element Galerkin operator for parabolic problems,* Numer. Funct. Anal. Optimiz. **4(4)**(1981-1982), 325-353.

[120] R. H. Nochetto, G. Savaré, and C. Verdi, A posteriori error estimates for variable time-step discretizations of nonlinear evolution equations, Comm. Pure Appl. Math. **53**(2000), 525-589.

[121] J. T. Oden and J. N. Reddy, *An Introduction to the Mathematical Theory of Finite Elements,* New York, John Wiley, 1976.

[122] A. Ostermann, *Stability of W-methods with applications to operator splitting and to geometric theory,* Appl. Numer. Math. **42**(2002), 353-366.

[123] A. Ostermann and M. Roche, *Runge-Kutta methods for partial differential equations and fractional orders of convergence,* Math. Comp. **59**(1992), 403-420.

[124] A. Ostermann and M. Roche, *Rosenbrock methods for partial differential equations and fractional orders of convergence,* SIAM J. Numer. Math. **30**(1992), 1084-1098.

[125] A. Ostermann and M. Thalhammer, *Convergence of Runge-Kutta methods for nonlinear parabolic equations,* Appl. Numer. Math. **42**(2002), 367-380.

[126] A. Ostermann, M. Thalhammer, and G. Kirlinger, *Stability of linear multistep methods and applications to nonlinear parabolic problems,* Appl. Numer. Math. **48**(2004), 389-407.

[127] C. Palencia, *A stability result for sectorial operators in Banach spaces,* SIAM J. Numer. Anal. **30**(1993), 1373-1384.

[128] C. Palencia, *On the stability of variable stepsize rational approximations of holomorphic semigroups,* Math. Comp. **62**(1994), 93-103.

[129] C. Palencia, *Stability of rational multstep approximations of holomorphic semigroups,* Math. Comp. **64**(1995), 591-599.

[130] D. Pathria, *The correct formulation of intermediate boundary conditions for Runge-Kutta time integration of initial boundary value problems,* SIAM J. Sci. Comput. **18**(1997), 1255-1266.

[131] A. Pazy, *Semigroups of Linear Operators and Applications to Partial Differential Equations,* Springer Verlag, New York, 1983.

[132] R. Rannacher, *Finite element solution of diffusion problems with irregular data,* Numer. Math. **43**(1984), 309-327.

[133] M. Reed and B. Simon, *Methods of Modern Mathematical Physics. II: Fourier Analysis, Self-Adjointness,* Academic Press, New York - San Francisco - London, 1975.

[134] M. Reed and B. Simon, *Methods of Modern Mathematical Physics. IV: Analysis of Operators,* Academic Press, New York - San Francisco - London, 1978.

[135] R. D. Richtmyer and K. W. Morton, *Difference Methods for Initial-Value Problems,* 2nd ed., New York, John Wiley, 1967.

[136] R. K. Ritt, *A condition that* $\lim_{n\to\infty} n^{-1}T^n = 0$, Proc. Amer. Math. Soc. **4**(1953), 898-899.

[137] E. B. Saff and R. S. Varga, *On the zeros and poles of Padé approximants to* $\exp z$, Numer. Math. **25**(1975), 1-14.

[138] N. Saito, *Remarks on the rational approximation of holomorphic semigroups with nonuniform partitions,* Japan J. Indust. Appl. Math. **21**(2004), 323-337.

[139] A. A. Samarskii, *Theory of Difference Schemes,* 2nd ed., Moscow, Nauka, 1983 (in Russian).

[140] J. M. Sanz-Serna, J. G. Verwer, and W. H. Hundsdorfer, *Convergence and order reduction of Runge-Kutta schemes applied to evolutionary problems in partial differential equations,* Numer. Math. **50**(1986), 405-418.

[141] A. H. Schatz, V. Thomée, and L. B. Wahlbin, *Maximum norm stability and error estimates in parabolic finite element equations,* Comm. Pure Appl. Math. **33**(1980), 265-304.

[142] S. I. Serdyukova, *The uniform stability with respect to initial data of a sixpoint symmetrical scheme for the heat conduction equation,* in: Numerical Methods for the Solution of Differential and Integral Equations and Quadrature Formulae, Nauka, Moscow, 1964, 212-216 (in Russian).

[143] S. I. Serdyukova, *On the C-stability of linear difference schemes with constant real-valued coefficients,* Zhurn. Vychisl. Mat. i Mat. Fiz. **6**(1966), 477-486 (in Russian).

[144] S. I. Serdyukova, *The uniform stability of a sixpoint scheme of increased order of accuracy for the heat equation,* Zhurn. Vychisl. Mat. i Mat. Fiz. **7**(1967), 214-218 (in Russian).

[145] S. I. Serdyukova, *On the maximum norm stability of systems of difference equations,* Zhurn. Vychisl. Mat. i Mat. Fiz. **7**(1967), 497-509 (in Russian).

[146] E. Sinestrari, *On the abstract Cauchy problem of parabolic type in spaces of continuous functions,* J. Math. Anal. Appl. **107**(1985), 16-66.

[147] I. H. Sloan and V. Thomée, *Time discretization of an integro-differential equation of parabolic type,* SIAM J. Numer. Anal. **23**(1986), 1052-1061.

[148] P. E. Sobolevskii, *On the equations of parabolic type in Banach space,* Annals Moscow Math. Soc. **10**(1961), 297-350 (in Russian).

[149] P. E. Sobolevskii, *On the stability and convergence of the Crank-Nicolson difference scheme,* Variat.-Difference Meths in Math. Phys., Novosibirsk, AN SSSR, Siber. Division, 1973, 146-151 (in Russian).

[150] P. E. Sobolevskii, *Fractional powers of coercive-positive sums of operators,* Sibirsk. Mat. Zhurn. **18**(1977), 637-657 (in Russian); Transl. into English.

[151] P. E. Sobolevskii, *Semigroup theory and stability of difference schemes,* in: Operator Theory in Function Spaces, G.P.Akilov, ed., Novosibirsk, SO Nauka, 1977, 304-337 (in Russian).

[152] P. E. Sobolevskii, *On the difference scheme of Crank-Nicolson for parabolic equations,* Nonlin. Oscil. Control Theory, Izhevsk, UdGU, 1978, 98-106 (in Russian).

[153] M. N. Spijker, *On a conjucture by Le Veque and Trefethen related to the Kreiss matrix theorem,* BIT **31**(1991), 551-555.

[154] M. N. Spijker and F. A. J. Straetemans, *Stability estimates for families of matrices of nonuniformly bounded order,* Linear Algebra Appl. **239**(1996), 77-102.

[155] H. J. Stetter, *Analysis of Discretization Methods for Ordinary Differential Equations,* Springer-Verlag, Berlin - Heidelberg - New York, 1973.

[156] H. B. Stewart, *Generation of analytic semigroups by strongly elliptic operators,* Trans. Amer. Math. Soc. **199**(1974), 141-162.

[157] E. Tadmor, *The equivalence of L_2-stability, the resolvent condition, and strict H-stability,* Linear Algebra Appl. **41**(1981), 151-159.

[158] E. Tadmor, *The resolvent condition and uniform power boundedness,* Linear Algebra Appl. **80**(1986), 250-252.

[159] H. Tanabe, *On the equations of evolution in a Banach space,* Osaka Math. J. **12**(1960), 363-376.

[160] A. E. Taylor, *Spectral theory of closed distributive operators,* Acta Math. **84**(1950), 189-224.

[161] V. Thomée, *Finite difference methods for linear parabolic equations,* in: Handbook of Numerical Analysis.I, P.G.Ciarlet and J.L.Lions, eds., North Holland, Amsterdam, 1990, 5-196.

[162] V. Thomée, *Galerkin Finite Element Methods for Parabolic Problems,* Springer Series in Computational Mathematics, Springer - Verlag, Berlin, 1997.

[163] V. Thomée and L. B. Wahlbin, *Maximum-norm stability and error estimates in Galerkin methods for parabolic equations in one space variable,* Numer. Math. **41**(1983), 345-371.

[164] H. Triebel, *Interpolation Theory. Function Spaces. Differential Operators,* VEB Deutscher Verlag, Berlin, 1978.

[165] P. Vitse, *Functional calculus under the Tadmor-Ritt condition, and free interpolation by polynomials of a given degree,* J. Funct. Anal. **210**(2004), 43-72.

[166] Y. Yan, *Smoothing properties and approximation of time derivatives for parabolic equations: variable time steps,* BIT **43**(2003), 647-669.

[167] Zh. Yuan, *On the Resolvent and Tauberian Conditions for Bounded Linear Operators,* Licentiate's Thesis, Helsinki University of Technology, January 2002.

[168] N.-Y. Zhang, *On fully discrete Galerkin approximations for partial integro-differential equations of parabolic type,* Math. Comp. **60**(1993), 133-166.

Index